WITHDRAWN FROM LIBRARY

This is a periodic table classified herein as subtype IIIC5–Ia (Mazurs 1967). The element symbols are in an order of *increasing atomic numbers* and *no interruptions* occur in the reading since it goes down series by series. The series are cut according to the *length of subshells s, p, d,* and *f* and are placed one below the other. With this arrangement the subshells are ordered according to the *Yeou Ta* (1946) *mathematical expression of the Periodic Law: t = n + l.* The drawing shows that the series of the table form inverted cones each of which represents a *period* according to the *new numeration*.

Since this table contains no numerations for vertical electron columns or for the family groups of related elements, the possibility of finding the related elements is achieved by *coloring the analogous elements*. The nonmetals are colored with cool colors (green, blue, and violet) and metals with warm colors (red, orange, yellow, brown, and gray) in agreement with the color legend given in the Conclusion of the book. Therefore, this *shell and subshell table* does not entirely lose the purpose assigned to the periodic table by D. I. Mendeleev in 1869, to show the element families and compare their properties.

This kind of series table represents in some ways the space curve type IIIA4–1 (Schaltenbrand 1920). Two other tables can be developed from this table simply by moving some series to the right: subtype IIIC5–2d (Mazurs 1958), which combines the advantages of chemical and electronic tables, or even type IIIC4–2 (Gardner 1930) which became very popular in recent times.

GRAPHIC REPRESENTATIONS OF THE PERIODIC SYSTEM DURING ONE HUNDRED YEARS

Motto:

"The failure of most chemistry teachers to exploit to the fullest the possibilities of modern theoretical principles is illustrated by their insistence on retaining the old Mendeleev form of the Periodic Chart even though the historical table is obsolete in the light of present day atomic theory. . . . a thorough, workable knowledge of the Periodic System and the relationship of this system to electron configurations of the various atoms is an absolute necessity for any student who aspires to attain any thorough understanding of chemical science."

Harry H. Sisler and Calvin A. Vanderwerf.
J. Chem. Educ. **20**, 479 a. 480 (**1943**).

Graphic Representations of the Periodic System During One Hundred Years

EDWARD G. MAZURS

THE UNIVERSITY OF ALABAMA PRESS
University, Alabama 35486

First Edition copyright © 1957 by
Edward G. Mazurs, under the title
   *Types of Graphic Representation of the
   Periodic System of Chemical Elements*

Revised (2nd) edition copyright © 1974 by
The University of Alabama Press

   ISBN 0-8173-3200-6
   Library of Congress Catalog Card Number—73- 8051

   All rights reserved

Manufactured in the United States of America

Contents

PREFACE	xi
1. PREHISTORY OF THE DISCOVERY OF THE PERIODIC SYSTEM	1
Law of Triads	1
Valence Tables	3
Law of Octaves	3
Forerunners of the Periodic Law	4
2. DISCOVERY OF THE PERIODIC LAW	6
D. I. Mendeleev and Lothar Meyer	6
Question of Priority	10
Portraits of Leading Men in Discovery	10
3. CLASSIFICATION OF PERIODIC TABLES ORIGINATED AFTER DISCOVERY OF THE LAW OF OCTAVES AND THE PERIODIC LAW	14
Basis of the Classification	14
DIVISION I. SHORT TABLES (WITH 8 COLUMNS)	18
Little periodic table: the lanthanides in two series	19
CLASS 1. TABLES WITH 8 GROUPS AND NO SUBGROUPS (LAW OF OCTAVES)	
Type 1: Tables where the Representative and Transition Elements are in the same Groups without Separation into Subgroups	20
Type 2: Tables where the Transition and the Inner Transition Elements are Separated from the Representative Elements in Special Tabulations	22
CLASS 2. TABLES WITH TWO SUBGROUPS "a" AND "b"	
Type 1: Symmetrical Chessboard-like Tables	26
Type 2: Symmetrical Tables with Bridge Elements	27
Type 3: Symmetrical Tables with the Elements of Short Periods Divided Symmetrically into Two Parts	30
Type 4: Unsymmetrical Tables with the Elements of Short Periods Placed in Subgroups "a" and "b"	31
Type 5: Unsymmetrical Tables with the Elements of the Short Periods Placed Exclusively in Subgroup "a"	34

CLASS 3. TABLES WITH THREE SUBGROUPS: "a," "b," AND "c"

    Type 1: Tables with the Inner Transition Elements in Two Series   38

    Type 2: Tables with the Inner Transition Elements in One Series   42

DIVISION II. MEDIUM TABLES (WITH 16 OR 18 COLUMNS)   46

CLASS 1. TABLES WITH 16 GROUPS

    Type 1: Helix and Spiral with Equal Revolutions, and Tables with Equal Series   48

    Type 2: Equal Lemniscate and Zigzag   50

CLASS 2. TABLES WITH DISPOSITION OF ELEMENTS: 2, 8, AND 18.

    Type 1: Step Tables with Group Zero on One Side of the Table   52

    Type 2: Lemniscates and Zigzags with Group Zero on One Side of the Table   55

    Type 3: Tables Symmetrical about a Vertical Line and with Group Zero on the Right Side of the Table   58

    Type 4: Tables with Interrupted Short Periods   60

    Type 5: Tables with Group Zero in the Middle of the Table   63

    Type 6: Step Tables with Group IIa on the Right Side of the Table   65

    Type 7: Mirror-Image Tables   66

DIVISION III. LONG TABLES (WITH 32 COLUMNS)   67

SUBDIVISION IIIA: CHEMICAL TABLES

CLASS 1. TABLES OF ONE REVOLUTION AND OF ONE ROW

    Type 1: Tables of One Revolution and of One Row   67

CLASS 2. TABLES WITH DISPOSITION OF ELEMENTS: 4, 16, 36, AND 64 (CYCLES)

    Type 1: Tables of Four Planes, Four Revolutions, or Four Cycles   72

    Type 2: Table of Four Lemniscates   74

CLASS 3. TABLES WITH DISPOSITION OF ELEMENTS: 2, 8, 18, AND 32 (PERIODS)

    Type 1: Step Tables with Group Zero on One Side of the Table   74

    Type 2: Lemniscates and Zigzags with Group Zero on One Side of the Table   79

    Type 3: Tables Symmetrical about a Vertical Line and with Group Zero on the Right Side of the Table   83

    Type 4: Tables with Interrupted Short and Medium Periods   87

    Type 5: Tables with Group Zero in the Middle of the Table   91

    Type 6: Step Tables with Group IIa on the Right Side of the Table   92

INTRODUCTION TO ELECTRONIC CONFIGURATION TABLES

    Basis in Atomic Structure   96
    Problem of Inner Transition Elements   105

SUBDIVISION IIIB: ELECTRONIC CONFIGURATION TABLES   108

CLASS 4. TABLES WITH ELECTRONIC CONFIGURATION DISPOSITION OF ELEMENTS: 2, 6, 10, AND 14 (BLOCKS OR SUBPERIODS)

    Type 1: Symmetrical Helices, Spirals, and Series Electronic Configuration Tables   108
    Type 2: Right-Side Electronic Configuration Tables   114
    Type 3: Left-Side Electronic Configuration Tables   119

CLASS 5. SHELL AND SUBSHELL TABLES WITH ELECTRONIC CONFIGURATION DISPOSITION OF ELEMENTS: 2, 6, 10, AND 14.

    Type 1: Symmetrical Tables of Concentric Circles and Parallel Lines   122
    Type 2: Right-Side Shell and Subshell Tables   128
    Type 3: Left-Side Shell and Subshell Tables   133

4. CONCLUSION   136

5. BIBLIOGRAPHY (IN ORDER OF TYPE NUMERATION; REFERENCES TO ARTICLES ABOUT TABLES NOT CLASSIFIED AND NOT FOUND IN LIBRARIES ARE INCLUDED)   144

APPENDICES   217

    I. Outline of the History of Discovery of the Periodic System and of the Classification of Types of Periodic Tables   217
    II. Element Blocks of which the Tables Consist   225
    III. Equations of the Periodic Table   226

ALPHABETICAL AUTHOR INDEX   228

Illustrations

TABLES OTHER THAN PERIODIC

1. Little Periodic Table	19
2. Authors who Originated Medium and Long Tables	46
3. Shells and Subshells in the Extranuclear Atomic Structure of a Hypothetical Element No. 120	98
4. Construction of the Periodic Table According to the Equation: $t = n + l$	100
5. Old Distribution of Subshells in Periods	102
6. Distribution of Subperiods in Periods According to the New Arrangement	103
7. The Lanthanide Completion with Electrons	106
8. The Actinide Completion with Electrons	107
9. Maximum Valence of Old VIIIb, Ib, and IIb Groups	107

PHOTOGRAPHS

1. J.B.A. Dumas (1800–1884)	11
2. A. É. Béguyer de Chancourtois (1820-1886)	11
3. W. Odling (1829–1921)	12
4. and 5. D. I. Mendeleev (1834–1907)	12
6. Ch. Janet (1849–1932)	141
7. L. M. Simmons (1905)	142
8. V. M. Klechkovskii (1900)	142

FIGURES

1. First Attempt of a Table	1
2-4. Law of Triads	1,2
5. Valence Table	2
6-8. Law of Octaves	3,4
9-11. Forerunners of Periodic Law	4,5
12-15. Mendeleev's Tables 1869	6,7
16. Lothar Meyer 1970	8
17,18. Mendeleev 1870 and 1871	9,10
19,20. Space and Plane Curves	15,16

SHORT TABLES

21-23. Tables of Law of Octaves	20,21
24-26. Octave Tables with Special Tabulations	23,24
27,28. Chessboard-like Tables	26,27
29-32. Tables with Bridge Elements	27-30
33. Symmetrical Table	30
34-36. Unsymmetrical Tables with Subgroups a and b for Short Periods	31-33

37-40.	Unsymmetrical Tables with Exclusively Subgroup *a* for Short Periods	34-37
41-45.	Tables with Three Subgroups and with Inner Transition Elements in Two Series	38-42
46-48.	Tables with Three Subgroups and with Inner Transition Elements in One Series	43-45

MEDIUM TABLES

49-51.	Tables with Equal Revolutions or Series with 16 Groups	48-50
52, 53.	Equal Lemniscate and Zigzag	51, 52
54-56.	Step Tables with Group Zero on One Side of the Table	53, 54
57-61.	Lemniscates and Zigzags with Group Zero on One Side of the Table	55-58
62-64.	Tables Symmetrical about a Vertical Line	58-60
65-67.	Tables with Interrupted Short Periods	61-63
68-70.	Tables with Group Zero in the Middle of the Table	64, 65
71.	Step Tables with Group IIa on the Right Side of the Table	65
72, 73.	Mirror-Image Tables	66

LONG TABLES

A. Chemical Tables

74-77.	Tables of One Revolution and of One Row	68-71
78-80.	Cycle Tables	72, 73
81.	Table of Four Lemniscates	74
82-85.	Step Tables with Group Zero on One Side of the Table	75-77
86-90.	Lemniscates and Zigzags with Group Zero on One Side of the Table	79-83
91-95.	Tables Symmetrical about a Vertical Line	83-87
96-98.	Tables with Interrupted Short and Medium Periods	88-90
99-102.	Tables with Group Zero in the Middle of the Table	91, 92
103-105.	Step Tables with Group IIa on the Right Side	93-95

B. Electron Configuration Tables

106, 107.	Schemes of Periodic Table	102
108-113.	Symmetrical Electronic Configuration Tables	109-113
114-118.	Right-Side Electronic Configuration Tables	114-118
119-122.	Left-Side Electronic Configuration Tables	119-121
123-129.	Symmetrical Shell and Subshell Tables	123-128
130-134.	Right-Side Shell and Subshell Tables	129-132
135-138.	Left-Side Shell and Subshell Tables	133-135

CONCLUSION

139.	Density Curve	138
140.	Melting Point Curve	139
141.	Atomic Radius Curve	139
142.	First Ionization Potential Curve	140
143.	Comparison of Periodic Tables	143

BACK MATTER

Colored periodic table subtype III C4–3c	*foldout*
Electronic energy level diagram	*foldout*
Colored periodic table subtype III C5–2d	*endpaper*

Preface

The importance of the Periodic System to science in general and to chemistry in particular is enormous. The gaps left in the first tables led directly to the discovery of many unknown elements: then there were sixty-three known elements, now there are one hundred four—forty-one new elements discovered in one hundred years.

Based on the gaps in the tables, Dmitrii Ivanovich Mendeleev, the discoverer of the Periodic System, predicted in 1871 the discovery of ten elements (Sc, Ga, Ge, Tc, Re, Po, Fr, Ra, Ac, and Pa) and fully described the properties of four of them (Sc, Ga, Ge, and Po). Five of them (Sc, Ga, Ge, Po, and Ac) were found in the nineteenth century. Mendeleev also alluded to the rare earth elements. Besides the elements predicted by Mendeleev, the eight lanthanides and six noble gases (He, Ne, Ar, Kr, Xe, and Rn) were also discovered in the nineteenth century.

The crowning achievement of the system is the continuation of the table due to the production, since 1940, of the twelve transuranium elements, eight of which have been synthesized by Glenn T. Seaborg and his associates. Some of these elements were important, of course, in the creation of the atomic bomb, but their greatest significance lies in their potential to produce energy for peaceful purposes.

The first edition of this book was published in 1957 as *Types of Graphic Representation of the Periodic System of Chemical Elements*. This second edition, under the title *Graphic Representations of the Periodic System During One Hundred Years*, is being published to commemorate the centennial of the discovery of the Periodic System by Mendeleev in 1869. Publication has been delayed, however, so that all the commemorative articles published in 1969 and 1970, and other recent items, could be examined and included. Many new tables have been published since the first edition, which also justifies the publication of the second edition since it was necessary to make some changes in the table classification.

The first two sections of this book—"Prehistory of the Discovery of the Periodic System" and "Discovery of the Periodic Law"—form the historical introduction. The third section, the main part of this book, is based on a survey and analysis of the approximately seven hundred periodic tables published during the past one hundred years. The number and variety of these charts represent the ability of the human mind to give disparate forms to the same body of matter. Often, however, the tables are similar or even identical. Classification of these corresponding tables as types and subtypes reduces the number to one hundred forty-six. The classification is arranged from the simpler, now obsolete, earlier charts to the more complex modern charts. It also shows the gradual evolution of the graphic presentation. The best developed tables are the electronic configuration tables. They are, so to speak, the center of gravity of this book. Therefore, there is a special introduction to Subdivision IIIB. The types and subtypes are designated by Roman numerals, capital letters, and Arabic numbers: IA1–1, for example. A more detailed explanation is given under Basis of Classification. One hundred sixteen drawings illustrate this section.

Although the previous edition of this book accepted the Gardner table as the best of the then published tables, our present Conclusion contains recommendations for three other tables that are more successful, including a newly constructed one. The physical property curves fit better into the frame of these tables; this property is published here for the first time.

The Bibliography contains lists of authors arranged under the same headings as the sections and the types and subtypes. These are the authors who worked with or discussed the particular table. Each section is numbered separately in chronological order with the numbers in parentheses: (1), (2), (3), etc. The originator of the table is designated as (1). Forerunners of the published tables are marked with zeros.

The Bibliography is as complete a listing as can be compiled. *Chemical Abstracts* and *Chemisches Zentralblatt* were used as sources, then the original articles were examined. Those articles that could not be found in libraries were not used. Chemistry textbooks were not considered unless they contain new or unusual tables. The Russian words are transliterated according to the rules accepted by the Library of Congress.

During the past hundred years, a number of scholars have made reviews of published periodic charts. A list of their contributions is included in the Bibliography section. The first man who did this kind of review was Venable (1) in the United States. His book, published in 1896, had an almost complete collection of the periodic tables published by that time. The next large collection was published in Russia in 1934 by Blokh (10) in order to honor the hundredth birthday of Mendeleev. Also in 1934, in the United States, Quam and Quam (11) published the collected charts in the *Journal of Chemical Education* and they tried to classify them. In the first edition of this book in 1957 (17) and in this second edition, I have made a more nearly complete collection of tables, classifying them on a different principle than Quam and Quam.

Here I would like to express my gratitude to Mr. Bertrand L. Chamberland (University of Pennsylvania, Philadelphia, Pa.) and Vasilii Ivanovich Semishin (Moscow, Russia), who provided me with new references; to the Academy of Sciences (Paris, France), the Russian Embassy (Washington, D.C.), Mr. Robert Janet (Neuilly s./S., France), and Dr. L. M. Simmons (Sydney, Australia), who kindly presented photographs; to the Library of Congress (Washington, D.C.) and the University of California at Santa Barbara, which offered me special services; and to chemist John R. Young (Westmont College, Santa Barbara, Calif.) for editing the language of the manuscript.

Santa Barbara, California
June, 1973.

E. MAZURS

**GRAPHIC REPRESENTATIONS
OF THE PERIODIC SYSTEM
DURING ONE HUNDRED YEARS**

1. Prehistory of the Discovery of the Periodic System

To our knowledge, the first attempt to draw a table of chemically simple substances was made in France by Louis Bernard Guyton de Morveau in 1782 (1). Five years later this table was repeated by a Commission of Chemical Nomenclature consisting of de Morveau, Antoine Laurent Lavoisier, Claude Louis Berthollet, and Antoine François de Fourcroy (2). The table is shown in Fig. 1, but the names of the substances have been changed to the present chemical symbols and formulas. Lavoisier used the same table in his textbook two years later (3).

Non Decomposable Substances.				
Simple Substances.	Simple Combustible Substances.	Metallic Substances.	Simple Earths.	Alkalies.
Light Heat O H	N C S P Cl B ·F etc.	As Fe Mo Sn W Pb Mn Cu Ni Hg Co Ag Bi Pt Sb Au Zn	SiO_2 Al_2O_3 BaO CaO MgO	NaOH KOH NH_4OH

FIG. 1. de Morveau 1782.

LAW OF TRIADS

The next step in the development of the tables of chemical elements was the discovery of the Law of Triads by Johann Wolfgang Döbereiner in 1829 (1a). He used the German word "Dreiheit" for the word "triad." Historians sometimes mention the year 1817 because in that year a letter by Wurzer about the work of Döbereiner and a letter from Döbereiner to Gilbert were published (1b), but

1	2	3	4	5
H	F	Se	P	B
	Cl	S	As	Si
	Br			
	I	O	N	C

FIG. 2. Dumas 1828.

neither contained an announcement of the triads of related elements. One year earlier, in 1828, Jean Baptiste Dumas had published a classification of related elements (0) but it included only the nonmetals (Fig. 2).

The triads of Döbereiner, which he described but did not draw, are illustrated in Fig. 3.

Li	Mg	Ca	Be	B	P	?	S	F	Cl
Na	?	Sr	Al	Si	As	Sb	Se	?	Br
K	?	Ba	?	?	?	Bi	Te	?	I
Y	Zr	Fe	Ni	Ru	Pt	Ag	?	?	?
Ce	Ti	Co	Cu	Rh	Ir	Pb	Sn	Au	W
?	Sn	Mn	Zn	Pd	Os	Hg	Cd	W	Ta

FIG. 3. Döbereiner 1829.

A well-known table of triads (Fig. 4), very similar to present tables, was published by Leopold Gmelin in 1843 (2). In this table the series are arranged downwards in order of increasing valence. On the left side of the table are the electronegative elements and on the right, the electropositive ones.

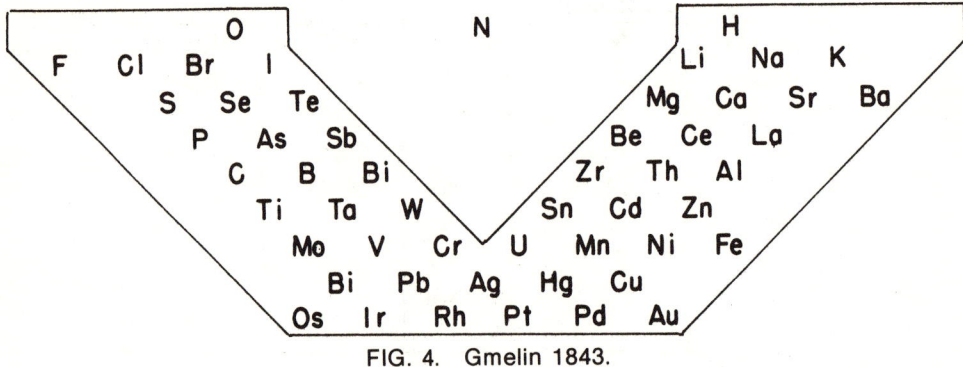

FIG. 4. Gmelin 1843.

Many authors have published articles with tables based on the Law of Triads (3–13). Among them Dumas excelled in popularizing the new ideas (5).

Valence:	4	6
	Ti	Mo
	Zr	V
	Ta	W

Valence:	4	3	2	1	1	2
Series.						
1.	-	-	-	-	Li	Be?
2.	C	N	O	F	Na	Mg
3.	Si	P	S	Cl	K	Ca
4.	-	As	Se	Br	Rb	Sr
5.	Sn	Sb	Te	I	Cs	Ba
6.	Pb	Bi	-	-	Tl?	-

Valence:	4	4	4	2	1
	Mn,Fe	Ni	Co	Zn	Cu
	Ru	Rh	Pd	Cd	Ag
	Pt	Ir	Os	Hg	Au

FIG. 5. Lothar Meyer 1864.

VALENCE TABLES

There have been authors who published tables of triads, tetrads, pentads (or more elements) with the chemical elements divided into groups according to their valences (1–7). The best of these tables was drawn by Lothar Meyer (1) (Fig. 5).

LAW OF OCTAVES

The tables demonstrating the Law of Octaves have 7 (or 8) vertical groups without separation into subgroups. A. É. Béguyer de Chancourtois, in 1862, was the first man to arrange the chemical elements in an order of increasing atomic weights and to find similar elements by this arrangement (1). De Chancourtois wrote also that the properties of substances depend upon the properties of numbers, but did not announce the Law of Octaves. De Chancourtois published many papers in *Comptes Rendus* in 1862 (1a), but no table was included. In the last published paper, he stated that a table had been lithographed and sent to the members of the Academy of Sciences of France. The original helix was published in 1863 (1b). There have been many adherents of de Chancourtois: the earlier ones published only a shortened version of his table (1c—1889, 1891, and 1895). In 1951, van Spronsen printed the whole original helix (1d). Actually, the table was drawn as the curve of an unrolled helix; and, therefore, the series of the table were oblique. The table is smaller if the series are drawn in horizontal lines (Fig. 6). On the abscissa of the table are the atomic weights from 1 to 16 (as is shown at the top of the table) or multiples of these figures. Seven main groups of similar elements are observable in this table.

1	2	3	4	5	6	7	8	9	10	11	12	13	14	15	16
H										Li		Be		B	C N
	F								Na Mg			Al Si			P S
	Cl								K Ca						Ti
			Cr						Mn Fe			Ni Co		Cu	
Zn			Zr									As		Br Se	
									Rb Sr			La Ce			Mo
		Di Y							Tl Rh			Pd Ag		Cd	
		Sn							Th U Sb			Cs		I	Te
									Ta W						
			Ir						Pt Au			Hg Ru			Os
Bi															

FIG. 6. de Chancourtois 1863.

1.	2.	3.	4.	5.	6.	7.
H 1	Li 2	Be 3	B 4	C 5	N 6	O 7
F 8	Na 9	Mg 10	Al 11	Si 12	P 13	S 14
Cl 15	K 16	Ca 17	Cr 19	Ti 18	Mn 20	Fe 21
Co,Ni 22	Cu 23	Zn 25	Y 24	In 26	As 27	Se 28
Br 29	Rb 30	Sr 31	Ce,La 33	Zr 32	Di,Mo 34	Rh,Ru 35
Pd 36	Ag 37	Cd 38	U 40	Sn 39	Sb 41	Te 43
I 42	Cs 44	Ba,V 45	Ta 46	W 47	Nb 48	Au 49
Pt,Ir 50	Tl 53	Pb 54	Th 56	Hg 52	Bi 55	Os 51

FIG. 7. Newlands 1865.

The discoverer of the Law of Octaves was John Alexander Reina Newlands. In 1864 he published a paper in which he wrote about octaves, but he did not give a complete table. Not until the next year, 1865, did Newlands announce his Law of Octaves (2) and present the table shown in Fig. 7. In this table, the elements are arranged in an order of increasing atomic weights, there are seven vertical groups of related elements, and the elements have ordinal numbers.

De Chancourtois was not very far from the discovery of the Law of Octaves. A comparison of the two tables (Fig. 6 and 7), after omitting the improperly placed symbols from both, shows the only difference to be that Newlands has more symbols in the right places than de Chancourtois (Fig. 8).

1	2	3	4	5	6	7
H	Li	Be	B	C	N	O
F	Na	Mg	Al	Si	P	S
Cl	K	Ca				
					As	Se
	Rb	Sr				Mo
						Te
Au						

1	2	3	4	5	6	7
H	Li	Be	B	C	N	O
F	Na	Mg	Al	Si	P	S
Cl	K	Ca		Ti		
	Cu	Zn			As	Se
Br	Rb	Sr		Zr		
	Ag	Cd			Sn	Sb Te
I	Cs	Ba				
					Bi	

FIG. 8. de Chancourtois 1863 and Newlands 1865.

FORERUNNERS OF THE PERIODIC LAW

The Periodic Law states that the properties of the elements repeat themselves periodically with increasing atomic weight (or number). The forerunners of the Periodic Law were the three authors who unknowingly drew tables according to this Law but did not announce the Law.

The first was William Odling (1) who, in 1864, drew an almost correct type of medium table with 18 vertical groups (Fig. 9). The elements are arranged in an order of increasing atomic weights.

Present groups:	8	1	2	1	2	3	4	5	6	7	1	2	4	-	6	7	8			1
		H	-	Li	Be	B	C	N	O	F	Na	Mg								
		-	-	-	-	Al	Si	P	S	Cl	K	Ca	Ti	-	Cr	Mn	Fe	Co	Ni	Cu
		-	Zn	-	-	-	-	As	Se	Br	Rb	Sr	Zr	Ce	Mo					
	Rh Ru Pd	Ag	Cd	-	-	U	Sn	Sb	Te	I	Cs	Ba	Ta	-	V	W				
	Pt Ir Os	Au	Hg	Tl	Pb	-	-	Bi	-	-	-	-	Th							

FIG. 9. Odling 1864.

PREHISTORY OF THE DISCOVERY OF THE PERIODIC SYSTEM 5

The second forerunner was Gustavus Detlef Hinrichs in the United States (2). His first table, published in 1867, was a spiral. The second table, drawn in the form of a series table and published in 1869, does not differ significantly from the spiral. This table is shown in Fig. 10 with arrows added to show the right order of element symbols. In this table, Hinrichs separated the nonmetals from the metals with a stepwise line, just as it is done today.

Pantoids	Kaloids	Calcoids	Cadmoids	Hydrargoids	Cuproids	Ferroids					Titanoids	Phosphoids	Sulphoids	Chloroids	Pantoids
															H
→ H	Li	-	-	-	-						C	N	O	F	
→ Na	-	Mg	-	-	Al					Si	P	S	Cl		
→ K	Ca														
		Zn	-	Cu	Co	Ni	Fe	Mn	Cr	Ti ←					
									→	As	Se	Br			
→ Rb	Sr														
		Cd	-	Ag		Rh				Pd ←					
								→	Sn	Sb	Te	I			
→	Ba														
		Pb	Hg	Au		Ir				Pt ←					
								→		Bi	-	-			

FIG. 10. Hinrichs (1867 and) 1869.

In 1868 the third forerunner, Lothar Meyer, drew a periodic table (Fig. 11) but unfortunately did not publish it. The table was published by Karl Seubert (3) only after Meyer's death in 1895. Since Meyer did not publish this table before or at the discovery of the Periodic Law in 1869 by Mendeleev, he cannot be regarded as a co-discoverer of the Periodic Law.

1	2	3	4	5	6	7	8	9	10	11	12	13	14	15
											Li	Be		
		Al	Al				C	N	O	F	Na	Mg		
							Si	P	S	Cl	K	Ca	Ti	Mo
Cr	Mn	Fe	Co	Ni	Cu	Zn	-	As	Se	Br	Rb	Sr	Zr	V
-	Ru	Rh	Pd	-	Ag	Cd	Sn	Sb	Te	I	Cs	Ba	Ta	W
-	Pt	Ir	Os	-	Au	Hg	Pb	Bi	-	-	Tl?			

FIG. 11. Lothar Meyer (drawn in 1868, published in 1895).

2. Discovery of the Periodic Law

D. I. MENDELEEV AND LOTHAR MEYER

The Periodic Law was discovered in 1869 by the Russian chemist Dmitrii Ivanovich Mendeleev. In March 1869, a paper by Mendeleev was read before the Russian Chemical Society in St. Petersburg by N.A. Menshutkin because Mendeleev himself was ill at the time. This paper was published in the *Journal of the Russian Chemical Society* in the same year (1a). Mendeleev repeated the paper and the drawing of the main table in two German journals (1b). Mendeleev's paper, in addition to the main table, contained suggestions and written fragments for three other types of tables. Paul Walden, a biographer of Mendeleev, mentioned in *Berichte* in 1908 (2) that Mendeleev had many kinds of tables; however, nobody tried to reconstruct them. In a recent publication, Kedrov explained a few of the different forms of tables discussed by Mendeleev (3). The following tables are an attempt to reconstruct tables from Mendeleev's fragments.

The main table (Fig. 12—1a, p. 70) belongs to the medium tables with 18 vertical columns, if the few elements written on the right side of the table are omitted; and to the long tables with 32 vertical columns, if the inner transition elements on the right side are written completely (types IIC2-6 and IIIC3-5).

1.								H					Li					
2.							Be	B	C	N	O	F	Na					
3.							Mg	Al	Si	P	S	Cl	K	Ca	-	Er?	Y?	In?
4.| Ti	V	Cr	Mn	Fe	Ni,Co	Cu	Zn	-	-	As	Se	Br	Rb	Sr	Ce	La	Di	Tb
5.| Zr	Nb	Mo	Rh	Ru	Pd	Ag	Cd	U	Sn	Sb	Te	I	Cs	Ba				
6.| -	Ta	W	Pt	Ir	Os	Hg	-	Au	-	Bi	-	-	Tl	Pb				

FIG. 12. Mendeleev 1869.

In a footnote (1a, p. 69), Mendeleev discussed a table that he said would not be a convenient one (Fig. 13). This table is particularly popular today (type IIC–4).

Li										Be	B	C	N	O	F
Na										Mg	Al	Si	P	S	Cl
K	Ca	-	Ti	V	Cr	Mn	Fe	Ni,Co	Cu	Zn	-	-	As	Se	Br
Rb	Sr	-	Zr	Nb	Mo	Rh	Ru	Pd	Ag	Cd	U	Sn	Sb	Te	I
Cs	Ba	-	-	Ta	W	Pt	Ir	Os	Hg	-	Au	-	Bi	-	-
Tl	Pb	-													

FIG. 13. Mendeleev 1869.

In another footnote (1a, p. 71), Mendeleev gave a description of the table shown in Fig. 14 (type IIC1-1). He completed this table himself in 1871 in the second volume of his textbook *Principles of Chemistry* (4).

Li	Be	B	C	N	O	F	Na	Mg	Al	Si	P	S	Cl
K	Ca	-	Ti	V	-	-	Cu	Zn	-	-	As	Se	Br
Rb	Sr	-	Zr	Nb	-	-	Ag	Cd	U	Sn	Sb	Te	I
Cs	Ba	-	-	Ta	-	-	Hg	-	Au	-	Bi	-	-

FIG. 14. Mendeleev 1869.

In a third footnote (1a, p. 70), Mendeleev suggested a helix or zigzag table, fragments of which he put down. The reconstructed table is presented in Fig. 15. The helix was published by B. K. Emerson in 1911 (type IIA2-1), the zigzag by Deeley in 1893 (subtype IIC2-2a).

1.	Li	Be	B	C	N	O	F			
2.	Na	Mg	Al	Si	P	S	Cl			
3.	K	Ca	-	Ti	V			Cr	Mn	Fe Ni,Co
4.	Cu	Zn	-	-	As	Se	Br			
5.	Rb	Sr	-	Zr	Nb			Mo	Rh	Ru Pd
6.	Ag	Cd	U	Sn	Sb	Te	I			
7.	Cs	Ba	-	-	-	-	-			
8.	-	-	-	-	Ta	W		Au	Pt	Os Ir,Hg?
9.	Tl	Pb	Bi?	-	-	-	-			

FIG. 15. Mendeleev 1869.

In connection with the last table, Mendeleev talked about odd and even series of elements. In effect, this table anticipated the short table with subgroups (type IC2-1), which was published the following year, 1870 (10c).

The publication of his article (1a) and table (Fig. 12) in 1869 established Mendeleev as the discoverer of the Periodic Law.

In 1869 Mendeleev read two more papers: one in August before the Congress of Naturalists and Physicians in Moscow, which was published in 1870 in the Transactions of this Congress (5); and the second in October before the Russian Chemical Society, also published in 1870 in the Journal of this Society (6a). However, in 1869 von Richter had referred to the former paper (then unpublished) in *Berichte* (6b). This first paper included a discussion of the atomic volumes of elements in the arrangement of the periodic table. In a remark in this paper, Mendeleev mentioned Lothar Meyer's article with the atomic volume curve, which had been published earlier in 1870 (8). Mendeleev's second paper discussed oxides and hydrogen compounds in connection with the vertical groups of the table. Both papers included the chart (Fig. 15) and emphasized the difference between the elements of even and odd series in the table.

Also in 1869, the first edition of Mendeleev's textbook *Principles of Chemistry* was published (7). This book contained the medium-size table with 18 columns published earlier in the *Journal of the Russian Chemical Society* (Fig. 12).

The forerunner of the Periodic System, Lothar Meyer, wrote an article in December 1869 which was published along with a table (Fig. 16) in 1870 (8).

Series.							
1.						Li	Be
2.	B	C	N	O	F	Na	Mg
3.	Al	Si	P	S	Cl	K	Ca
4.	-	Ti	V	Cr	Mn Fe Co,Ni	Cu	Zn
5.	-	-	As	Se	Br	Rb	Sr
6.	-	Zr	Nb	Mo	Ru Rh Pd	Ag	Cd
7.	In	Sn	Sb	Te	I	Cs	Ba
8.	-	-	Ta	W	Os Ir Pt	Au	Hg
9.	Tl	Pb	Bi	-	-	-	-

FIG. 16. Lothar Meyer 1870.

In this table, the element symbols of the even series are moved to one side, although no real arrangement into subgroups is shown. This table can be regarded as a prototype of Mendeleev's short table with 8 groups, published by von Richter in December 1870 (10c). In this connection, some scientists regard Meyer as a co-discoverer of the Periodic System, although at that time Meyer did not feel himself to be a co-discoverer. This is clear from the following quotation from his article (8, p. 355):

> Recently Mendeleev showed [*Z. f. Chem.* **12**, 405 (**1869**)] that this kind of arrangement is obtained if the atomic weights of all elements are ordered in one series not by some arbitrary choice, but simply according to the values of their figures. This series is then divided into sections and these sections are added to each other in unchanged order. The following table is essentially identical to that given by Mendeleev.°

Actually Meyer's table belongs to a very original subtype (IC2–5a) that is related to the modern tables. It was in this article that Meyer gave the well-known curve of atomic volumes.

In 1870 another author, Heinrich Baumhauer, wrote a book (9) about the connections between the atomic weights and properties of chemical elements. He gave a spiral table belonging to the type IB2–4. He mentioned both Mendeleev's first article of 1869 (1a) and Meyer's article of 1870 (8).

In November 1870, Mendeleev wrote an article that was published the next year in the *Bulletin of the Russian Academy of Sciences* in St. Petersburg in German

°Vor kurzem hat Mendelejeff gezeigt [*Z. f. Chem.* **12**, 405 (**1869**)],dass man eine solche Anordnung schon dadurch erhält, dass man die Atomgewichte aller Elemente ohne willkürliche Auswahl einfach nach der Grösse ihrer Zahlenwerthe in eine einzige Reihe ordnet, diese Reihe in Abschnitte zerlegt und diese in ungeänderter Folge an einander fügt. Die nachstehende Tabelle ist im Wesentlichen identisch mit der von Mendelejeff gegebenen.

(10a). This publication was reprinted in the same language in another journal of this Academy in 1872 (10b). In this article, Mendeleev discussed the position of In, Ce, and U in the first periodic table, suggested a change of their atomic weights, and arranged them in their correct positions in the enclosed table (Fig. 17). However, Mendeleev's table had been published earlier, in December 1870, in the German *Berichte* by von Richter (10c).

Groups:	I	II	III	IV	V	VI	VII	VIII
Typical elements:	H Li	Be	B	C	N	O	F	
Periods / Series								
1. — 1.	Na	Mg	Al	Si	P	S	Cl	
1. — 2.	K	Ca	-	Ti	V	Cr	Mn	Fe Co Ni Cu
2. — 3.	(Cu)	Zn	-	-	As	Se	Br	
2. — 4.	Rb	Sr	Y?	Zr	Nb	Mo	-	Ru Rh Pd Ag
3. — 5.	(Ag)	Cd	In	Sn	Sb	Te	I	
3. — 6.	Cs	Ba	-	Ce	-	-	-	- - - -
4. — 7.	-	-	-	-	-	-	-	
4. — 8.	-	-	-	-	Ta	W	-	Os Ir Pt Au
5. — 9.	(Au)	Hg	Tl	Pb	Bi	-	-	
5. — 10.	-	-	-	Th	-	U	-	

FIG. 17. Mendeleev 1870.

Also in December 1870, Mendeleev read his third paper before the Russian Chemical Society, and this paper was published in the *Journal of the Russian Chemical Society* the following year (11). Here the same table was presented (Fig. 17). This table is the classical short Mendeleev table (type IC2–1) containing 8 vertical groups with subgroups, chessboard-like. In this table, the elements of the zero series are the prototypical elements (Mendeleev named them "typical"). The periods, all equal (with 17 elements), begin with the elements Na, Cu, Ag, and Au, and end with iron and platinum metals (group VIII). The period numeration is different from that used now. Although Mendeleev later changed the numeration, it still does not agree with that currently accepted. The short table was popular for a long time; it was changed in different ways and is still used in a changed form. In this paper (11), Mendeleev pointed out that by means of the periodic table it is possible to correct atomic weights. For the first time, he predicted the discovery of many elements and described their possible properties and position in the periodic table. The elements later discovered were Sc, Ga, Ge, Po, Fr, Ra, Ac, and Pa. He also suspected the possibility of a series of elements that were later found as rare earth elements.

Mendeleev published the fundamental article about the Periodic System with the explanation of its possible application in 1871 in German (12a). Here Mendeleev presented two tables: one medium size (Fig. 18) and the other short size (Fig. 17).

									H							
									Li	Be	B	C	N	O	F	
									Na	Mg	Al	Si	P	S	Cl	
K	Ca	-	Ti	V	Cr	Mn	Fe	Co	Ni	Cu	Zn	-	-	As	Se	Br
Rb	Sr	Y?	Zr	Nb	Mo	-	Ru	Rh	Pd	Ag	Cd	In	Sn	Sb	Te	I
Cs	Ba	Di?	Ce	-	-	-	-	-	-	-	-	-	-	-	-	-
-	-	Er?	La?	Ta	W	-	Os	Ir	Pt	Au	Hg	Tl	Pb	Bi	-	-
-	-	-	Th	-	U	-	-	-	-	-	-	-	-	-	-	-

FIG. 18. Mendeleev 1871.

This medium-size table (Fig. 18) differs from the first one published in 1869 (Fig. 12). The table is divided into short and medium periods, each period starting with an alkali metal and ending with a nonmetal (type IIC2–1). This important article by Mendeleev was published later in English, shortened in 1877 (12b) and complete in 1879/80 (12c), and in French in 1879 (12d).

As has been shown in this historical survey, the most significant contributions to the discovery of the Periodic System were made by Jean Baptiste Dumas and Dmitrii Ivanovich Mendeleev. Dumas popularized the idea of the relationship of chemical elements in papers in 1828, 1851, 1857, and 1859. Even Mendeleev was acquainted with and inspired by Dumas' ideas. The particular importance of Mendeleev was that he not only defined clearly the periodicity in relationship of elements and predicted new elements, but also originated many types of periodic tables: in 1869—IIIC3–5 (Mendeleev's main table Fig. 99), IIC2–6 (Fig. 71), IIC2–4 (Fig. 67), and IIC1–1 (Fig. 51); in 1870—IC2–1 (Fig. 28); in 1871—IIC2–1 (Fig. 56); and in 1889—IIIC1–1a (Fig. 75). Lothar Meyer, the forerunner of the discovery of the Periodic System, originated and published only one type, IC2–5a (Fig. 40), in 1870. Type IIC2–6, originated by him in 1868, was not published, but was originated and published by Mendeleev in the following year.

QUESTION OF PRIORITY

The question of priority caused polemics between Meyer and Mendeleev. Newlands also claimed priority. The main articles of the polemics are listed in the Bibliography section.

PORTRAITS OF LEADING MEN IN DISCOVERY

The following are portraits of the four men who were the first to formulate new concepts. The first to publish the triads of chemical elements was Jean Baptiste André Dumas in 1828; the first to find the octaves of elements was Alexandre Émile Béguyer de Chancourtois in 1863; the first forerunner of the Periodic Law, with an almost complete periodic table, was William Odling in 1864; and, finally, the first scientist to announce the Periodic Law was Dmitrii Ivanovich Mendeleev in 1869.

Jean Baptiste André Dumas (1800–1884) began his studies in Geneva and continued in Paris. He was a professor of chemistry at the Athenaeum and the Sorbonne in Paris.

Alexandre Émile Béguyer de Chancourtois (1820–1886) was a professor of geology at the École des Mines in Paris.

12 GRAPHIC REPRESENTATIONS OF THE PERIODIC SYSTEM

William Odling (1829–1921) was born in England. He studied medicine as well as chemistry and from 1850–1868 taught chemistry in Hospitals. He was a professor of chemistry at the Royal Institute from 1868–1872, and at Oxford from 1872–1912.

Mendeleev at the discovery of the Periodic System.

Mendeleev in later years.

Dmitrii Ivanovich Mendeleev (1834–1907) was a professor of chemistry at the University of St. Petersburg (1857–1890). He was the author of the chemistry textbook "Osnovy Khimii" (*Principles of Chemistry*), published in eight editions (1869–1905). In 1890, he resigned as a professor, and the last years of his life were spent as a director of the Main Institution of Measures and Weights.

3. Classification of Periodic Tables Originated after Discovery of the Law of Octaves and the Periodic Law

BASIS OF THE CLASSIFICATION

The Periodic System of chemical elements was discovered by Mendeleev when he arranged the elements in an order of increasing atomic weights. In 1913, H.G.J. Moseley, analysing the atomic x-ray spectra, found the atomic numbers, which are used with better success than the atomic weights. The now generally accepted knowledge that the properties of the elements depend on the extranuclear electron configuration of their atoms has established the modern electronic configuration periodic tables, which are described in class 4 of Division III, as the most accurate reflectors of the nature of the chemical elements in a graphic representation.

As a law of nature, the Periodic System is a single entity and does not change with time; but the tables of the System, which try to reflect this Law, have changed as scientists developed them into better forms. In this study, all the tables from the discovery of the Law of Octaves to the present are reviewed.

Since many tables are structurally identical or at least similar, the total number of collected tables can be reduced to 146 different forms of periodic charts. These different forms of tables are classified according to the method described in the following paragraphs.

The Periodic System contains periods of different length: besides a foreperiod of 2 elements (H and He), there are 2 short periods with 8 elements, 2 medium periods with 18 elements, and 2 long periods with 32 elements. In recent times, some authors have combined the two periods with equal lengths into one cycle and, therefore, have obtained four cycles. The first cycle contains the zero period, with "e" (electron) and "n" (neutron), and the first period, with H and He. The second cycle contains the short periods (2nd and 3rd); the third cycle, the medium periods (4th and 5th); and the fourth cycle, the long periods (6th and 7th).

According to this, the tables are divided first into three large divisions by their length (the proper word would be width): short, medium, and long tables. The short tables possess 8 vertical groups (columns) of related elements, each group usually being divided into two subgroups, "a" and "b." The medium tables consist of 18 vertical groups without separation into subgroups. Neither the short nor the medium tables contain the inner transition elements (lanthanides and actinides). Short tables containing these elements have been published only recently. The long tables have 32 vertical groups with the inner transition elements included.

Second, the divisions are divided into classes of tables related by structure.

Third, the classes are divided into types that combine those individual tables having almost no structural differences. Tables are listed as subtypes when their structural difference in comparison with the basic type is small but noticeable.

Besides classification according to length, types of tables may be classified according to their dimensional graphic representation: 1) as a space curve, 2) as a top view of this space curve, or 3) as an elevation of this space curve.

1) The *space curves* in three dimensions (Fig. 19) usually are: helices on a cylinder, like a "screw" line (A); helices on a cone, like a "snail" line (B); space lemniscates (C); space concentric circles (D), and also space squares. The symbols of the elements are placed on the three-dimensional curves as points, and the generating lines of the cylinder or cone connect the related elements. In series tables, these are the elements of the same vertical group.

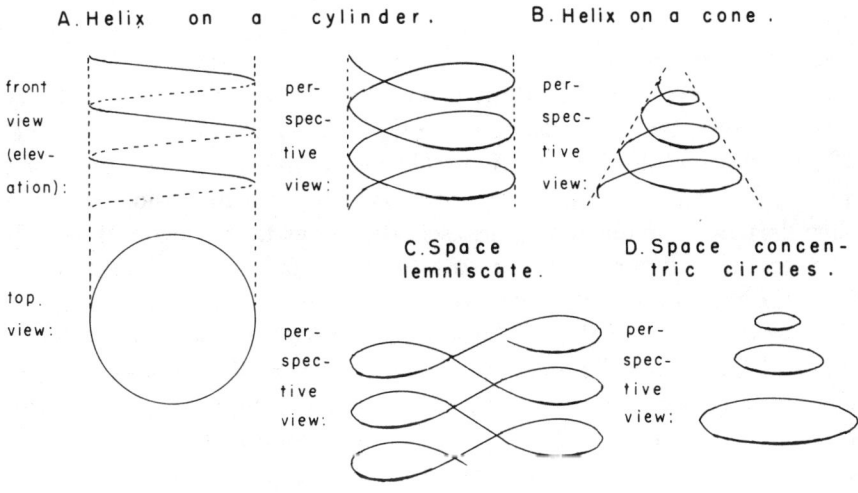

FIG. 19. Space Curves.

2) The *top views* (projections) of the three-dimensional curves are two-dimensional curves on a plane (Fig. 20). The top view of a helix on a cylinder gives only a circle; but drawn like a polar projection, it gives a spiral (A). The usual projection of a helix on a cone is also a spiral. The projection of a space lemniscate is a plane lemniscate, which could be drawn like a polar projection (B). The top view of space concentric circles gives plane concentric circles (C). The generating lines of cylinders or cone turn into radii of the spirals. On these radii, the related elements lie.

16 GRAPHIC REPRESENTATIONS OF THE PERIODIC SYSTEM

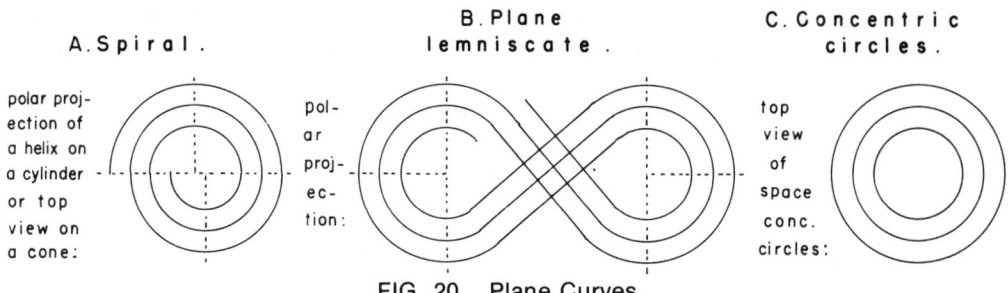

FIG. 20. Plane Curves.

3) The *elevations* of the space curves actually give zigzag lines. In zigzag tables, the symbols of elements have to be read in both directions, but if the zigzags are broken and ordered so that the symbols on the lines can be read only in one direction, then the usual series tables are obtained. Another method of obtaining a typical series table from a space curve would be to cut the space curve (e.g. helix) along the generating line of the body (e.g. cylinder) on which the curve is wound, and then the surface of the body is flattened. In the case of a series table, the related elements lie in vertical columns (groups).

The space curves in three dimensions best represent in graphs the types of the tables because the symbols of elements usually are arranged on the curve at equal intervals without interruptions. Therefore, the space curves can be considered the basis of a type from which the curves in two dimensions and the series tables are derived. For teaching purposes, the series tables are preferred because the three- and two-dimensional curves often are harder to read and to draw.

Thus, the tables are classified into types, the types are combined into classes, and the classes are combined into divisions. The three divisions of the tables—short, medium, and long—are denoted by means of the Roman figures I, II, and III, respectively. The dimensional representation of the type, either a space curve (helix) or a plane curve (spiral) or a series table, is denoted by means of the capital letters A, B, and C, respectively. The classes of tables are denoted by means of Arabic figures. The types of tables are also denoted by Arabic figures after a dash. The subtypes are denoted by small letters. For example, in the denotation "IIIC4-3c," the "III" means the division of long tables, "C" means that the table is drawn as a series table, "4" means the class of tables with a certain disposition of elements (2, 6, 10, and 14), "3" means the type of left-side electronic configuration table, and, finally, "c" means that the table is constructed according to the "differentiating" electrons of the elements. This table is shown in Fig. 122 and also as a separate color enclosure.

As was mentioned above, 146 different forms of periodic charts can be distinguished. This is the case if the charts originated in different dimensions are considered as separate ones. Very often all three dimensional representations (helix, spiral, and series table) can be found in the literature, usually originated by different authors. But at other times, the series table can be found, but no helix nor spiral. If the charts constructed on the same basis, although in different dimensions, are considered as one type, then we obtain only 33 different types. Deriving from these basic types the so-called subtypes, tables with little structural differences, we obtain 66 subtypes more. Altogether we have 99 types and subtypes. (See Appendix I: "Outlines of the Classification of the Types of Periodic Tables.")

The tables, especially the long ones and some of the medium tables, are divided into large vertical blocks. These blocks are not included as unities in the classification; however, they are mentioned throughout the book. They are:
1) the block of representative elements (the alkali metals, alkaline earth metals, nonmetals, and metals of nonmetallic groups) with 8 elements in each series found in all periods;
2) the block of transition (or related) elements (metals) with 10 elements in each series found in medium and long periods;
3) the block of inner transition (or similar) elements with 14 elements in each series.

The inner transition elements are located only in the long periods (6th and 7th): they are the lanthanides (the former rare earth elements) and the actinides (the former radioactive elements). Often, four blocks of elements are distinguishable when the block of representative elements is split into two blocks consisting of the "alkaline metals" and the nonmetal-group elements.

For comparative purposes, drawings of the different types of tables are presented, sometimes in all three dimensions. Drawings of some subtypes are not included if little difference exists between the subtype and the basic type. For the purpose of comparison, all these drawings of types and subtypes give all the elements known up to this time and are done according to the present status of the science. The position of the lanthanides and actinides is discussed in the Introduction to Electronic Configuration Tables (Subdivision IIIB). For better comparison, each table that was vertical in the original drawing is changed to a horizontal one so that each table can be read from left to right, but not from top to bottom. Since the drawings represent the types and subtypes of the periodic tables, the drawings are not intended to match exactly the originators' or other authors' drawings. The peculiarity and purpose of this classification is to combine similar tables suggested by different authors into one type, even if they differ slightly. Of course, such a coalescence of drawings reflects subjective judgments. Also, a criticism of some published tables is included under the headings of the types and subtypes.

Division I. Short Tables

Short tables are of three classes:
1) tables without subgroups (tables drawn according to the Law of Octaves).
2) tables with two subgroups.
3) tables with three subgroups.

The tables of *class 1* represent the Law of Octaves; therefore, they are obsolete. An exception is type 2 of this class, which shows the transition and inner transition elements in special tabulations. Therefore, the tables of this type can be used, and Blanshard's table (type IC1–2) is important.

To *class 2* belong five types of tables that differ among themselves.

Type 1 is similar to a chessboard and was originated by Mendeleev in 1870. Here, Na is placed in one subgroup with Cu, Ag, and Au, an arrangement not accepted today.

To type 2 belong tables with "bridge" elements, meaning that the elements of the first two short periods are placed between the "a" and "b" subgroups of the medium periods. This assumes that their properties are intermediate between these subgroups, a supposition known to be false.

In the type 3 table, the elements of the short periods are placed symmetrically. This arrangement does not provide a good table.

To types 4 and 5 belong the tables that have been used until recent times. The arrangement of the tables in these two types has caused disagreements among scientists centering around the question of which subgroups should be used to place the elements of short periods. The principle of arrangement of type 4 is that the first ten elements of the medium periods (K, Ca, . . . Fe, Co, Ni) are placed in subgroups "a" and the next eight elements (Cu, Zn, . . . Se, Br, Kr) in subgroups "b." In this case, the placement of the short period elements is split: I and II groups are placed in subgroups "a" and III–VIII in subgroups "b." Actually, these kinds of tables do not have a sound basis of arrangement and should be dropped, but many scientists still use them.

In my opinion, only type 5 (IC2–5) has the proper arrangement of subgroups. The basis of this arrangement is that all the short period elements are representative elements and must be placed in subgroups "a". Naturally, this causes the representative elements of the medium periods also to be placed in subgroups "a" while the transition elements are placed in subgroups "b." The table originated by Lothar Meyer in 1870 belongs to this type (subtype IC2–5a).

Tables of *class 3* with three subgroups have been developed in recent times. By using the short tables with two subgroups, the elements of subgroups "a" and "b" can be compared; i.e., the representative and transition elements. However, the properties of these elements are not always similar, especially when compared to the subgroups of group I—Li, Na, K, Rb, Cs, Fr with Cu, Ag, Au; or group VII—F, Cl, Br, I, At with Mn, Tc, Re. Only by basing the arrangement on valence is it possible to place these elements in the same groups. The transition elements are arranged in the same groups as the representative elements only conventionally, since a similarity in the properties of the elements does not always exist. Nevertheless, a third subgroup in the short tables is arranged where the inner transition elements are placed.

Two types of tables of this class with three subgroups are known. Type 1 contains the inner transition elements arranged in two series and type 2 arranged in one series. Murashov's table belonging to type 2 (subtype IC3–2a) is a good one.

Little periodic table: the lanthanides in two series.

The arrangement of the lanthanides into two series (the "little" periodic table) arises from Brauner (2) and was popularized especially by Klemm (9) (Table 1).

Table 1.

La	Ce	Pr	Nd	Pm	Sm	Eu
Gd	Tb	Dy	Ho	Er	Tm	Yb

20 GRAPHIC REPRESENTATIONS OF THE PERIODIC SYSTEM

CLASS 1. TABLES WITH 8 GROUPS AND NO SUBGROUPS (LAW OF OCTAVES)

Type 1: Tables where the Representative and Transition Elements are in the same Groups without Separation into Subgroups

Type IA1–1

Helix drawn on one cylinder (Fig. 21). This type was originated by de Chancourtois in 1863 (1). The helix is obsolete as is the Law of Octaves itself. Although Elena Bogdan (4) placed her helix on a space figure similar to a paraboloid, in actuality it belongs to this type of helix.

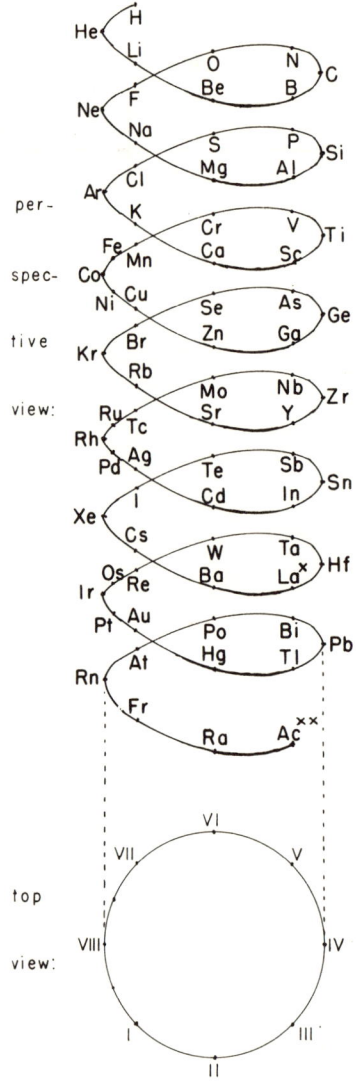

FIG. 21. de Chancourtois 1863. Type IA1–1.

DIVISION I. SHORT TABLES   21

Type IB1-1

Spiral with 8 radii. This spiral (Fig. 22), originated by Wallin in 1926 (1), is obsolete since it belongs to the Law of Octaves. One year earlier, Kurbatov (0) had made a double spiral in two directions; the turning point was at the lanthanides.

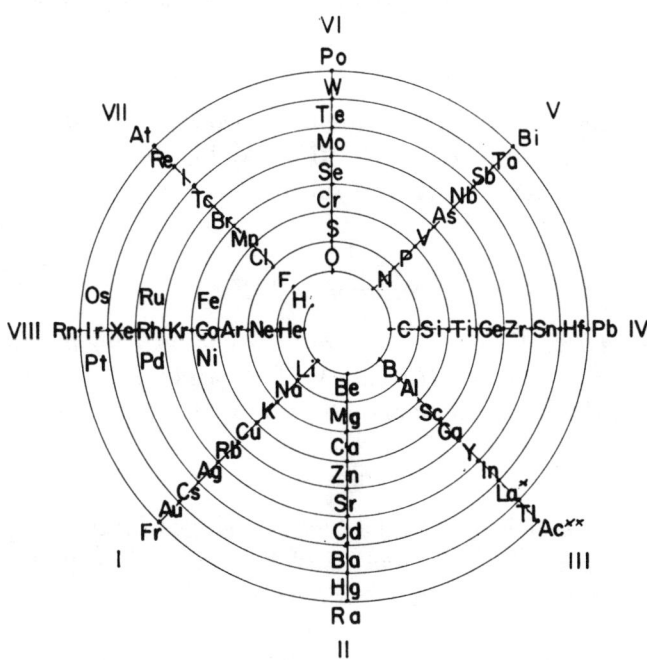

FIG. 22.   Wallin 1926.   Type IB1-1.

Type IC1-1

Series table containing 8 groups and no subgroups and beginning with group I (Fig. 23). The originator was Newlands in 1865 (1). This series table is also obsolete since it belongs to the Law of Octaves.

Groups / Series:	I	II	III	IV	V	VI	VII	VIII
1.							H	He
2.	Li	Be	B	C	N	O	F	Ne
3.	Na	Mg	Al	Si	P	S	Cl	Ar
4.	K	Ca	Sc	Ti	V	Cr	Mn	Fe Co Ni
5.	Cu	Zn	Ga	Ge	As	Se	Br	Kr
6.	Rb	Sr	Y	Zr	Nb	Mo	Tc	Ru Rh Pd
7.	Ag	Cd	In	Sn	Sb	Te	I	Xe
8.	Cs	Ba	La^x	Hf	Ta	W	Re	Os Ir Pt
9.	Au	Hg	Tl	Pb	Bi	Po	At	Rn
10.	Fr	Ra	Ac^{xx}					

La^x = lanthanides;   Ac^{xx} = actinides.

FIG. 23.   Newlands 1865.   Type IC1-1.

Subtype IC1-1a

Series table beginning with group IV. Instead of group numerations, the valences are shown starting with the valence −4. This table was originated by Lewis R. Gibbes in 1875 (1). Kohlweiler (3) drew this table in three dimensions, but three dimensions do not play a role in the construction of the table (it is only a form of presenting the table); therefore, his table belongs to this subtype.

Subtype IA1-1b

Helix which represents approximately the subtype IC1-1b. The elements of group VIII are placed into groups II, III, and IV. No advantage is gained by this change in the Octave Law. The subtype was originated by Oppegaard in 1948 (1).

Subtype IC1-1b

Series table where the Law of Octaves is changed by placing the elements of group VIII into groups II, III, and IV. The originator was Reed in 1885 (1). Preyer (2) placed the same elements into groups II, IV, and VI.

Subtype IC1-1c

Zigzag table in which the Octave Law is changed by dividing one series into two sections of one zigzag. One section of the zigzag has groups I, II, III, and IV. The other section uses the numbering of the first section but in descending order, namely, III, II, I, and 0 for the groups V, VI, VII, and VIII. The table was originated by Nechaev (1) in 1894. Goldhammer (2) presented a similar idea, although a little different from Nechaev's.

Subtype IC1-1d

Series table according to the Law of Octaves where the inner transition elements are arranged in the same groups as the representative and transition elements. This table was published by H. Biltz and W. Biltz in 1928 (1).

Type 2: Tables where the Transition and Inner Transition Elements are Separated from the Representative Elements in Special Tabulations.

Type IC1-2

Series table containing only the representative elements and beginning with group I, with supplementary tabulations for the transition and inner transition elements (Fig. 24). The originator of this type was Blanshard in 1895 (1), and this type is still usable. Petrovici (7) drew the inner transition elements in two rows in a zigzag manner. Sanderson (8) used lines to connect the transition elements with the representative elements and the inner transition elements with the transition elements.

DIVISION I. SHORT TABLES

Representative Elements

Groups\Periods	I a	II a		III a	IV a	V a	VI a	VII a	VIII a
1.	H								He
2.	Li	Be		B	C	N	O	F	Ne
3.	Na	Mg		Al	Si	P	S	Cl	Ar
4.	K	Ca		Ga	Ge	As	Se	Br	Kr
5.	Rb	Sr	1) {	In	Sn	Sb	Te	I	Xe
6.	Cs	Ba	2)	Tl	Pb	Bi	Po	At	Rn
7.	Fr	Ra		113	114	115	116	117	118

1) Transition Elements

Groups\Periods	III b	IV b	V b	VI b	VII b	VIII b			I b	II b
4.	Sc	Ti	V	Cr	Mn	Fe	Co	Ni	Cu	Zn
5.	Y	Zr	Nb	Mo	Tc	Ru	Rh	Pd	Ag	Cd
6.	Lu	Hf	Ta	W	Re	Os	Ir	Pt	Au	Hg
7.	Lw	Ku	105	106	107	108	109	110	111	112

2) Inner Transition Elements

Periods														
6.	La	Ce	Pr	Nd	Pm	Sm	Eu	Gd	Tb	Dy	Ho	Er	Tm	Yb
7.	Ac	Th	Pa	U	Np	Pu	Am	Cm	Bk	Cf	Es	Fm	Md	No

FIG. 24. Blanshard 1895. Type IC1–2.

Subtype IC1–2a

Series table containing 8 groups with special flaps for the transition and inner transition elements. This kind of table has two separate sheets of paper attached by one end to the basic table, usually between the groups II and III. The transition elements are printed on both sides of one sheet; similarly, the inner transition elements are printed on the second sheet. By flipping these sheets back and forth, the aforementioned elements can be compared to the representative elements. In Fig. 25, the transition element flap is turned to the right, and the inner transition element flap to the left. This kind of table was originated by Masson (1a) and Ramsay (1b) in 1896. Masson did not include the inner transition elements. Ramsay used one flap and placed the inner transition elements into the conventional groups. McCutchon's (2) use of two flaps for the arrangement of this table was more successful. Mathur (3) arranged his table with one flap on which he printed the transition elements in one line on both sides and the inner transition elements in two lines on one side. Satou (4) used the same kind of table as McCutchon, except that his table must be read from bottom to the top.

24   GRAPHIC REPRESENTATIONS OF THE PERIODIC SYSTEM

Periods \ Groups:	I	II	III	IV	V	VI	VII	VIII
1.	H							He
2.	Li	Be	B	C	N	O	F	Ne
3.	Na	Mg	Al	Si	P	S	Cl	Ar
4.	K	Ca	Sc	Ti	V	Cr	Mn	Kr
5.	Rb	Sr	Y	Zr	Nb	Mo	Tc	Xe
Gd Tb Dy Ho Er	Tm	Yb	Lu	Hf	Ta	W	Re	Rn
Cm Bk Cf Es Fm	Md	No	Lw	Ku	105	106	107	118

FIG. 25.   Masson; Ramsay 1896.   Subtype IC1–2a.

Subtype IB1–2b

Spiral with 8 radii (Fig. 26). The representative elements were placed on these radii, but the transition and the inner transition elements were simply marked and placed between the radii of groups III and IV. In Fig. 26 these elements are placed between groups II and III. This table corresponds to subtype IC1–2b. This spiral was originated by Lyon in 1928 (1).

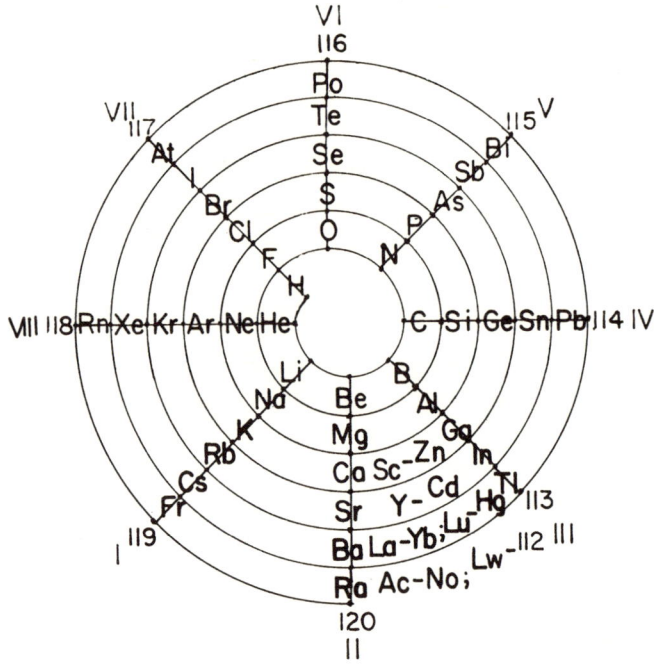

FIG. 26.   Lyon 1928.   Subtype IB1–2b.

Subtype IC1–2b

Series table containing the representative elements with the position of the transition and inner transition elements only marked, the so-called "gekürzte Tabelle" in German. The table was published by E. Rabinowitsch and E. Thilo in 1930 (1). This table corresponds to the spiral subtype IB1-2b. The construction of French's table (2) was a little different because he used oblique vertical groups.

Subtype IC1-2c

Series table similar to the preceding table, beginning with group III. This table differs from the previous table in that the marked positions of the transition and inner transition elements are not inserted in the middle of the table but are located on the far right side of it. This is similar to types IIC2-5 and IIIC3-5a. Lecoq de Boisbaudran originated this table in 1895 (1).

Class 2. Tables with Two Subgroups "a" and "b"

Type 1: Symmetrical Chessboard-like Tables

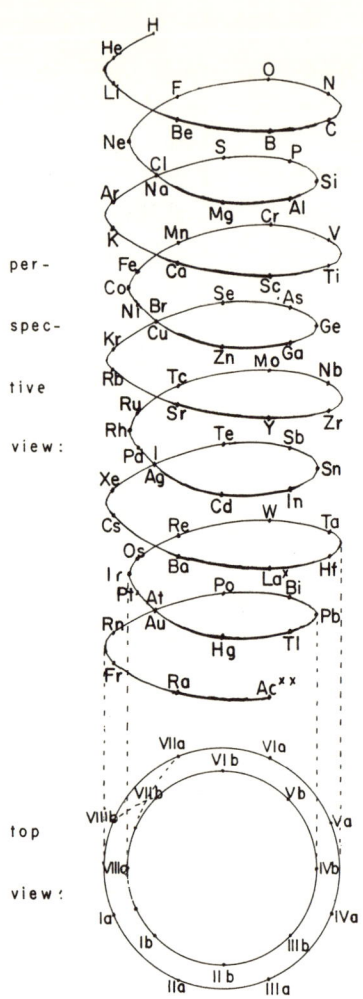

Type IA2-1

Helix drawn alternately on two sizes of cylinders that are placed one into the other concentrically (Fig. 27). The originator was Alfred Stewart in 1919 (1).

FIG. 27.   A. W. Stewart 1919.   Type IA2-1.

Type IC2-1

Symmetrical series table, chessboard-like (Fig. 28). The table was originated by Mendeleev in 1870 (1). (See also Fig. 17.) Before the origination, he published three articles in which he discussed the odd and even series in a table (0). The table, placing the elements of the short periods (periods 2 and 3) into different subgroups "a" and "b," does not agree with present views; therefore, this table is obsolete.

DIVISION I. SHORT TABLES

Groups Periods	I a | b	II a | b	III a | b	IV a | b	V a | b	VI a | b	VII a | b	VIII a | b
1.	H							He
2.	Li	Be	B	C	N	O	F	Ne
3.	Na	Mg	Al	Si	P	S	Cl	Ar
4.	K	Ca	Sc	Ti	V	Cr	Mn	Fe Co Ni
	Cu	Zn	Ga	Ge	As	Se	Br	Kr
5.	Rb	Sr	Y	Zr	Nb	Mo	Tc	Ru Rh Pd
	Ag	Cd	In	Sn	Sb	Te	I	Xe
6.	Cs	Ba	La^x	Hf	Ta	W	Re	Os Ir Pt
	Au	Hg	Tl	Pb	Bi	Po	At	Rn
7.	Fr	Ra	Ac^{xx}					

FIG. 28.   Mendeleev 1870.   Type IC2-1.

Type 2: Symmetrical Tables with Bridge Elements

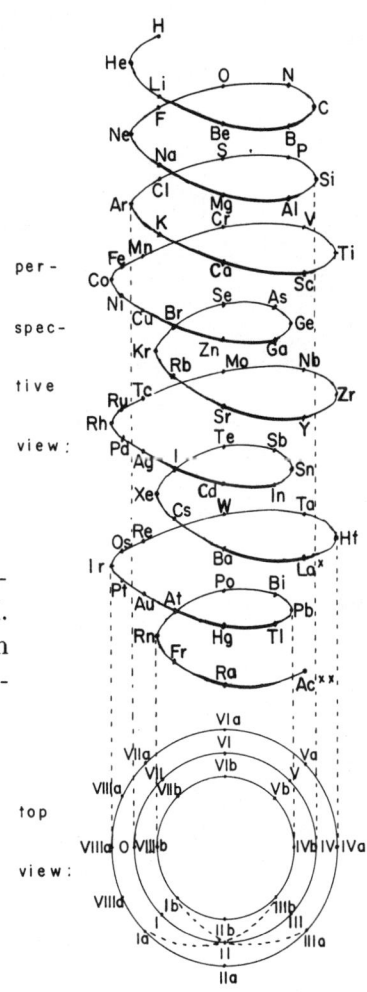

Type IA2-2

Helix drawn on three sizes of concentric cylinders. (Fig. 29). The originator of this type was O. I. Stewart in 1928 (1); although he drew the helix on a cone, the principle was the same. This helix corresponds to the types IB2-2 and IC2-2.

FIG. 29.   O. I. Stewart 1928.   Type IA2-2.

Type IB2-2

Spiral with 8 radii (Fig. 30). The elements of the short periods are placed on the radii. The subgroups of the medium periods "a" and "b" are distinguished by writing the symbols of the elements on both sides of the radii. The originator was O. I. Stewart in 1928 (1). The spiral is obsolete because the properties of the elements of the short periods are not intermediate between the properties of the elements of different subgroups of the medium periods.

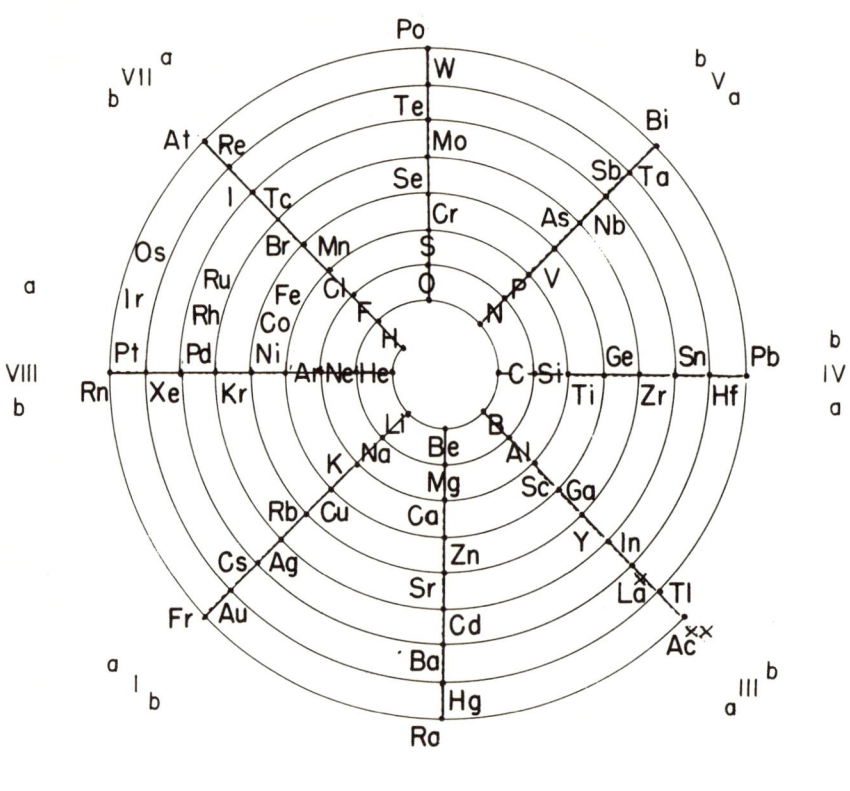

FIG. 30. O. I. Stewart 1928. Type IB2-2.

Type IC2-2

Symmetrical series table with "bridge" elements (Fig. 31). The elements of the short periods are placed midway between the subgroups "a" and "b"; they are the so-called bridge elements. The originator was von Richter in 1889(?) (1). The table is obsolete because the relationship between the elements of the short and medium periods is different from that shown in this table. Friend (13) placed the elements on the surface of a sphere. However, he did not have a space curve around the surface of the sphere, but placed the element symbols, as is usual, in squares. Therefore, his arrangement cannot be accepted as a three-dimensional one. The two circles representing the projections of the sphere should be judged only as decorative. Something similar could be said about Heribert's table (15): he drew the table in the form of a slide rule. A little change in Bjerrum's (18) table obtains a table of this type. Shchukarev (21) included the lanthanides and actinides in the table, but did not arrange them in definite places. Furthermore, he placed the first half of the transition elements in "b" subgroups and the second half in "a" subgroups.

Groups Periods	I a | b	II a | b	III a | b	IV a | b	V a | b	VI a | b	VII a | b	VIII a | b
1.	H							He
2.	Li	Be	B	C	N	O	F	Ne
3.	Na	Mg	Al	Si	P	S	Cl	Ar
4.	K	Ca	Sc	Ti	V	Cr	Mn	Fe Co Ni
	Cu	Zn	Ga	Ge	As	Se	Br	Kr
5.	Rb	Sr	Y	Zr	Nb	Mo	Tc	Ru Rh Pd
	Ag	Cd	In	Sn	Sb	Te	I	Xe
6.	Cs	Ba	La*	Hf	Ta	W	Re	Os Ir Pt
	Au	Hg	Tl	Pb	Bi	Po	At	Rn
7.	Fr	Ra	Ac**					

FIG. 31. von Richter 1889. Type IC2-2.

Subtype IC2-2a

Series table similar to the previous table with the difference that the elements of group VIII are placed in group VII. Originated by H. Biltz in 1902 (1) this is not a successful subtype. Sborgi (4) placed the elements of groups VIIIb and VIIb in group VIb.

Subtype IC2-2b

Series table similar to the basic type IC2-2, the difference being that only the elements of the first series of the short periods are placed as "bridge" elements (Fig. 32). The originator was Preyer in 1892 (1). His table did not have a group VIII, but the elements of this group were placed into the next row as the elements of groups II, III, and IV. Sears (4) drew the second short period differently than did the other authors. This is not a successful subtype.

30 GRAPHIC REPRESENTATIONS OF THE PERIODIC SYSTEM

Groups\Periods	I a | b	II a | b	III a | b	IV a | b	V a | b	VI a | b	VII a | b	VIII a | b
1.	H |							| He
2.	Li |	Be |	| B	C |	| N	| O	| F	| Ne
3.	| Na	| Mg	| Al	| Si	| P	| S	| Cl	| Ar
4.	K |	Ca |	Sc |	Ti |	V |	Cr |	Mn |	Fe Co Ni |
	| Cu	| Zn	| Ga	| Ge	| As	| Se	| Br	| Kr
5.	Rb |	Sr |	Y |	Zr |	Nb |	Mo |	Tc |	Ru Rh Pd |
	| Ag	| Cd	| In	| Sn	| Sb	| Te	| I	| Xe
6.	Cs |	Ba |	La˟ |	Hf |	Ta |	W |	Re |	Os Ir Pt |
	| Au	| Hg	| Tl	| Pb	| Bi	| Po	| At	| Rn
7.	Fr |	Ra |	Ac˟˟ |					

FIG. 32. Preyer 1892. Subtype IC2-2b.

Type 3: Symmetrical Tables with the Elements of Short Periods Divided Symmetrically into Two Parts

Type IC-3

Series table with symmetrically divided short periods (Fig. 33). The periods begin with group I. The elements of the short periods are placed symmetrically: the elements of groups I to III in subgroups "a," of groups V to VIII in subgroups "b," but the elements of group IV are placed midway between subgroups "a" and "b." Benedicks was the originator (1). He and Meissner (9) both placed group III midway between these two subgroups. This type can not be used because the elements of the short periods are not placed according to their relationship, but only to be symmetrical.

Groups\Periods	I a | b	II a | b	III a | b	IV a | b	V a | b	VI a | b	VII a | b	VIII a | b
1.	H |							| He
2.	Li |	Be |	B |	C	| N	| O	| F	| Ne
3.	Na |	Mg |	Al |	Si	| P	| S	| Cl	| Ar
4.	K |	Ca |	Sc |	Ti |	V |	Cr |	Mn |	Fe Co Ni |
	| Cu	| Zn	| Ga	| Ge	| As	| Se	| Br	| Kr
5.	Rb |	Sr |	Y |	Zr |	Nb |	Mo |	Tc |	Ru Rh Pd |
	| Ag	| Cd	| In	| Sn	| Sb	| Te	| I	| Xe
6.	Cs |	Ba |	La˟ |	Hf |	Ta |	W |	Re |	Os Ir Pt |
	| Au	| Hg	| Tl	| Pb	| Bi	| Po	| At	| Rn
7.	Fr |	Ra |	Ac˟˟ |					

FIG. 33. Benedicks 1904. Type IC2-3.

Type 4: Unsymmetrical Tables with the Elements of Short Periods Placed in Subgroups "a" and "b"

Type IB2–4

Spiral with 8 radii (Fig. 34). The elements of subgroups "a" are placed on the radii and the elements of subgroups "b" are placed in the spaces between the radii. The originator was Baumhauer in 1870 (1). Tansley (3) and Saccardo (4) started their spirals from the outside; the last element U was in the center. This type of spiral is not a very good one; the reason for this is explained under type IC2–4.

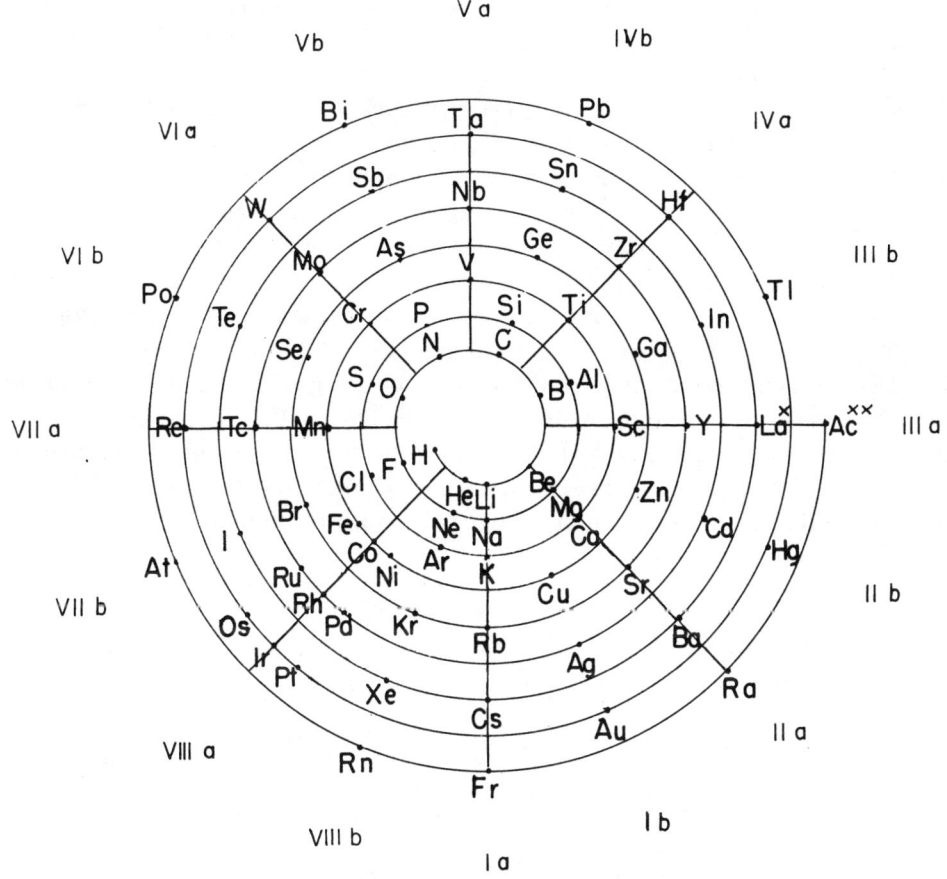

FIG. 34. Baumhauer 1870. Type IB2–4.

Type IC2–4

Unsymmetrical series table with the elements of short periods in different subgroups (Fig. 35). The 10 elements of the first series of a medium-size period are placed in subgroups "a," and the 8 elements of the second series in subgroups "b." This is the same way that Mendeleev in 1870 drew his original short type

IC2-1—simply to have a chessboard-like table. The elements of the short periods are arranged in those subgroups where the related elements of the medium periods already are: the elements of groups I and II into subgroups "a" and the elements of groups III to VIII into subgroups "b." Therefore, an interruption occurs in the short periods between groups II and III. In earlier times, there was a different concept of the relationship of the elements, and the elements of short periods were sometimes placed differently than they are in Fig. 35. The originators of this table were Gretschel and Bornemann in 1883 (1). This type is the most popular one among the short tables, and this table is printed in many textbooks that are not listed here. However, this is not a good method of arranging the elements into subgroups. The elements of the two short periods should be drawn as belonging to the same block of elements, as is done in the electron configuration tables (class 4 of Division III). This block is called the block of representative elements, and the elements of the short periods are sometimes considered as prototypes of other elements. These elements should belong to the same subgroups. However, some scientists are of a different opinion, namely, that the s and p elements of the representative block should be separated. What could be said about the unity of the block of representative elements could also be said about the block of transition elements, the properties of which have something common. However, the transition elements in this table are distributed in different subgroups "a" and "b." Rydberg (4), Butler (11), Fillinger (15), Hopkins (17), Syrkin (19), and Höltje (22) placed the short period elements of groups II, III, and IV inbetween the "a" and "b" subgroups, while Paneth (8) and Bolin (9) did the same with the elements of groups I and II. Loung (44) arranged the subgroups according to this type, although the naming of them was similar to the type IC2-5. Trifonov (48) added the 8th period, which brought the total of elements to 168.

Groups Periods	I		II		III		IV		V		VI		VII		VIII	
	a	b	a	b	a	b	a	b	a	b	a	b	a	b	a	b
1.														H		He
2.	Li		Be		B		C		N		O		F			Ne
3.	Na		Mg		Al		Si		P		S		Cl			Ar
4.	K		Ca		Sc		Ti		V		Cr		Mn		Fe Co Ni	
		Cu		Zn		Ga		Ge		As		Se		Br		Kr
5.	Rb		Sr		Y		Zr		Nb		Mo		Tc		Ru Rh Pd	
		Ag		Cd		In		Sn		Sb		Te		I		Xe
6.	Cs		Ba		La̅		Hf		Ta		W		Re		Os Ir Pt	
		Au		Hg		Tl		Pb		Bi		Po		At		Rn
7.	Fr		Ra		Ac̽											

FIG. 35. Gretschel a. Bornemann 1883. Type IC2-4.

Subtype IC2-4a

Series table similar to the preceding table with the elements of groups VIIa and VIIIa placed into group VIa. This table originated by Dauvillier in 1921 (1) is not a successful subtype.

Subtype IC2-4b

Series table similar to the basic type IC2-4 with the difference that all the elements of the short periods are placed in subgroups "b." The table was originated by Martin (1a) and Molinari (1b) in 1905. The naming of the subgroups in Martin's table was reversed in comparison with the basic type IC2-4.

Subtype IC2-4c

Series table similar to the preceding table with the elements of the short periods placed in subgroups "a." This is not a successful subtype. It was first published by Herz in 1912 (1).

Subtype IC2-4d

Series table where the series start with group III (Fig. 36). In Morette's table (1), the names of the subgroups were changed. However, in principle, this table does not differ from the main type IC2-4.

III		IV		V		VI		VII		VIII		I		II		
a	b	a	b	a	b	a	b	a	b	a	b	a	b	a	b	
											n	H				
										He	Li		Be			
	B		C		N		O		F		Ne	Na		Mg		
	Al		Si		P		S		Cl		Ar	K		Ca		
Sc		Ti		V		Cr		Mn		Fe Co Ni			Cu		Zn	
	Ga		Ge		As		Se		Br		Kr	Rb		Sr		
Y		Zr		Nb		Mo		Tc		Ru Rh Pd			Ag		Cd	
	In		Sn		Sb		Te		I		Xe	Cs		Ba		La×
Lu		Hf		Ta		W		Re		Os Ir Pt			Au		Hg	
	Tl		Pb		Bi		Po		At		Rn	Fr		Ra		Ac××
Lw		Ku		105		106		107		108 109 110			111		112	
	113		114		115		116		117		118	119		120		

FIG. 36. Morette 1941. Subtype IC2-4d.

Type 5: Unsymmetrical Tables with the Elements of the Short Periods Placed Exclusively in Subgroups "a"

Type IA2–5

Helix on two cylinders of different sizes that have a common generating line on which group III is placed (Fig. 37). The two different revolutions have 8 and 10 elements. The originators were Harkins and Hall in 1916 (1). This helix represents the types IB2–5 and IC2–5.

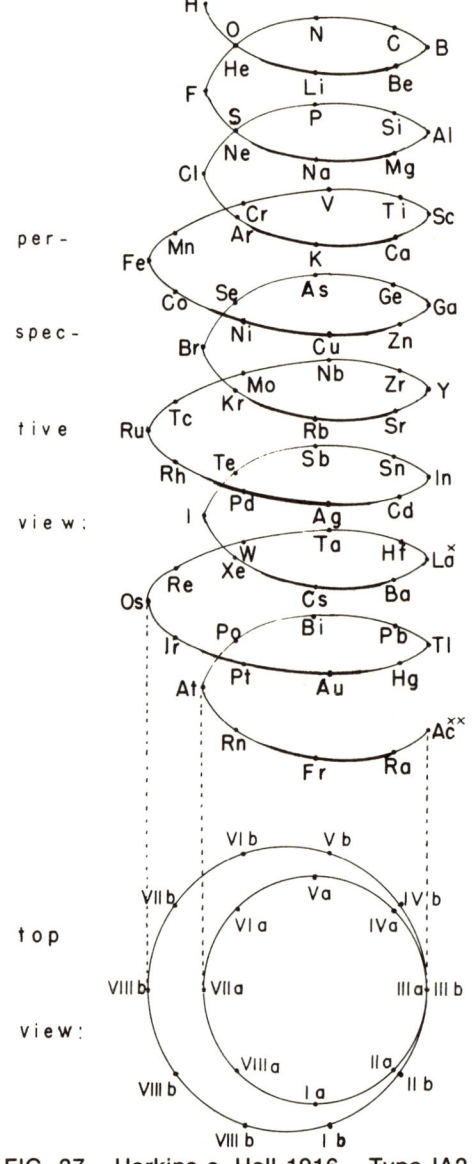

FIG. 37. Harkins a. Hall 1916. Type IA2–5.

DIVISION I. SHORT TABLES 35

Type IB2–5

Spiral with 8 radii (Fig. 38). The elements of the short periods and of subgroups "a" of the medium periods (in agreement with the next series table) are placed on the radii and the elements of subgroups "b" in the spaces between the radii. Hinrichs' spiral (0) was incomplete, and it belongs to the tables of the forerunners of the Periodic Law. Huth in 1884 (1) must be acknowledged as the originator, although his table was not perfect. Huth (1) and von Wolff (5) did not place the symbols on the radii and between them, but they marked the symbols differently to distinguish the subgroups. Only Wells (2) gave a good table of this type. This type of spiral is the best of the spirals drawn thus far.

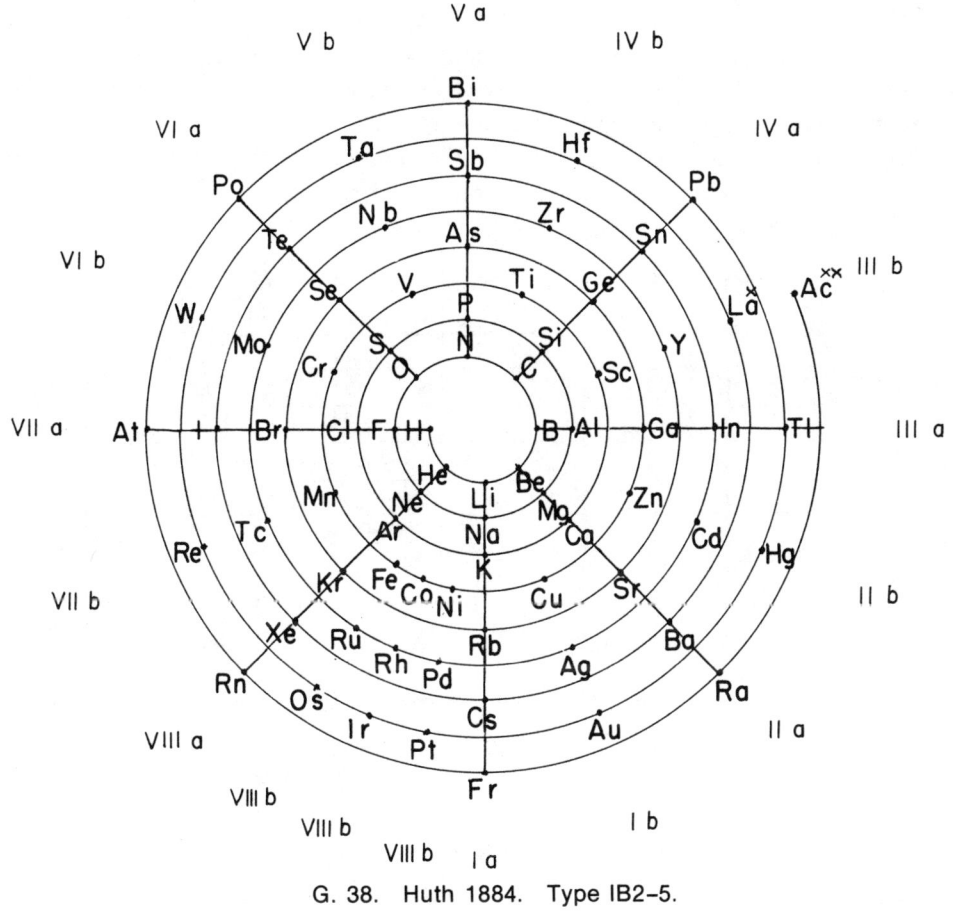

G. 38. Huth 1884. Type IB2–5.

Type IC2–5

Unsymmetrical series table with the elements of the short periods exclusively in subgroups "a" (Fig. 39). The elements of the medium periods, which consist of two kinds of elements, the representative and the transition elements, are arranged by considering the existing arrangement of the short periods. The representative elements, as the more important ones, are placed in subgroups "a" and the transition elements, as the less characteristic ones, in subgroups "b." Therefore, in the medium periods, a transition exists between groups II and III from the representative elements to the transition elements and back. The originator of this table was Arnold in 1885 (1). This is the best table among the short tables with two subgroups. Hereafter, the names of the subgroups in the tables that follow will be in agreement with this table. Huth's table (0) belongs to this type in only some respects. The arrangement of Morozov's table (2) was not very clear. Mazzucchelli (9) marked the "a" subgroups as s and p and the "b" subgroups as d. Bokii (14) gave the table the shape of type IC2–4, but regarded the representative elements as belonging to the main subgroups and the transition elements to the subordinate subgroups. Kvasnik (15) and Banerjee (17) named the "a" subgroups as "b" and vice versa. Some authors, Sheringa (4), Bauer (5), Mazzucchelli (9), Bakker (12), and Sanderson (23), marked or colored the transition elements to distinguish them from the representative elements. The first man to publish a more elaborate table was Centnerszwer (8). The type IC2–5 table was adopted as a standard table by the Chemical Society of France (13).

Groups Periods	I		II		III		IV		V		VI		VII		VIII	
	a	b	a	b	a	b	a	b	a	b	a	b	a	b	a	b
1.	H														He	
2.	Li		Be		B		C		N		O		F		Ne	
3.	Na		Mg		Al		Si		P		S		Cl		Ar	
4.	K		Ca			Sc		Ti		V		Cr		Mn	Fe Co Ni	
		Cu		Zn	Ga		Ge		As		Se		Br		Kr	
5.	Rb		Sr			Y		Zr		Nb		Mo		Tc	Ru Rh Pd	
		Ag		Cd	In		Sn		Sb		Te		I		Xe	
6.	Cs		Ba			La^x		Hf		Ta		W		Re	Os Ir Pt	
		Au		Hg	Tl		Pb		Bi		Po		At		Rn	
7.	Fr		Ra			Ac^{xx}		Ku								

FIG. 39.   Arnold 1885.   Type IC2–5.

Subtype IC2–5a

Series table similar to the preceding table with the difference that each series begins with group III (Fig. 40). Here no abrupt transition exists from subgroups "a" to "b" in a series of the table. The originator of this table was Lothar Meyer (1) in 1870 (Fig. 16). Meyer had no followers in support of this table, although his table is a good one and is related to the modern tables, such as the Janet series table type IIIC3–6 (Fig. 103). It permits the use of a new division of periods, which is explained in the Introduction to the Electronic Configuration Tables.

DIVISION I. SHORT TABLES 37

New periods	III		IV		V		VI		VII		VIII			I		II	
	a	b	a	b	a	b	a	b	a	b	a		b	a	b	a	b
1.a.2.									H		He			Li		Be	
3.	B		C		N		O		F		Ne			Na		Mg	
4.	Al		Si		P		S		Cl		Ar			K		Ca	
5.		Sc		Ti		V		Cr		Mn		Fe Co Ni			Cu		Zn
	Ga		Ge		As		Se		Br		Kr			Rb		Sr	
6.		Y		Zr		Nb		Mo		Tc		Ru Rh Pd			Ag		Cd
	In		Sn		Sb		Te		I		Xe			Cs		Ba	
7.		Ld		Hf		Ta		W		Re		Os Ir Pt			Au		Hg
	Tl		Pb		Bi		Po		At		Rn			Fr		Ra	
8.	Ac		Ku														

FIG. 40. Lothar Meyer 1870. Subtype IC2-5a.

Class 3. Tables with Three Subgroups: "a," "b," and "c"

Type 1: Tables with the Inner Transition Elements in Two Series

Type IA3–1

Helix with three sizes of revolutions for 8, 10, and 7 elements (Fig. 41). 8 elements are the representative elements, 10 are the transition elements, and two revolutions with 7 elements in each are the inner transition elements. Bassett (1) suggested this kind of helix in 1892, although he did not draw it. Horie's table (3) was constructed on a different basis: the atomic volumes of the elements were taken as the radii of the revolutions. This helix corresponds to types IB3–1 and IC3–1.

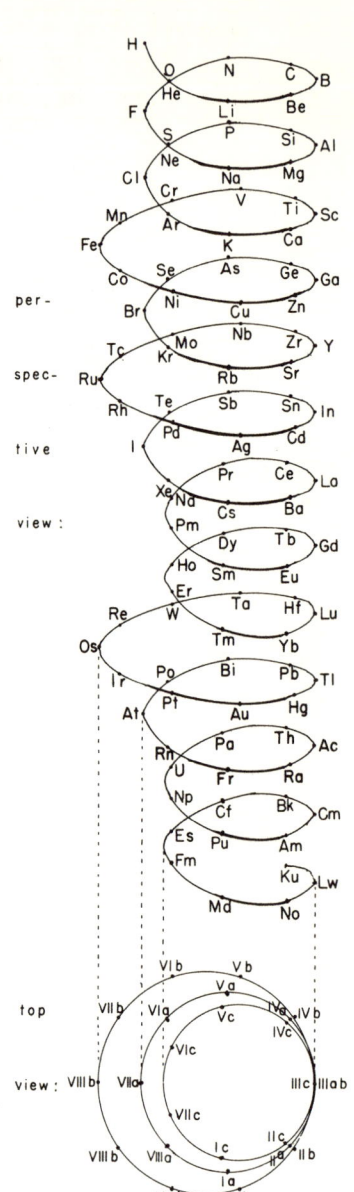

FIG. 41. Bassett 1892. Type IA3–1.

DIVISION I. SHORT TABLES   39

Type IB3-1

Spiral with 8 radii for subgroups "a," 10 supplementary radii for subgroups "b," and 7 supplementary radii for subgroups "c" (Fig. 42). The supplementary radii are shorter and placed between the main radii. This type corresponds with type IC3-1 where some of the inner transition elements are placed in groups I and II, as is shown in this spiral. This spiral is hard to read as a table. Green and Jackson in 1950 were the originators (1), but their concept was a little different from that of the table in Fig. 42. Earlier, in 1918, Steinmetz (0) drew approximately similar spirals, starting from a different point of view, i.e., using the Riemann surfaces.

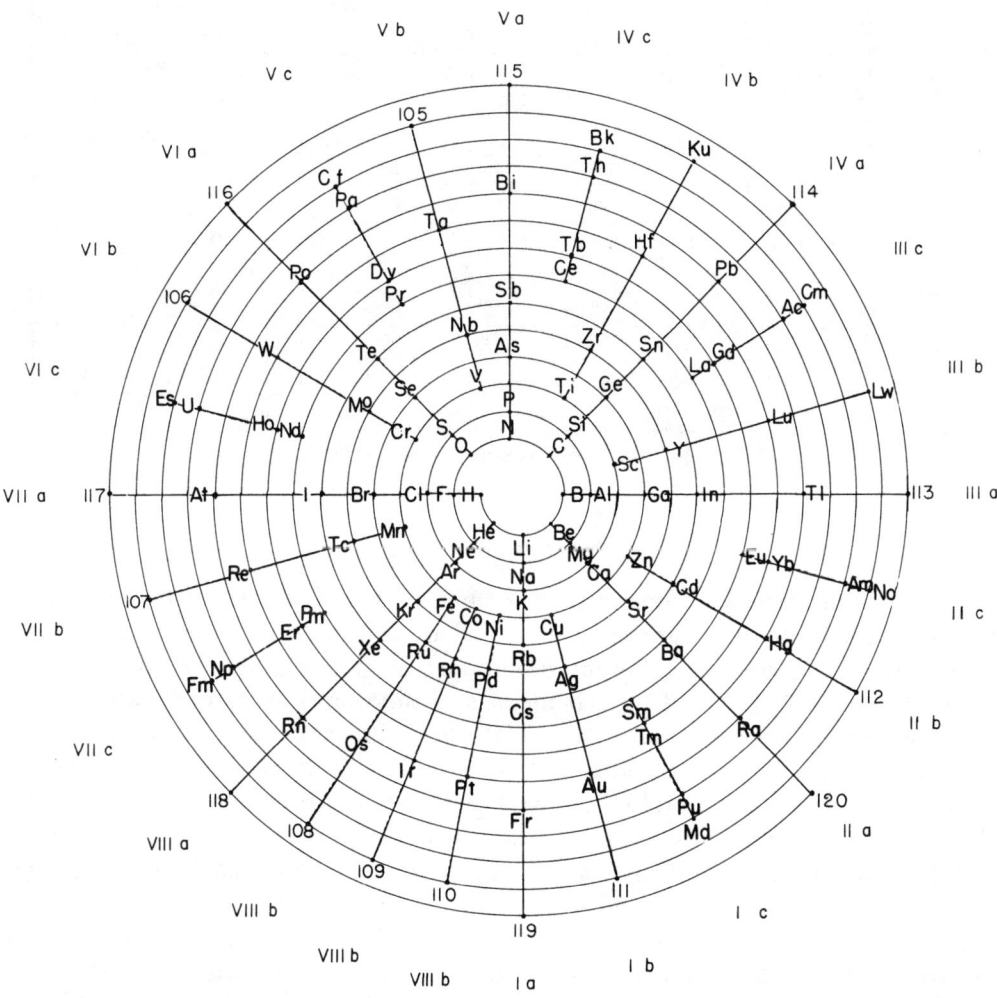

FIG. 42. Green a. Jackson 1950. Type IB3-1.

Type IC3–1

Series table with 8 groups and 3 subgroups "a," "b," and "c" with the inner transition elements placed in two series (Fig. 43). In 1949, Schenk (1a) and Faustov (1b) were the first two authors to publish this table. Grigorovich (6) placed the elements of group Ic into subgroup VIIIc. This type of table is not perfect because the positioning of the inner transition elements does not fit them into the groups according to their relationship. Ali (8) did not designate the subgroups.

Groups → Periods ↓	I a	I b	I c	II a	II b	II c	III a	III b	III c	IV a	IV b	IV c	V a	V b	V c	VI a	VI b	VI c	VII a	VII b	VII c	VIII a	VIII b
1.	H																					He	
2.	Li			Be			B			C			N			O			F			Ne	
3.	Na			Mg			Al			Si			P			S			Cl			Ar	
4.	K			Ca					Sc			Ti			V			Cr			Mn	Fe Co Ni	
4.		Cu			Zn		Ga			Ge			As			Se			Br			Kr	
5.	Rb			Sr					Y			Zr			Nb			Mo			Tc	Ru Rh Pd	
5.		Ag			Cd		In			Sn			Sb			Te			I			Xe	
6.	Cs			Ba					La			Ce			Pr			Nd			Pm		
6.		Sm			Eu				Gd			Tb			Dy			Ho			Er		
6.		Tm			Yb				Lu			Hf			Ta			W			Re	Os Ir Pt	
6.			Au			Hg	Tl			Pb			Bi			Po			At			Rn	
7.	Fr			Ra					Ac			Th			Pa			U			Np		
7.		Pu			Am				Cm			Bk			Cf			Es			Fm		
7.		Md			No				Lw			Ku			105			106			107	108 109 110	
7.			111			112	113			114			115			116			117			118	

FIG. 43. Schenk; Faustov 1949. Type IC3–1.

Subtype IC3-1a

Series table having 8 groups with 4 subgroups included in each group (Fig. 44). The inner transition groups are placed in two series, but the elements of each series are placed in different subgroups. In groups I and II, the order of subgroups is 1, 2, 3, and 4, which mean s, f (first half), f (second half), and d elements respectively. In groups III to VIII, the order of the subgroups is different, namely, 4, 3, 2, and 1, which mean f (first half), f (second half), d, and p elements. By this arrangement, a trend of physical property values is obtained among the elements of the subgroups belonging to the same group and period. This kind of table was arranged by Chistiakov in 1964 (1), and he called these subgroups "microperiods." These appeared in the 4th and 5th periods as two elements in groups I–VII and in the 6th and 7th periods as four elements in all groups. In his article, Chistiakov gave tables of values for ionization potential, oxidation-reduction potential, and atomic radius. By these schemes, he predicted unknown values for constants.

DIVISION I. SHORT TABLES

Diads	Periods	Series	Groups and subgroups																													
			I				II				III				IV				V				VI				VII				VIII			
			1	2	3	4	1	2	3	4	4	3	2	1	4	3	2	1	4	3	2	1	4	3	2	1	4	3	2	1	4	3	2	1
I.	0	1	n																															
	1	2	H				He																											
2.	2	3	Li				Be							B				C				N				O				F				Ne
	3	4	Na				Mg							Al				Si				P				S				Cl				Ar
3.	4	5	K				Ca						Sc				Ti				V				Cr				Mn				Fe Co Ni
		6			Cu				Zn				Ga				Ge				As				Se				Br					Kr
	5	7	Rb				Sr						Y				Zr				Nb				Mo				Tc				Ru Rh Pd
		8				Ag			Cd				In				Sn				Sb				Te				I					Xe
4.	6	9	Cs				Ba				La				Ce				Pr				Nd				Pm						
		10	Sm				Eu				Gd				Tb				Dy				Ho				Er						
		11		Tm				Yb				Lu				Hf				Ta				W				Re				Os Ir Pt
		12			Au				Hg				Tl				Pb				Bi				Po				At					Rn
	7	13	Fr				Ra				Ac				Th				Pa				U				Np						
		14	Pu				Am				Cm				Bk				Cf				Es				Fm						
		15		Md				No				Lw				Ku				105				106				107				108 109 110
		16			111				112				113				114				115				116				117					118
Subperiods and electrons			s^1 f^6 f^{13} d^9				s^2 f^7 f^{14} d^{10}				f^1 f^8 d^1 p^1				f^2 f^9 d^2 p^2				f^3 f^{10} d^3 p^3				f^4 f^{11} d^4 p^4				f^5 f^{12} d^5 p^5				d^6 d^7 d^8 p^6			

FIG. 44. Chistiakov 1964. Subtype IC3–1a.

Subtype IB3–1b

The same kind of spiral as in the basic type IB3–1 with the difference that those inner transition elements which were in groups I and II are placed in group VIII. Therefore this subtype corresponds to subtype IC3–1b. Agafoshin originated this subtype in 1952 (1).

Subtype IC3–1b

Series table with 8 groups, 3 subgroups "a," "b," and "c," and with a transition group (Fig. 45). Groups I and II have exclusively one subgroup, "a"; groups III–VIII have 3 subgroups, "a," "b," and "c." The elements which are not placed into groups I–VIII belong to a special transition group. The inner transition elements are placed in subgroups "c" in two lines, exactly as in the little periodic table. The originator of this table was Shemiakin in 1932 (1). This table also is not perfect because of the position of the inner transition elements, although it is better than the previously discussed type IC3–1. Mashentsev's table (2) was a little different.

42 GRAPHIC REPRESENTATIONS OF THE PERIODIC SYSTEM

Cycles	Periods	I a	II a	III a	III b	III c	IV a	IV b	IV c	V a	V b	V c	VI a	VI b	VI c	VII a	VII b	VII c	VIII a	VIII b	VIII c	transition b	transition c	transition b	transition b	transition b
1.	1.	H																	He							
2.	2.	Li	Be	B			C			N			O			F			Ne							
	3.	Na	Mg	Al			Si			P			S			Cl			Ar							
3.	4.	K	Ca	Sc			Ti			V			Cr			Mn			Fe	Co		Ni	Cu	Zn		
				Ga			Ge			As			Se			Br			Kr							
	5.	Rb	Sr	Y			Zr			Nb			Mo			Tc			Ru	Rh		Pd	Ag	Cd		
				In			Sn			Sb			Te			I			Xe							
4.	6.	Cs	Ba	La			Ce			Pr			Nd			Pm			Sm	Eu						
				Gd			Tb			Dy			Ho			Er			Tm	Yb						
				Lu			Hf			Ta			W			Re			Os	Ir		Pt	Au	Hg		
				Tl			Pb			Bi			Po			At			Rn							
	7.	Fr	Ra	Ac			Th			Pa			U			Np			Pu	Am						
				Cm			Bk			Cf			Es			Fm			Md	No						
				Lw			Ku			105			106			107			108	109		110	111	112		
				113			114			115			116			117			118							

FIG. 45. Shemiakin 1932. Subtype IC3–1b.

Type 2: Tables with the Inner Transition Elements in One Series

Type IA3-2

Helix with three sizes of revolutions for 8 representative, 10 transition, and 14 inner transition elements (Fig. 46). The originator was Vogel in 1918 (1), who did not draw the helix, but gave only the top view of the helix. Thix helix is analogous to type IC3-2.

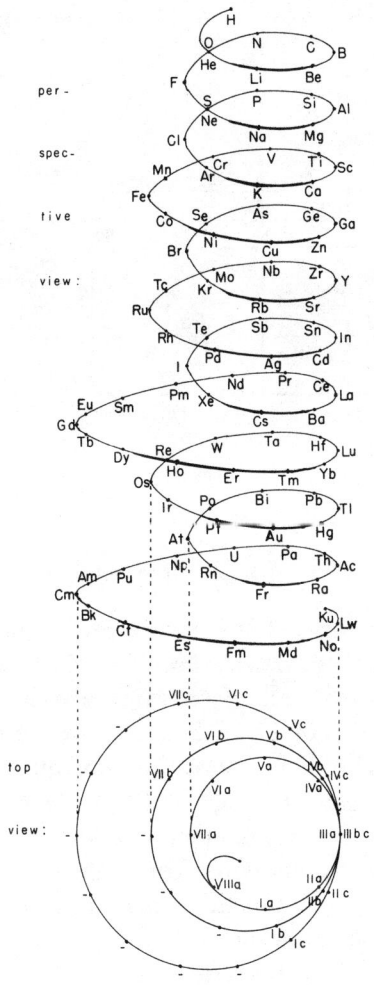

FIG. 46. Vogel 1918. Type IA3-2.

Type IC3–2

Series table with 8 groups and 3 subgroups "a," "b," and "c," in which the inner transition elements are placed in one series (Fig. 47). Instead of using group VIII, the noble gases, the iron-platinum elements, and seven inner transition elements are placed in a transition group in subgroups "a," "b," and "c" respectively. This kind of transition group was introduced by Bury in 1921 in type IIIC4–3 (00) and used by LeRoy in 1927 in type IIIC3–6 (2), Stareck in 1932 in subtype IIIC3–1d (1), John D. Clark in 1950 in type IIIB3–1 (4), and by myself in 1955 in this type IC3–2 (1) and in subtype IIIC3–6b (1). Clifford (2) did not place the inner transition elements into groups I and II; therefore, he had 9 elements in the "c" subgroup of the transition group instead of 7.

Groups:	I			II			III			IV			V			VI			VII			transition		
Periods	a	b	c	a	b	c	a	b	c	a	b	c	a	b	c	a	b	c	a	b	c	a	b	c
1.	H																					He		
2.	Li			Be			B			C			N			O			F			Ne		
3.	Na			Mg			Al			Si			P			S			Cl			Ar		
4.	K			Ca				Sc			Ti			V			Cr			Mn		Fe Co Ni		
		Cu			Zn		Ga			Ge			As			Se			Br			Kr		
5.	Rb			Sr				Y			Zr			Nb			Mo			Tc		Ru Rh Pd		
		Ag			Cd		In			Sn			Sb			Te			I			Xe		
6.	Cs			Ba				La			Ce		Pr				Nd			Pm				Sm Eu Gd Tb Dy Ho Er
		Tm			Yb		Lu			Hf			Ta			W			Re		Os Ir Pt		
		Au			Hg		Tl			Pb			Bi			Po			At			Rn		
7.	Fr			Ra				Ac			Th			Pa			U			Np				Pu Am Cm Bk Cf Es Fm
		Md			No		Lw			Ku			105			106			107		108 109 110		
			111			112	113			114			115			116			117			118		

FIG. 47. Mazurs 1955. Type IC3–2.

Subtype IC3–2a

Series table with 8 groups, 3 subgroups, and a transition group, and with the inner transition elements in one series. (Fig. 48). This table differs from the previous table in that groups I and II contain only the alkali and alkaline earth elements that possess valences of 1 and 2 in accordance with their group number. Groups III to VIII contain elements with the maximum valence according to the group number in all subgroups "a," "b," and "c." Exceptions exist for the lanthanides. The transition group contains elements with descending valences; this is especially true of the transition elements. The originator of the table was Murashov in 1949 (1). This table is better than the other short tables with three subgroups. It is "elegant" because the marking of the shells, subshells, and electrons can be done in a regular order; and, therefore, it could also be used as an electronic configuration table.

Groups:	I	II	III			IV			V			VI			VII			VIII			transition								
Periods.Shells	a	a	a	b	c	a	b	c	a	b	c	a	b	c	a	b	c	a	b	c	b	c	b	c	b	c	b	c	c	c	c	c
1.	1. H	He																														
2.	2. Li	Be	B			C			N			O			F			Ne														
3.	3. Na	Mg	Al			Si			P			S			Cl			Ar														
4.	4. K	Ca	3. Sc			Ti			V			Cr			Mn			Fe			Co		Ni		Cu		Zn					
			4. Ga			Ge			As			Se			Br			Kr														
5.	5. Rb	Sr	4. Y			Zr			Nb			Mo			Tc			Ru			Rh		Pd		Ag		Cd					
			5. In			Sn			Sb			Te			I			Xe														
6.	6. Cs	Ba	4. La			Ce			Pr			Nd			Pm			Sm			Eu		Gd		Tb		Dy		Ho	Er	Tm	Yb
			5. Lu			Hf			Ta			W			Re			Os			Ir		Pt		Au		Hg					
			6. Tl			Pb			Bi			Po			At			Rn														
7.	7. Fr	Ra	5. Ac			Th			Pa			U			Np			Pu			Am		Cm		Bk		Cf		Es	Fm	Md	No
			6. Lw			Ku			105			106			107			108			109		110		111		112					
			7. 113			114			115			116			117			118														
Subshells and electrons:	s^1	s^2	d^1 f^1	p^1		d^2 f^2	p^2		d^3 f^3	p^3		d^4 f^4	p^4		d^5 f^5	p^5		d^6 f^6	p^6		d^7 f^7		d^8 f^8		d^9 f^9		d^{10} f^{10}		f^{11}	f^{12}	f^{13}	f^{14}

FIG. 48. Murashov 1949. Subtype IC3-2a.

Subtype IC3-2b

Series table similar to the basic type with the difference that the inner transition elements are placed in an unusual order: one element each in groups I, II, III, and V, VI, VII; 3 elements in group IV; and 5 elements in group VIII. This subtype, suggested by Fritz Scheele in 1949 (1), is not successful. Lebedev in 1972 (2) suggested a similar table: in his table groups III, IV, V, VI, and VII contained each one element; group VIII, 4 elements; group I, 3 elements; and group II, 2 elements.

Division II. Medium Tables

Medium tables are of two classes:
1) tables with 16 vertical groups.
2) tables with disposition of elements: 2, 8, and 18.

The tables of *class 1* are similar in construction to the tables of the Law of Octaves. The difference is that the medium periods with 16 groups are taken as the basis for construction instead of the short periods with 8 groups. These tables are obsolete today.

Type 1 of this class contains three different series tables.

Type 2 contains a lemniscate and a zigzag.

The tables of *class 2* are divided into seven types. These types, except number 7, are analogous to the corresponding types of long tables. The only difference is that the medium tables do not contain the inner transition elements. By reason of analogy, the medium tables seemingly could be omitted in order to simplify our classification; but to do so would not be feasible because the medium and the long tables were originated by different authors (Table 2).

Table 2. Authors who Originated Medium and Long Tables

	Medium Tables			Long Tables	
IIA2–1	Emerson	1911	IIIA3–1	Stintzing	1916
IIB2–1	Emerson	1911	IIIB3–1	Hackh	1914
IIC2–1	Mendeleev	1871	IIIC3–1	Bassett	1892
IIC2–1a	Loring	1913	IIIC3–1e	Verschoyle	1908
IIA2–2	Soddy	1914	IIIA3–2	Schirmeisen	1900
			IIIA3–2a	Gooch and Walker	1905
IIB2–2	Kipp	1942	IIIB3–2	Janet	1928
IIC2–2	Woodiwiss	1906	IIIC3–2	Saz	1922
IIB2–2c	Bindel and Blickle	1952	IIIB3–2a	Rinck and Feschotte	1962
IIA2–3	Tomkeieff	1954	IIIA3–3	Aucken	1951
IIC2–3	Carnelley	1886	IIIC3–3	Bayley	1882
IIC2–3a	von Antropoff	1925	IIIC3–3b	Hackh	1914
IIA2–4	von Stackelberg	1911	IIIA3–4	Zmaczynski	1937
IIB2–4	Tocher	1910	IIIB3–4	Janet	1928
IIC2–4	Mendeleev	1869	IIIC3–4	Werner	1905
IIC2–5	Schmidt	1918	IIIC3–5a	Schmidt	1911
			IIIC3–5c	Sheehan	1961
IIC2–5a	van den Broek	1911	IIIC3–5b	Mazurs	1957
IIC2–6	Mendeleev	1869	IIIC3–6	Janet	1927

The tables of class 2 have a disposition of elements 2, 8, and 18—according to the number of elements in the foreperiod, the short period, and the medium period.

Type 1 contains stepwise tables with group 0 on the right side of the table. This arrangement provides poorly constructed tables.

To type 2 belong lemniscates and a zigzag table with the same basis of arrangement as type 1.

Type 3 tables are symmetrical about a vertical line, and they also have group 0 on the right side. These are good tables, and they are popular, although the long tables of a similar type are used more.

Type 4 has tables with group 0 on one side like types 1 and 2, but, since the short periods are rearranged so that they are interrupted, these tables are acceptable. Mendeleev's table (type IIC2–4) is a very popular one.

Type 5 tables have group 0 in the middle of the table. These tables are acceptable, although they are not very popular.

Type 6 contains a stepwise table with group IIa on the right side of the table. This table is related to a very good long table originated by Janet which should be generally accepted as a modern one.

Type 7 contains mirror-image tables where the representative elements are opposite the transition elements. This type of table does not exist as a long table and could be classified even as a type of short table.

CLASS 1. TABLES WITH 16 GROUPS

Type 1: Helix and Spiral with Equal Revolutions, and Tables with Equal Series

Type IIA1-1

Helix with one size revolution with 16 or 18 elements on it (Fig. 49). The helix was originated by Hackh in 1910 (1). This type is obsolete.

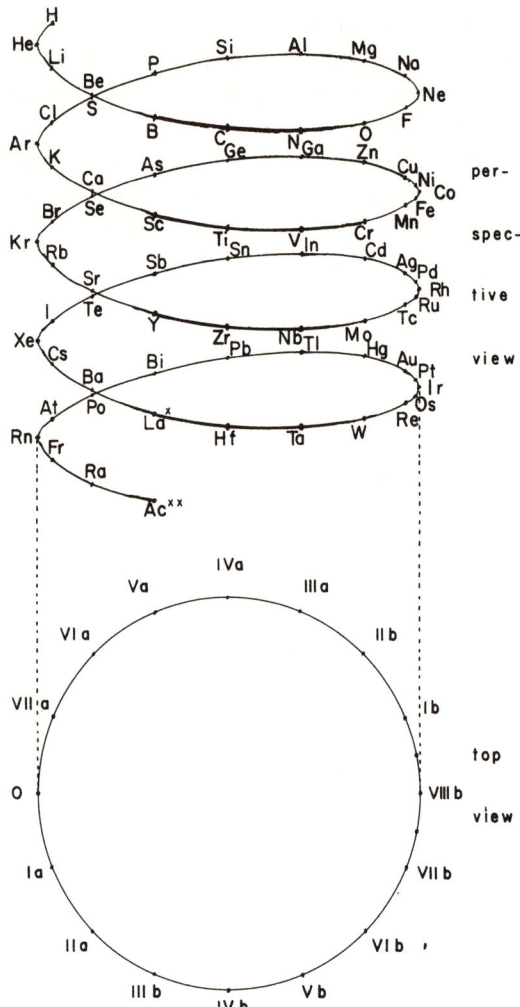

FIG. 49. Hackh 1910. Type IIA1-1.

Type IIB1-1

Spiral with 16 radii (Fig. 50). The originator of this spiral was Flavitskii in 1887 (1), although he did not draw the spiral itself, but wrote the symbols of the elements around a circle in the order of a spiral. This type is obsolete (see next type IIC1-1).

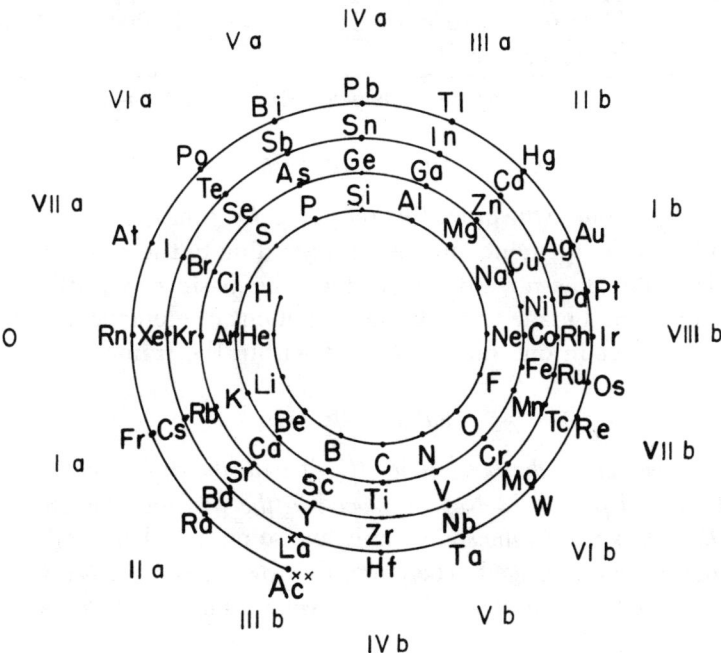

FIG. 50. Flavitskii 1887. Type IIB1-1.

Type IIC1-1

Series table with 16 vertical groups (Fig. 51). The first series consists of two short periods. The originator was Mendeleev in 1869 (1, p. 71), when he gave the fragments of this table in a footnote (see also Fig. 14). Later, Mendeleev wrote that this was the best and most convenient table. This table cannot be adopted now because no relationship exists between the elements in the middle of the first series (B, C, N, O, F, Ne, and Na), which are representative elements, and the elements in the middle of the next second series (Sc, Ti, V, Cr, Mn, Fe, Co, Ni, and Cu), which are transition elements.

Groups\Periods	I a	II a	III b	IV b	V b	VI b	VII b	VIII b	I b	II b	III a	IV a	V a	VI a	VII a	O
1.															H	He
2. 3.	Li	Be	B	C	N	O	F	Ne	Na	Mg	Al	Si	P	S	Cl	Ar
4.	K	Ca	Sc	Ti	V	Cr	Mn	Fe Co Ni	Cu	Zn	Ga	Ge	As	Se	Br	Kr
5.	Rb	Sr	Y	Zr	Nb	Mo	Tc	Ru Rh Pd	Ag	Cd	In	Sn	Sb	Te	I	Xe
6.	Cs	Ba	La˟	Hf	Ta	W	Re	Os Ir Pt	Au	Hg	Tl	Pb	Bi	Po	At	Rn
7.	Fr	Ra	Ac˟˟													

FIG. 51. Mendeleev 1869. Type IIC1–1.

Subtype IIC1–1a

Series table similar to the preceding table, but beginning with group III. Vaisman in 1948 (1) placed the elements in a series according to their isotope numbers: I = = N–Z. This table has no advantage over the basic type, IIC1–1. Earlier, in 1920, Kirchhof (0) had made a similar arrangement of elements according to their isotope numbers, but in one line instead of within the frame of a table.

Subtype IIC1–1b

Series table similar to the basic type IIC1–1 with the difference that the 1st period (H, He) and the 2nd period (Li–Ne) are placed in the first row, the 3rd period (Na–Ar) and half of the 4th period in the second row, and so on. In other words, the two halves of the drawing are interchanged. Therefore, the rows start with Na, Cu, Ag, and Au. The originator was Loew in 1897 (1). Dobrocvetov's table (3) belongs to this subtype only in some respects.

Type 2: Equal Lemniscate and Zigzag

Type IIA1–2

Space lemniscate (Fig. 52). The lemniscate is a curve that, like the numeral "8," consists of two loops. Two short periods are placed on the first lemniscate, each on one loop. The medium period elements are arranged on the following lemniscates: the first 10 elements on one loop and the remaining 8 elements on the second loop. The lemniscate and zigzag tables of this type, and also in the next types of lemniscates and zigzags, have a peculiarity that the Periodic Law of Mendeleev does not have. In these tables, the elements with opposite valences, such as $+1$ and -1, $+2$ and -2 etc., can be compared. Some examples are the elements Na and Cl, Mg and S, Al and P, and so on. However, this comparison of opposite valences is successful only for the representative elements, not for the transition elements. The originator of the space lemniscate was Sir Crookes in 1898 (1). The type is obsolete.

DIVISION II. MEDIUM TABLES 51

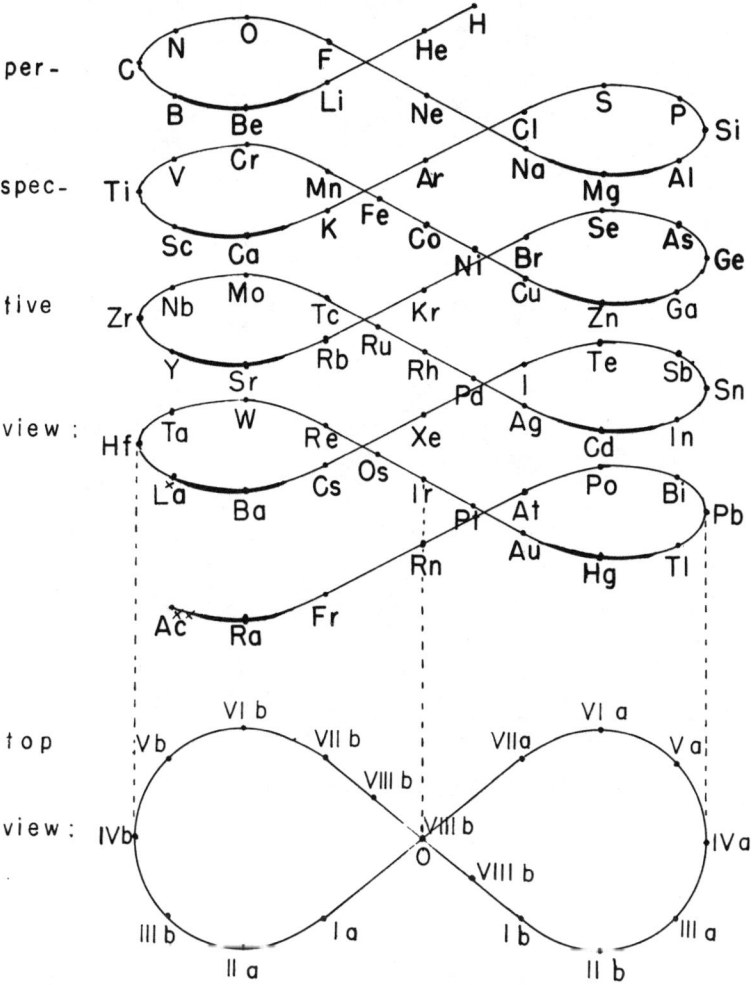

FIG. 52. Crookes 1898. Type IIA1–2.

Type IIC1–2

Zigzag table (Fig. 53). The principle of construction of the zigzag table and the previous lemniscate table is the same. The only difference is in the kind of drawing. This table was originated by Reynolds in 1886 (1), earlier than the preceding Crookes' lemniscate (in 1898). Crookes (2) referred to the Reynold's table (1). This table is obsolete.

52 GRAPHIC REPRESENTATIONS OF THE PERIODIC SYSTEM

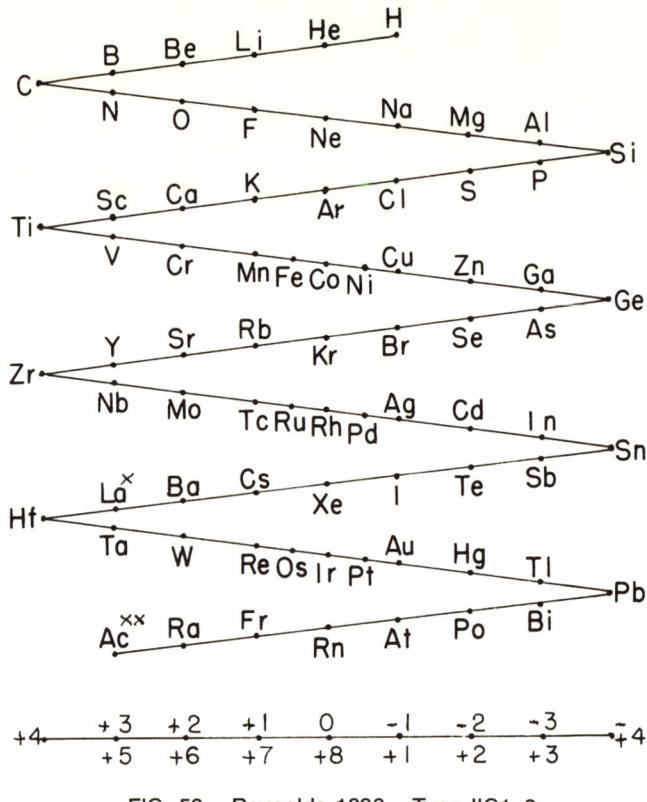

FIG. 53. Reynolds 1886. Type IIC1-2.

CLASS 2. TABLES WITH DISPOSITION OF ELEMENTS: 2, 8, AND 18

Type 1: Step Tables with Group Zero on One Side of the Table

Type IIA2-1

Helix wound on two main stepwise cylinders placed laterally with a common generating line at group 0 (Fig. 54). The originator of this table was Emerson in 1911 (1). Mendeleev suggested this helix and set down fragments of it in 1869 (00), but did not draw the helix itself. Ramsay (0) also suggested this kind of helix, but he did not draw it either. This type is not a convenient one.

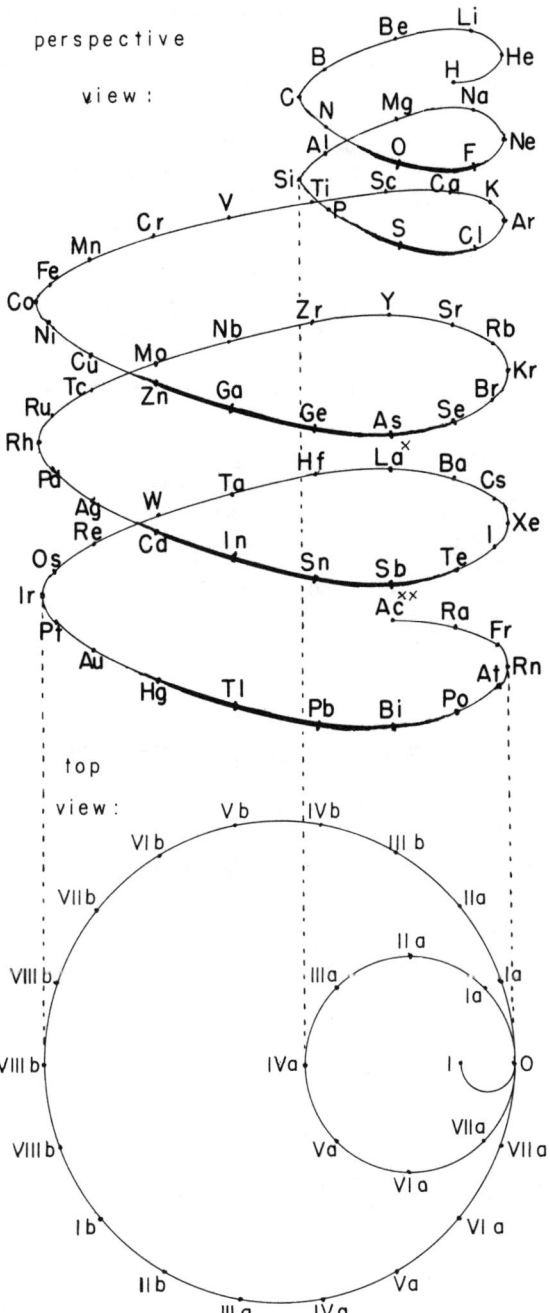

FIG. 54. B. K. Emerson 1911. Type IIA2-1.

Type IIB2-1

Spirals with two main sizes of revolutions for 8 and 18 elements (Fig. 55). The revolutions are not placed concentrically. Emerson was the originator in 1911

54   GRAPHIC REPRESENTATIONS OF THE PERIODIC SYSTEM

(1). Kunz (2) referred to Emerson. This type is not a convenient one.

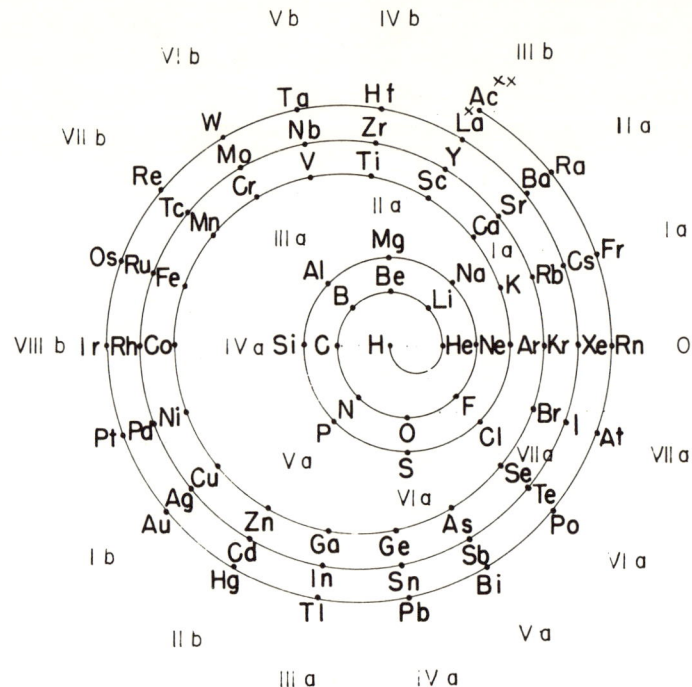

FIG. 55.   B. K. Emerson 1911.   Type IIB2-1.

Type IIC2-1

Left-step series table with group 0 on the right side of the table (Fig. 56). In this table, two elements of group Ia (Li and Na) are placed above the elements of group Ib—Cu, Ag, and Au. Nowadays, this is not acceptable because the aforementioned elements (Li, Na) belong to group Ia (above K, Rb, Cs, and Fr). The same can be said about the elements Be and Mg. This table was originated by Mendeleev in 1871 (see also Fig. 18). Scarpa (3) drew the elements of groups from Ia to VIIIb in reversed order.

Groups Periods:	I a	II a	III b	IV b	V b	VI b	VII b	VIII b	I b	II b	III a	IV a	V a	VI a	VII a	O
1.															H	He
2.									Li	Be	B	C	N	O	F	Ne
3.									Na	Mg	Al	Si	P	S	Cl	Ar
4.	K	Ca	Sc	Ti	V	Cr	Mn	Fe Co Ni	Cu	Zn	Ga	Ge	As	Se	Br	Kr
5.	Rb	Sr	Y	Zr	Nb	Mo	Tc	Ru Rh Pd	Ag	Cd	In	Sn	Sb	Te	I	Xe
6.	Cs	Ba	La	Hf	Ta	W	Re	Os Ir Pt	Au	Hg	Tl	Pb	Bi	Po	At	Rn
7.	Fr	Ra	Ac[xx]													

FIG. 56.   Mendeleev 1871.   Type IIC2-1.

Subtype IIC2–1a

Right-step series table with group 0 on the left side of the table. This table is not acceptable at all because the elements of most of the groups, III–VII, are not arranged according to their relationship. The table was suggested by Loring in 1913 (1). This table could be made acceptable by introducing connecting lines between the representative elements of the different periods, as in the table of type IIC2–3.

Type 2: Lemniscates and Zigzags with Group 0 on One Side of the Table

Type IIA2–2

Complex of helix and space lemniscate (Fig. 57). The helix is drawn for the short periods and the lemniscate for the medium periods. Because the representative elements lie one above the other on similar loops of the curves (as do the transition elements), it is easy to make comparisons among them. Also, the representative elements with opposite valences (+ and −) can be compared. The originator of this table was Soddy in 1914 (1). This type is better than the lemniscate type IIA1–2 (Fig. 52).

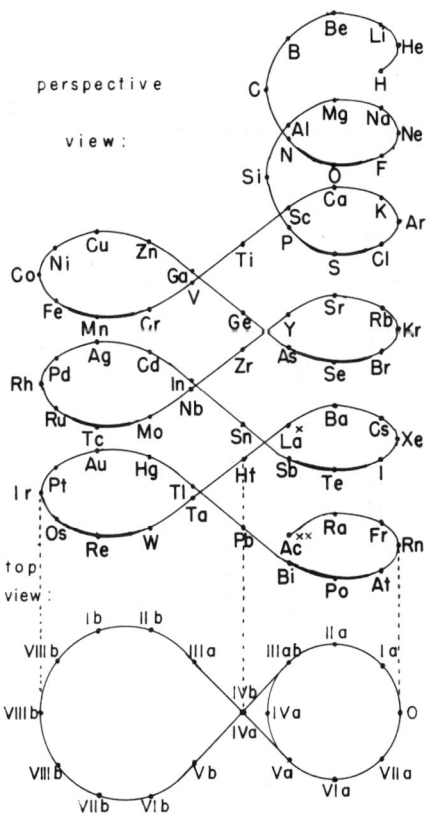

FIG. 57. Soddy 1914. Type IIA2–2.

Type IIB2-2

Complex of spiral and plane lemniscate (Fig. 58). The spiral and one-half of the lemniscate are drawn together and are reserved for the representative elements. The originator of this spiral-lemniscate was Kipp in 1942 (1). This type is acceptable as a medium-size table and can be used.

FIG. 58. Kipp 1942. Type IIB2-2.

Type IIC2-2

Two size zigzag table (Fig. 59). In principle, this zigzag table is no different from the helix-lemniscate of Soddy (Fig. 57). This zigzag table was originated earlier, by Woodiwiss in 1906 (1). The zigzag drawings of the originator and other cited authors had a step to the right. In Fig. 59, the drawing has been changed so that the step is to the left. This was done so that the zigzag could be compared with the lemniscate. This type is not a convenient one.

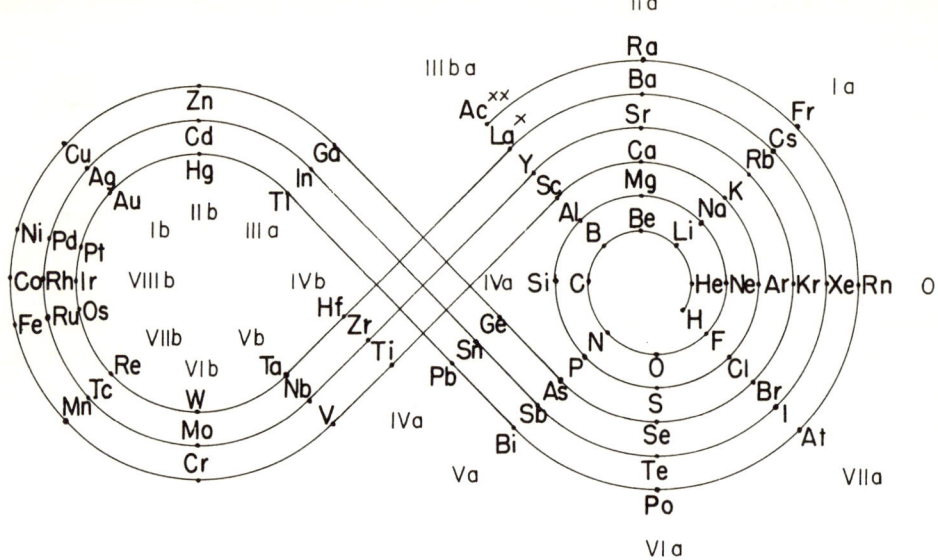

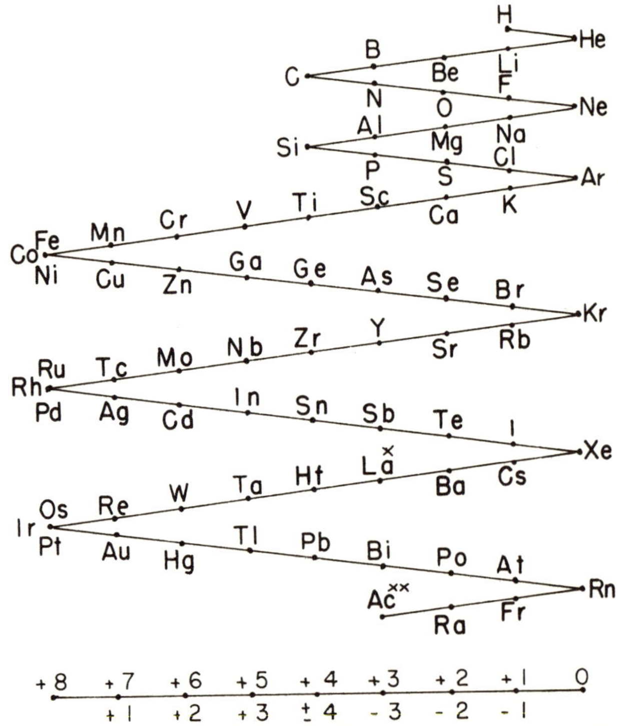

FIG. 59. Woodiwiss 1906. Type IIC2-2

Subtype IIC2-2a

Zigzag table as before, except that the short periods are in one series. This subtype has no advantages. Deeley (1) did not draw the zigzag line, but the table was built as a zigzag table because the series had to be read in different directions. Mendeleev (0) drew only a fragment of this table.

Subtype IIC2-2b

Partial zigzag table where the short periods are in one series and the medium periods in a triple zigzag. In the middle part of the triple zigzag, the elements of group VIII are located opposite groups VI, IV, and II. This table was originated by Spring in 1881 (1). Earlier, in 1869, the forerunner Hinrichs (0) published a table (see Fig. 10) that resembles this subtype except that located on the middle part of the triple zigzag were only those transition elements known at the time. At the present time this kind of table is obsolete. However, Shchukarev in 1965 (3) used the idea of this type and located the second half of the transition elements on the middle part of the zigzag line. With this kind of table, he acknowledges the retreat of valences in this part of the transition elements; and, therefore, the numeration of the groups occurs in reversed order on the appropriate line of the zigzag. His table (Fig. 60) is also acceptable as a short table.

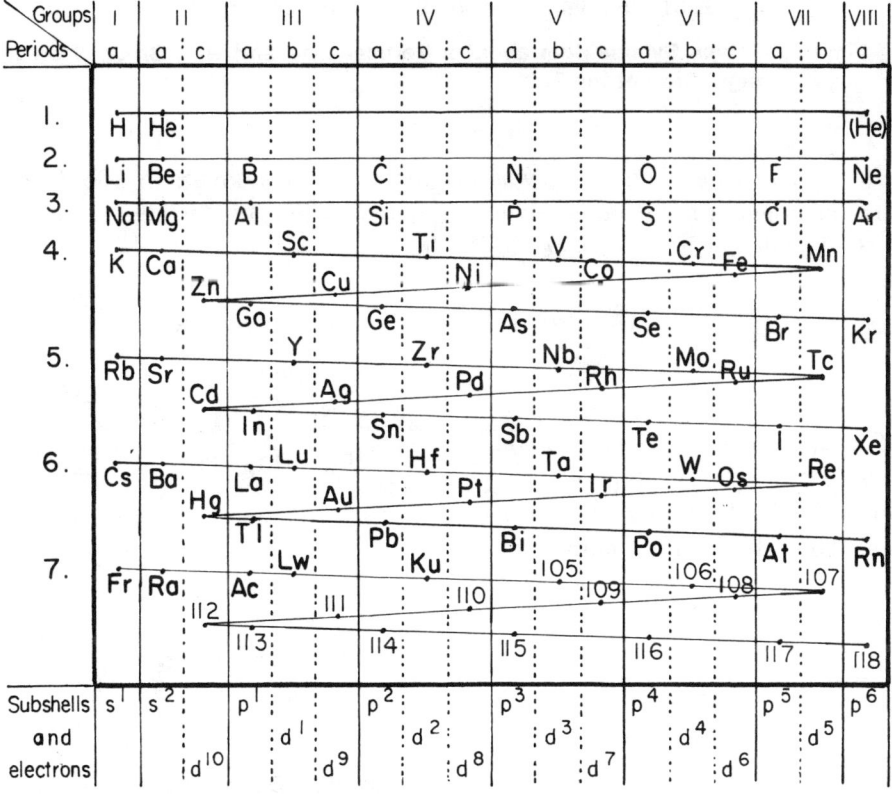

FIG. 60. Shchukarev 1965. Subtype IIC2-2b.

Subtype IIB2-2c

Complex of sinusoids and partial lemniscates (Fig. 61). Actually, drawing these curves represents an attempt to put the space lemniscate of Soddy (type IIA2-2) on a plane. The comments about that table apply to this one. In this table, the valence increases from 0 to 8 from the bottom to the top, excepting the second half of the transition elements (Co, Ni, Cu, Zn, etc.). This table was suggested by Bindel and Blickle in 1952 (1).

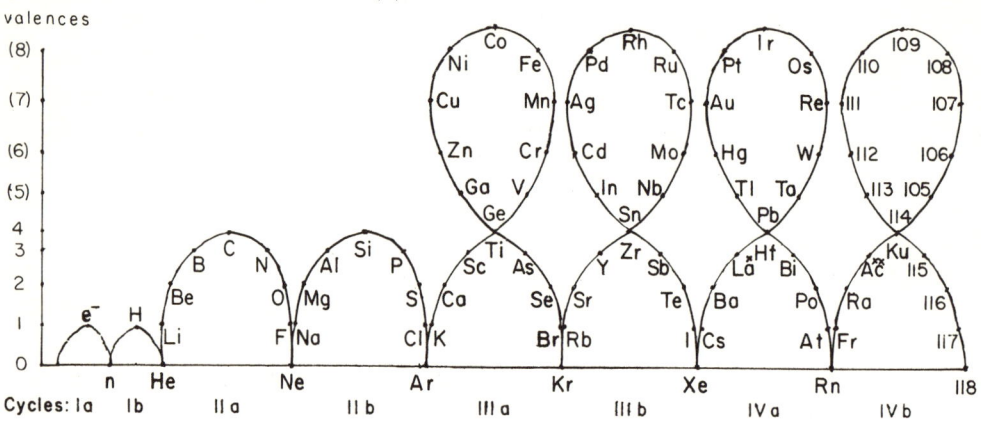

FIG. 61. Bindel a. Blickle 1952. Subtype IIB2-2c.

Type 3: Tables Symmetrical about a Vertical Line and with Group 0 on the Right Side of the Table

Type IIA2-3

Helix on a cone. This helix was originated by Tomkeieff in 1954 (1). The long size analogue, the "Pagoda," was originated earlier, by Aucken in 1951 (type IIIA3-3). This is not a very successful table.

Type IIB2-3

Two sizes of a spiral placed concentrically (Fig. 62). This type was suggested by Oddo (1), who gave only its scheme. The table is not a convenient one.

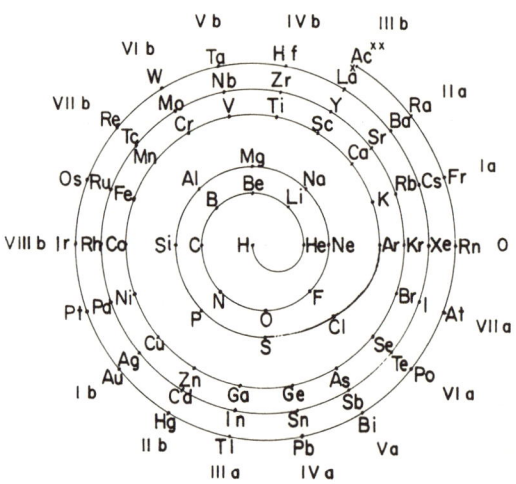

FIG. 62. Oddo 1925. Type IIB2-3.

Type IIC2–3

Pyramidal symmetrical series table with centered steps (Fig. 63). The originator of this type was Carnelley in 1886 (1). This table is a successful one. The analogous type of long table IIIC3–3 (Fig. 92) was originated by Bayley in 1882. Carnelley refers to him. (The table originated by Bayley is the so-called Thomsen-Bohr table.) Some authors—von Richter (2), Nernst (3), Stefan Meyer (5), Roscoe and Schorlemmer (6), Rudorf (7), Hopkins (11), Emerson (12), and Butler (15)—did not draw the connecting lines between the elements of the short and medium periods.

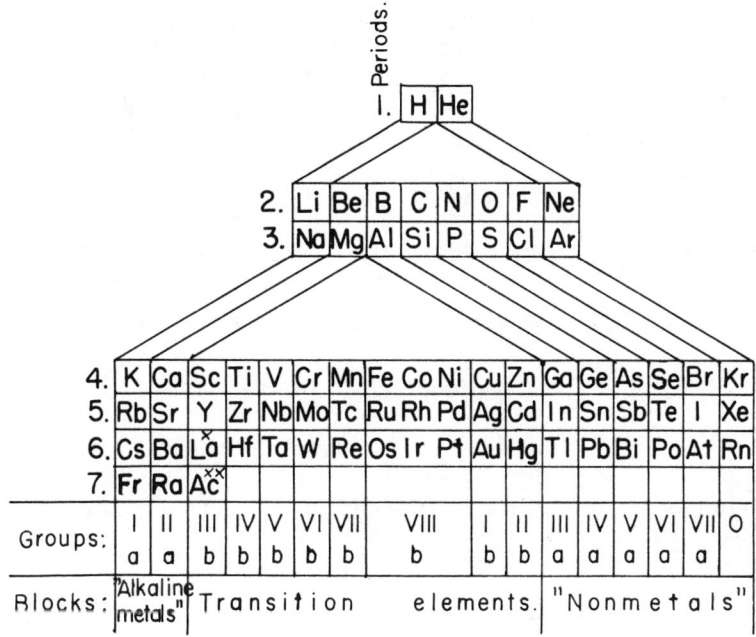

FIG. 63. Carnelley 1886. Type IIC2–3.

Subtype IIC2–3a

Symmetrical series table constructed in the form of a quadrangle (Fig. 64). This table does not have steps because the short periods are "stretched" in a horizontal direction. The table was originated by von Antropoff, but published by Mark von Stackelberg in 1925 (1), one year earlier than von Antropoff himself published it. This subtype is a successful one. In 1909, Stefan Meyer (0) published an imperfect table of this type without the connecting lines. Its analogue among the long tables is the subtype IIIC3–3b (Fig. 94), which was originated earlier, by Hackh in 1914.

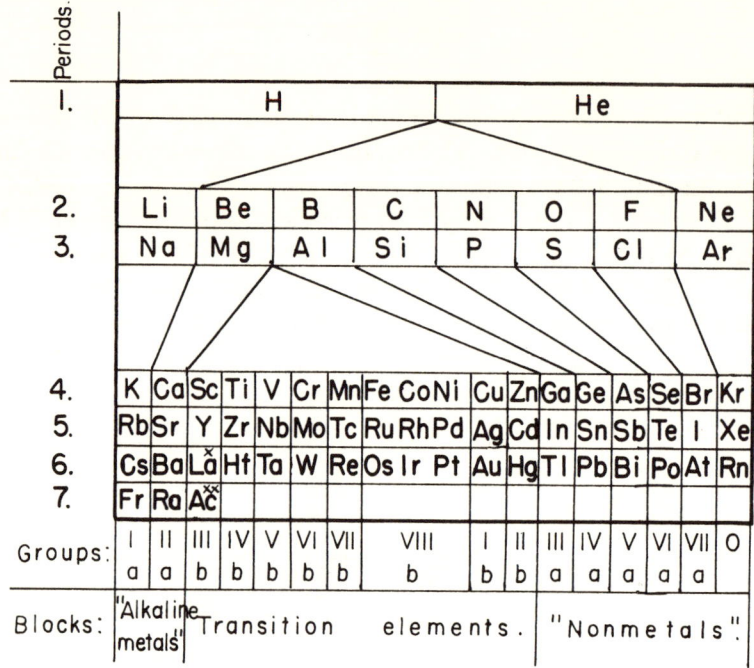

FIG. 64. von Antropoff 1925. Subtype IIC2–3a.

Subtype IIC2–3b

Symmetrical series table with centered steps similar to the basic type IIC2†3. The difference is that the inner transition elements are arranged in two rows as subgroup elements in groups IIa–VIIIb. Grigorovich (1) suggested this subtype in 1963.

Subtype IIC2–3c

Symmetrical series table as a quadrangle similar to the subtype IIC2–3a. The difference is that the inner transition elements are inserted in one row as subgroup elements in groups IIIb–VIIb, Ib–VIIa, and in Ia and IIa. This table, suggested by Schenk (1) in 1951, is not a successful one.

Type 4: Tables with Interrupted Short Periods

Type IIA2–4

Helix wound on one cylinder with interruptions in the placing of the elements of the short periods on the curve (Fig. 65). Stackelberg suggested this kind of helix in 1911 (1), but did not draw it. This type is a good one and could be used.

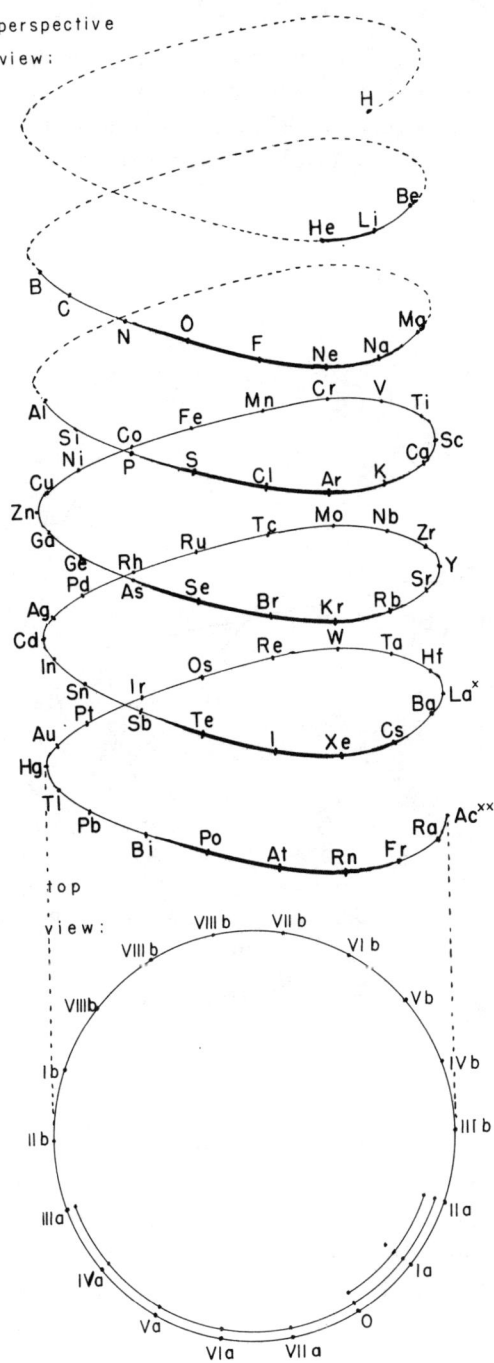

FIG. 65. E. von Stackelberg 1911. Type IIA2-4.

Type IIB2–4

Spiral with 16 radii with interruptions in the placing of the elements of the short periods. (Fig. 66). The originator of this spiral was Tocher in 1910 (1). This type could be used. Caswell (5), Bilibin (6), Emerson (7), and Sokoloff (9) had 18 radii on the spiral. Hackh's spiral (3) was reversed—it started from the outside; element U was placed in the center.

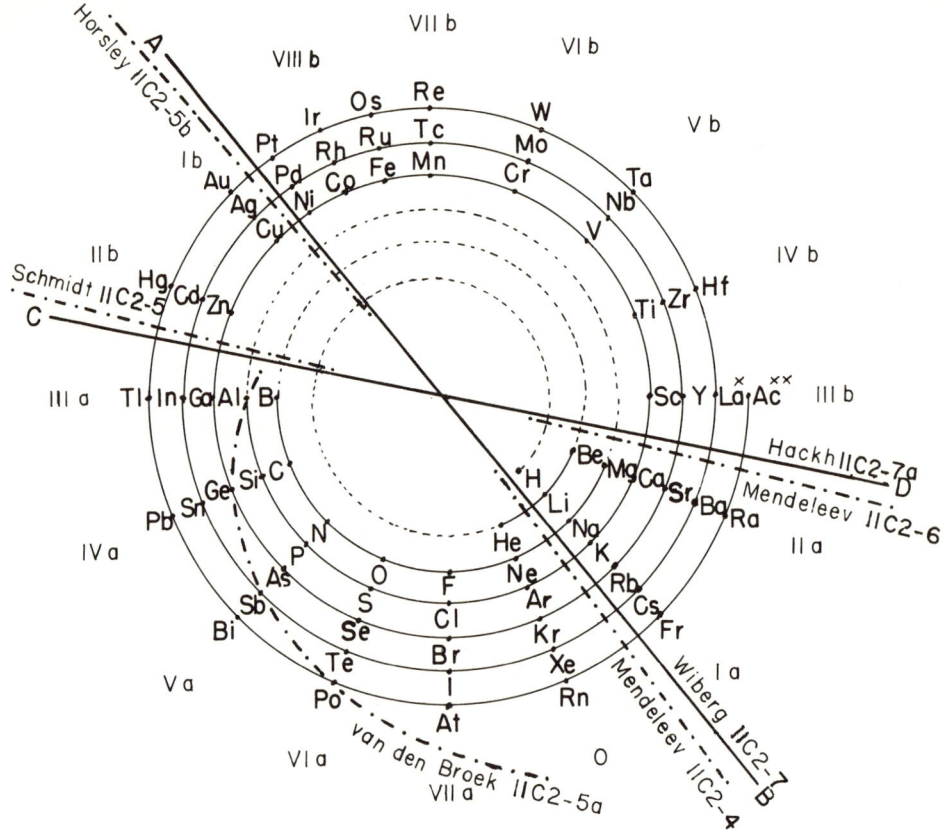

FIG. 66. Tocher 1910. Type IIB2–4.

The helix of this type, IIA2–4 (Fig. 65), and the spiral, IIB2–4 (Fig. 66), are prototypes not only for the series table IIC2–4, but also for the next series tables; types IIC2–5, –5a, –5b, –6, –7, and –7a. All these series tables could be obtained from the helix or spiral by cutting them in different places and flattening the surface of the cylinder on which the helix is wound or straightening the revolutions of the spiral. The various ways of cutting are shown on the spiral in Fig. 66.

Type IIC2–4

Series table with interrupted short periods (Fig. 67). This table can be obtained by cutting the mentioned spiral (Fig. 66) between groups 0 and Ia. The table was originated by Mendeleev in 1869 (1) (see also Fig. 13). Since Mendeleev did not like interrupted series, he did not draw the table but gave only fragments of it. This is the most popular table among the medium-size tables and is a good one. The phenomenon that element properties change from nonmetallic to metallic from top to bottom in vertical groups and from metallic to nonmetallic from left to right in horizontal rows is best reflected by this table and by Werner's table, type IIIC3–4. Wrigley, Mast, and McCutcheon (49) printed the element symbols on different levels of a model by the use of lamina closer to or further from the base in agreement with their membership in different shells. Seaborg (44, p. 149)), in his article of 1969, included, besides the lanthanides and actinides, the superactinides for the 8th period as a footnote; therefore, the total number of elements in this table is 168. Mojos and Incolla (64) and Nebel (66) also used superactinides and had a table with a total of 168 elements. Heist (63) constructed a folding table that enabled him to obtain from this type a table including only the representative elements. Allen (65, p. 23) drew the horizontal rows obliquely by diagonals on his table while the vertical columns remained as they were. On his second table (65, p. 24), he drew both the horizontal rows and the vertical columns by diagonals.

Groups / Periods	I a	II a	III b	IV b	V b	VI b	VII b	VIII b	I b	II b	III a	IV a	V a	VI a	VII a	O
1.	H															He
2.	Li	Be									B	C	N	O	F	Ne
3.	Na	Mg									Al	Si	P	S	Cl	Ar
4.	K	Ca	Sc	Ti	V	Cr	Mn	Fe Co Ni	Cu	Zn	Ga	Ge	As	Se	Br	Kr
5.	Rb	Sr	Y	Zr	Nb	Mo	Tc	Ru Rh Pd	Ag	Cd	In	Sn	Sb	Te	I	Xe
6.	Cs	Ba	La˟	Hf	Ta	W	Re	Os Ir Pt	Au	Hg	Tl	Pb	Bi	Po	At	Rn
7.	Fr	Ra	Ac˟˟													
Blocks:	"Alkali metals"		Transition elements.								"Nonmetals."					

FIG. 67. Mendeleev 1869. Type IIC2–4.

Subtype IIC2–4a

Series table similar to the basic type IIC2–4. The difference is that the inner transition elements are arranged in subgroups "c" so that the symbols are placed in groups IIIb to VIIIb and in IIb in two rows. This kind of table was arranged by Grigorovich (1). Earlier, in 1922, a similar attempt was made by Norrish (0). The inner transition elements are not well fitted into their places.

Type 5: Tables with Group Zero in the Middle of the Table

Type IIC2–5

Series table with group 0 in the middle of the table (Fig. 68). Each series starts with group IIIa. The representative elements are on the left side of the

64 GRAPHIC REPRESENTATIONS OF THE PERIODIC SYSTEM

table and the transition elements on the right. This table can be obtained by cutting the spiral (Fig. 66) between groups IIb and IIIa. The originator of this table was Schmidt in 1918 (1). The table can be used as a medium-size table. Grimm (2) substituted valence designations (positive and negative) for group names and arranged the nonmetals not only on the left but also on the right side of the table. Remy (4) suggested establishing a new numeration of periods: the nonmetals should be before the noble gases and the alkaline metals after them.

Groups:	III a	IV a	V a	VI a	VII a	O	I	II	III b	IV b	V b	VI b	VII b	VIII b		I b	II b
							H	He	Li	Be								
	B	C	N	O	F	Ne	Na	Mg									
	Al	Si	P	S	Cl	Ar	K	Ca	Sc	Ti	V	Cr	Mn	Fe	Co	Ni	Cu	Zn
	Ga	Ge	As	Se	Br	Kr	Rb	Sr	Y	Zr	Nb	Mo	Tc	Ru	Rh	Pd	Ag	Cd
	In	Sn	Sb	Te	I	Xe	Cs	Ba	La	Hf	Ta	W	Re	Os	Ir	Pt	Au	Hg
	Tl	Pb	Bi	Po	At	Rn	Fr	Ra	Ac**									
Blocks:	Represent. elements.								Transition elements.									

FIG. 68. Schmidt 1918. Type IIC2–5.

Subtype IIC2–5a

Series table similar to the previous table with the difference that each series begins with a nonmetal (Fig. 69). This table is obtained by cutting the spiral (Fig. 66) by means of a curved line, as shown in that figure. Van den Broek originated the table in 1911 (1). This subtype is even better than the preceding table because the elements are approximately in the order in which the elements would be taught in an elementary chemistry course.

Groups:	III a	IV a	V a	VI a	VII a	O	I	II	III b	IV b	V b	VI b	VII b	VIII b		I b	II b	III a	IV a	V a	VI a	VII a
							H	He	Li	Be												
	B	C	N	O	F	Ne	Na	Mg	Al													
		Si	P	S	Cl	Ar	K	Ca	Sc	Ti	V	Cr	Mn	Fe	Co	Ni	Cu	Zn	Ga	Ge		
			As	Se	Br	Kr	Rb	Sr	Y	Zr	Nb	Mo	Tc	Ru	Rh	Pd	Ag	Cd	In	Sn	Sb	
				Te	I	Xe	Cs	Ba	La	Hf	Ta	W	Re	Os	Ir	Pt	Au	Hg	Tl	Pb	Bi	Po
					At	Rn	Fr	Ra	Ac**	Ku	105	106	107	108	109	110	111	112	113	114	115	116	117
Blocks:	Nonmetals.						"Alkaline metals"		Transition elements.									Metals of nonmet.groups.				

FIG. 69. van den Broek 1911. Subtype IIC2–5a.

Subtype IIC2–5b

Series table similar to the basic type with the difference that each series begins with the group Ib (Fig. 70). The series of the short periods begin with group IIb.

The table can be obtained by cutting the spiral (Fig. 66) between groups VIIIb and Ib. The originator was Horsley in 1893 (1). This table is related to subtype IIC1–1b. Sommerfeldt's table (7) can be classified as belonging to either of these two subtypes. This table is not as good as the basic type.

Groups:	I b	II b	III a	IV a	V a	VI a	VII a	O	I a	II a	III b	IV b	V b	VI b	VII b	VIII b
								H	He	Li						
		Be	B	C	N	O	F	Ne	Na							
		Mg	Al	Si	P	S	Cl	Ar	K	Ca	Sc	Ti	V	Cr	Mn	Fe Co Ni
	Cu	Zn	Ga	Ge	As	Se	Br	Kr	Rb	Sr	Y	Zr	Nb	Mo	Tc	Ru Rh Pd
	Ag	Cd	In	Sn	Sb	Te	I	Xe	Cs	Ba	La˟	Hf	Ta	W	Re	Os Ir Pt
	Au	Hg	Tl	Pb	Bi	Po	At	Rn	Fr	Ra Ac˟˟						

FIG. 70. Horsley 1893. Subtype IIC2–5b.

Type 6: Step Tables with Group IIa on the Right Side of the Table

Type IIC2–6

Series table with steps to the left and with group IIa on the right side of the table (Fig. 71). This table is obtained by cutting the spiral (Fig. 66) between groups IIa and IIIb. Mendeleev was the originator of the table in 1869 (1) (see also Fig. 12). Two of the forerunners of the Periodic Law, William Odling (00) in 1864 and Lothar Meyer (0) in 1868, originated similar periodic tables. Meyer did not publish his table at this time; it was published after his death by Karl Seubert in 1895. Silbermann (2) and Pascal (6) had slightly different tables. This table is a good one and is related to the long table, type IIIC3–6, originated by Janet. Fig. 71 shows the adjustment of the new periods to this table.

New periods \ Groups	III b	IV b	V b	VI b	VII b	VIII b	I b	II b	III a	IV a	V a	VI a	VII a	O	I a	II a
1.															H	He
2.															Li	Be
3.									B	C	N	O	F	Ne	Na	Mg
4.									Al	Si	P	S	Cl	Ar	K	Ca
5.	Sc	Ti	V	Cr	Mn	Fe Co Ni	Cu	Zn	Ga	Ge	As	Se	Br	Kr	Rb	Sr
6.	Y	Zr	Nb	Mo	Tc	Ru Rh Pd	Ag	Cd	In	Sn	Sb	Te	I	Xe	Cs	Ba
7.	La˟	Hf	Ta	W	Re	Os Ir Pt	Au	Hg	Tl	Pb	Bi	Po	At	Rn	Fr	Ra
8.	Ac˟˟															
Blocks:	Transition elements.								Represent. elem.							

FIG. 71. Mendeleev 1869. Type IIC2–6.

Type 7: Mirror-Image Tables

Type IIC2-7

Series table consisting of two parts which are the mirror images of each other (Fig. 72). The spiral (Fig. 66) is cut in two places by means of the diameter AB, and groups Ib and IIb are replaced by Ia and IIa and vice versa. In addition, the whole section that contains the transition elements is turned so that the subgroups "b" are in the same vertical columns as the corresponding subgroups "a." Each series begins with group I. Wiberg originated this type in 1936 (1). He placed the transition elements below the representative elements to accomplish the comparison of these two kinds of elements instead of using the subgroup method, but Fig. 72 contains the transition elements on the top of the representative elements to obtain a method of mirror-image comparison. The table is more like a short-size table, but it is not a very convenient one. Woodiwiss (0) had a similar table.

Periods									Blocks
6b 7b	Au	Hg	Ac^xx						Trans. elem.
5b 6b	Ag	Cd	La^x	Hf	Ta	W	Re	Os Ir Pt	
4b 5b	Cu	Zn	Y	Zr	Nb	Mo	Tc	Ru Rh Pd	
4b			Sc	Ti	V	Cr	Mn	Fe Co Ni	
Groups:	I	II	III	IV	V	VI	VII	VIII	
1.	H							He	Representat. elem.
2.	Li	Be	B	C	N	O	F	Ne	
3.	Na	Mg	Al	Si	P	S	Cl	Ar	
4a	K	Ca	Ga	Ge	As	Se	Br	Kr	
5a	Rb	Sr	In	Sn	Sb	Te	I	Xe	
6a	Cs	Ba	Tl	Pb	Bi	Po	At	Rn	
7a	Fr	Ra							

FIG. 72. Wiberg 1936. Type IIC2-7

Subtype IIC2-7a

Series table similar to the preceding table with the difference that each series begins with group III (Fig. 73). The cut in the spiral (Fig. 66) is made in two places by means of the diameter CD. The originator was Hackh in 1914 (1). This table cannot be judged a convenient one. Hackh tried to arrange the inner transition elements into the table: later he abandoned this attempt.

Periods										Blocks
7 b	Lw	Ku								Transit. elem.
6 b	Lu	Hf	Ta	W	Re	Os Ir Pt	Au	Hg		
5 b	Y	Zr	Nb	Mo	Tc	Ru Rh Pd	Ag	Cd		
4 b	Sc	Ti	V	Cr	Mn	Fe Co Ni	Cu	Zn		
Groups:	III	IV	V	VI	VII	VIII	I	II		
1. 2.					H	He	Li	Be		Repres. elem.
2. 3.	B	C	N	O	F	Ne	Na	Mg		
3. 4a	Al	Si	P	S	Cl	Ar	K	Ca		
4a 5a	Ga	Ge	As	Se	Br	Kr	Rb	Sr		
5a 6a	In	Sn	Sb	Te	I	Xe	Cs	Ba	La^x	
6a 7a	Tl	Pb	Bi	Po	At	Rn	Fr	Ra	Ac^xx	

FIG. 73. Hackh 1914. Subtype IIC2-7a.

Division III. Long Tables

In the long tables, the inner transition elements are arranged as a substantial part of the table. Therefore, the long tables are better than the medium tables or the short tables without the third subgroup because the long tables reflect the true picture of nature.

Long tables are divided as Chemical Tables and Electronic Configuration Tables.

SUBDIVISION IIIA: CHEMICAL TABLES

Chemical Tables are of three classes:
1) tables of one revolution and of one row.
2) tables with a disposition of elements: 4, 16, 36, and 64 (cycles).
3) tables with a disposition of elements: 2, 8, 18, and 32 (periods).

Class 1 has more theoretical interest than practical importance. Interesting tables in this class are the valence table of Mendeleev, the so-called "periodic chain" (subtype IIIC1-1a), and the tables where the elements are described by means of electrons (subtype IIIB1-1b and IIIC1-1b).

Class 2 is important because of the disposition of elements in cycles. Type 2 of this class contains the lemniscates.

Class 3 is analogous to class 2 of the medium tables. The disposition of the elements is 2, 8, 18, and 32. It has six types.

The tables of type 1 are stepwise tables with group 0 usually on the right side of the table. These are not good tables.

Type 2 contains lemniscate and zigzag tables with the same basis of arrangement as type 1.

Tables belonging to type 3 are symmetrical about a vertical line. Here belong 1) the famous so-called Thomsen-Bohr table, actually originated by Bailey (type IIIC3-3), 2) Hackh's quadrangle table (subtype IIIC3-3b), and, finally, 3) the very interesting triangle table by Wagner and Booth (subtype IIIC3-3c).

Type 4 has interruptions in the short and medium periods. The famous Werner's table (type IIIC3-4) belongs to this type.

Type 5 contains tables with group 0 in the middle of the tables.

To type 6 belong left step tables with group IIa on the right side. One of these tables is the very important Janet table (type IIIC3-6).

Class 1. Tables of One Revolution and One Row

Type 1: Tables of One Revolution and of One Row.

68 GRAPHIC REPRESENTATIONS OF THE PERIODIC SYSTEM

Type IIIB1-1

One revolution of an Archimedes spiral for all elements (Fig. 74). Opolonick originated this spiral in 1935 (1). It has no advantage over other tables. Earlier, Loew (0) had made a similar attempt by placing the elements on 2½ revolutions of the Archimedes spiral. Opolonick had 90 elements on one revolution; the angle θ between two elements was 4°. As is shown on the figure, putting 120 elements on one revolution is more successful; in this case, the angle θ between the two elements is 3°. The concentric circles separate areas where the elements belonging to different periods (according to the new division) are located.

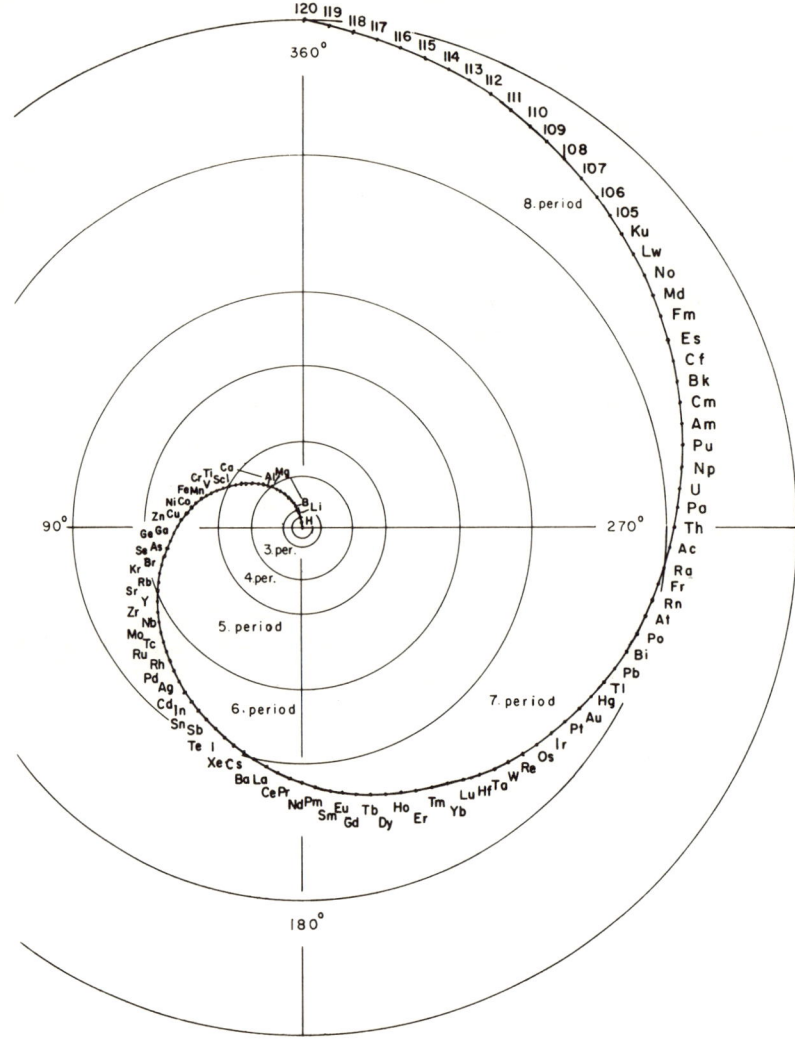

FIG. 74. Opolonick 1935. Type IIIB1-1.

Type IIIC1-1

Table of one series. Lea (1) placed the colorless and colored elements on two different lines of a single series. By this arrangement, he obtained a table where the representative elements were on the lower line and the transition and inner transition elements on the upper line. Von Antropoff (2) gave a special numeration for each period.

Subtype IIIC1-1a

Table of one series with valence numbers for each element, so-called "periodic chain" (Fig. 75.) Mendeleev originated the table in 1889 (1). This is not a bad table since it is naturally empirical. Rinck and Feschotte (7) improved the table by dividing it and writing the cycles separately.

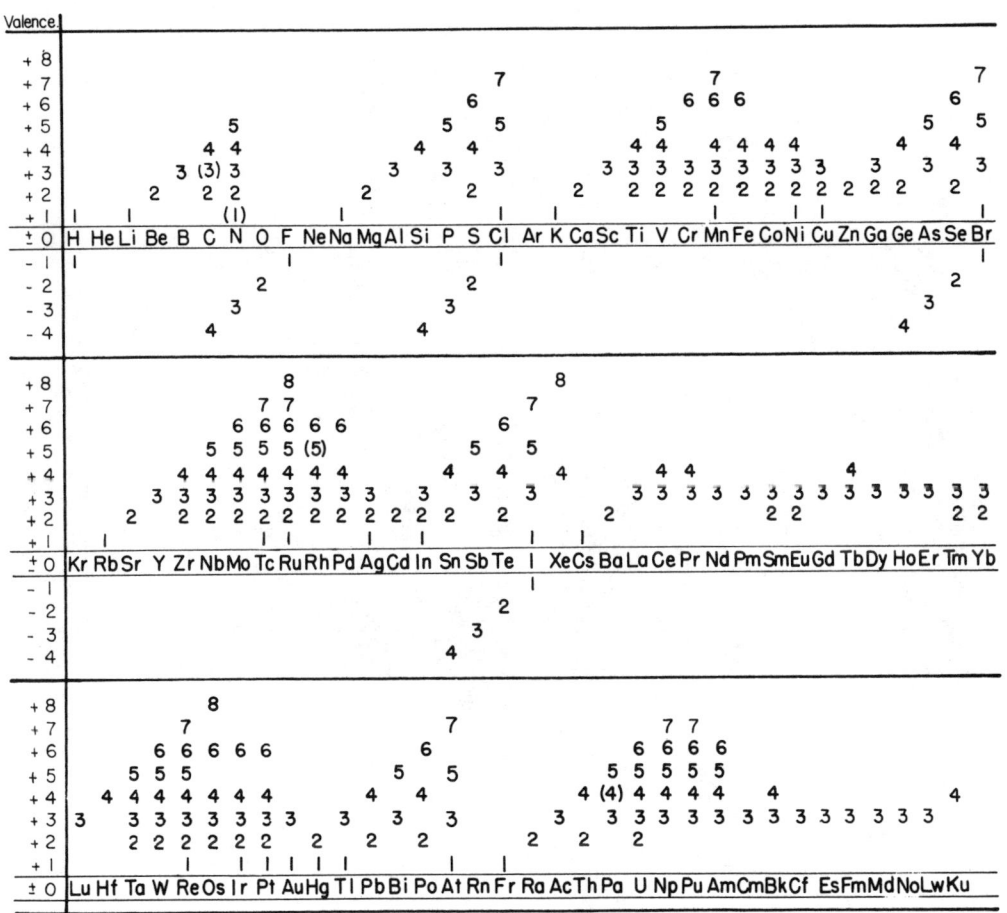

FIG. 75. Mendeleev 1889. Subtype IIIC1–1a.

Subtype IIIB1–1b

Table of one circle around a disc. The atomic numbers and the element symbols are placed around the outside circle of a disc. The center of the disc contains the positive nucleus. Around the nucleus the shells and subshells are located in the same order as is shown in Table 3 (see p. 98). The completion of the subshells with electrons occurs according to the Table 4 (see p. 100). The subshells are separated from each other by concentric circles. The disc is divided also by radii at angles of 3° into 120 sectors, which are determined for 120 elements: 104 known and 16 hypothetical. Both the concentric circles and the radii form small quadrangles similar to trapezoids. Each trapezoid represents one subshell. In these trapezoids are located as many electrons as are required for a certain element. For subshells s the electrons are shown as ·, for subshells p as ^{x}, for subshells d as $_-$, and for subshells as ^{1}. By this arrangement, the whole picture of electron configuration for all elements is represented in one drawing.

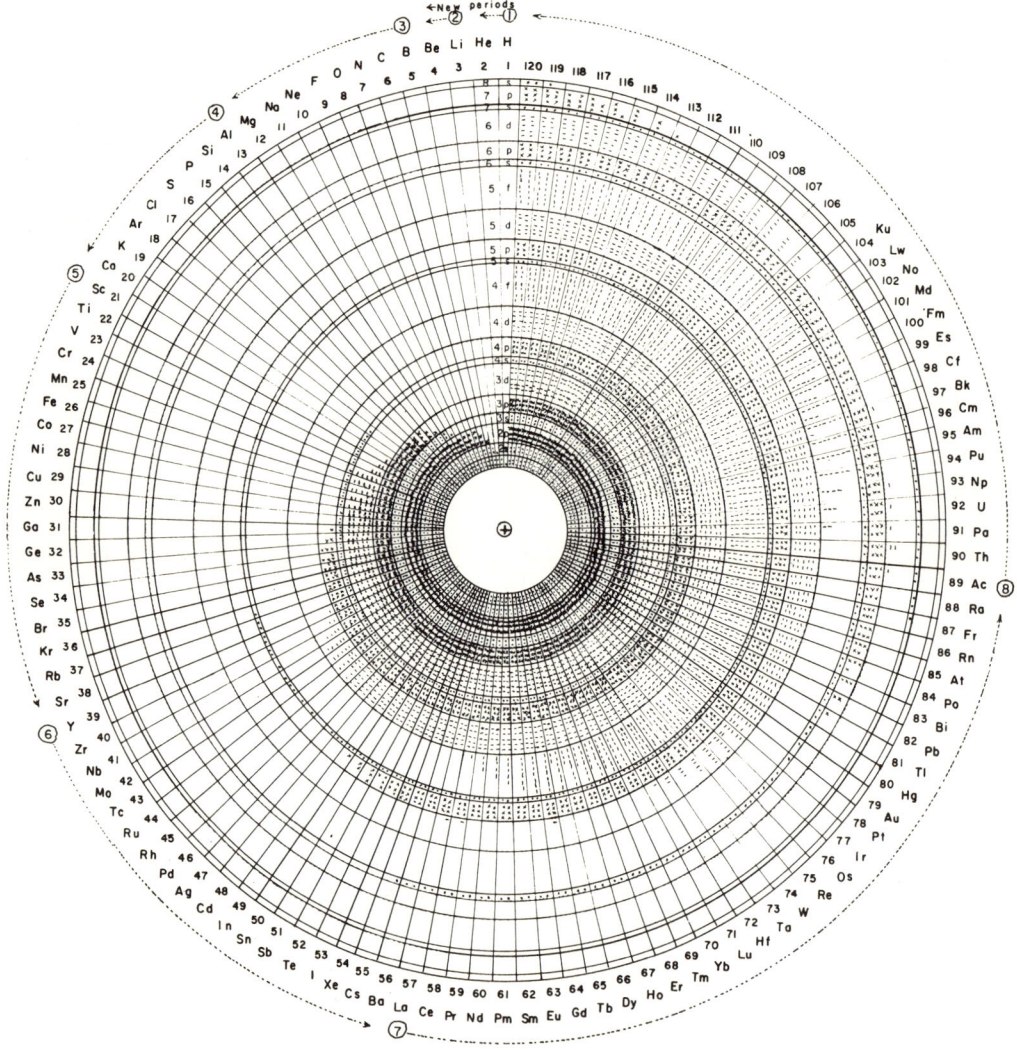

FIG. 76. Stoye 1954. Subtype IIIB1–1b.

Subtype IIIC1–1b

Table of one vertical column, where the atomic numbers and the element symbols are placed from the top to the bottom (Fig. 77). On the table itself, the number of valence electrons of the elements are shown by dots. Since the electrons

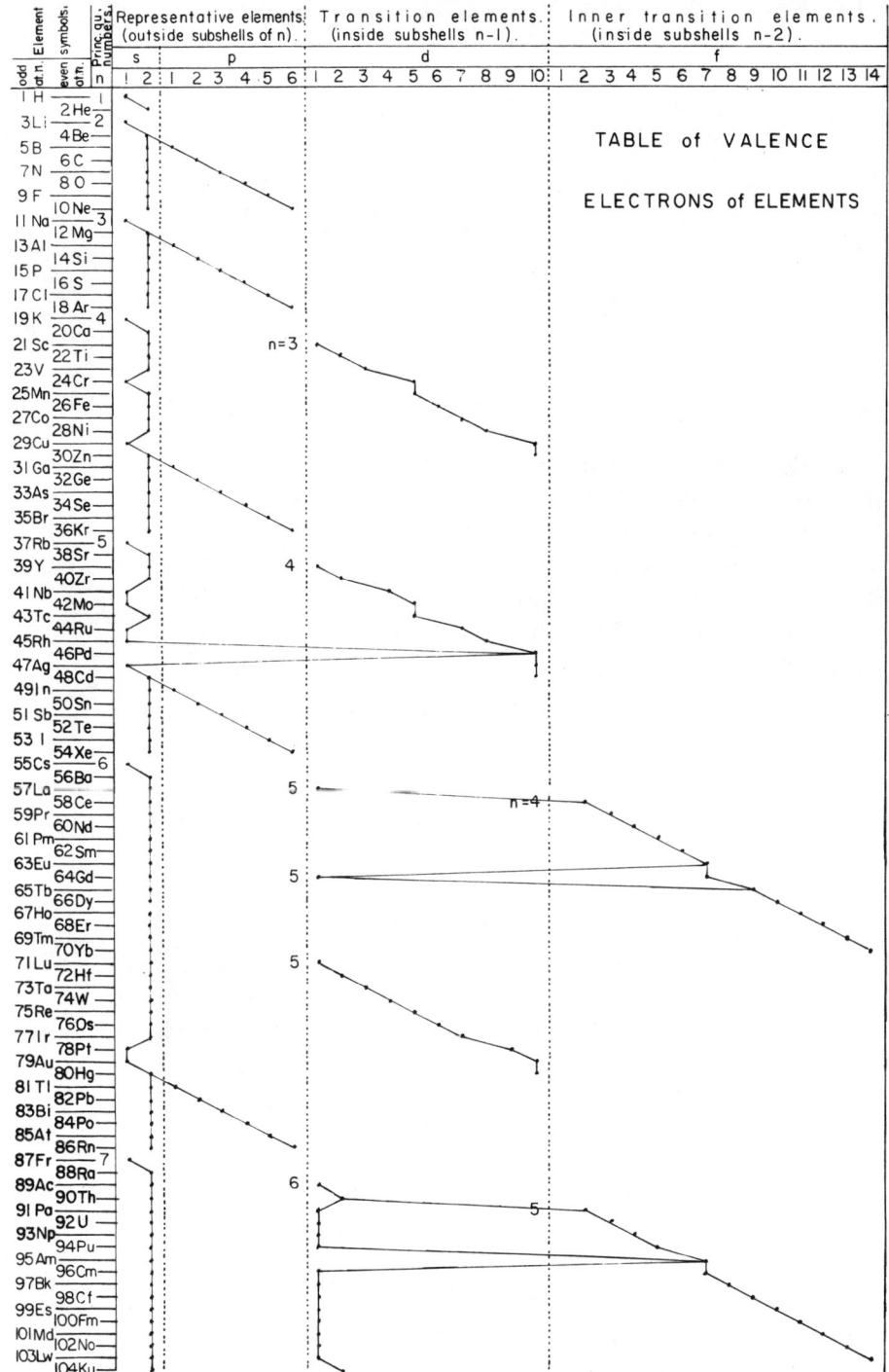

FIG. 77. Vincent 1969. Subtype IIIC1–1b.

often appear in different subshells, they should be summed. In some sections the electrons are located irregularly; this is shown with additional lines. This series table represents the circle table of subtype IIIB1-1b, but only for the valence electrons. This interesting table, originated by Vincent in 1969 (1), is reminiscent the Gardner table (type IIIC4-2).

CLASS 2. TABLES WITH DISPOSITION OF ELEMENTS: 4, 16, 36, AND 64 (CYCLES)

Type 1: Tables of Four Planes, Four Revolutions, or Four Cycles.

Type IIIA2-1

Four planes in space with 4, 16, 36, and 64 elements on each plane (Fig. 78). Each plane contains two periods. The originator was Kapustinskii in 1953 (1). Bindel (2), in 1958, gave only a scheme for this kind of table. This is not a convenient table.

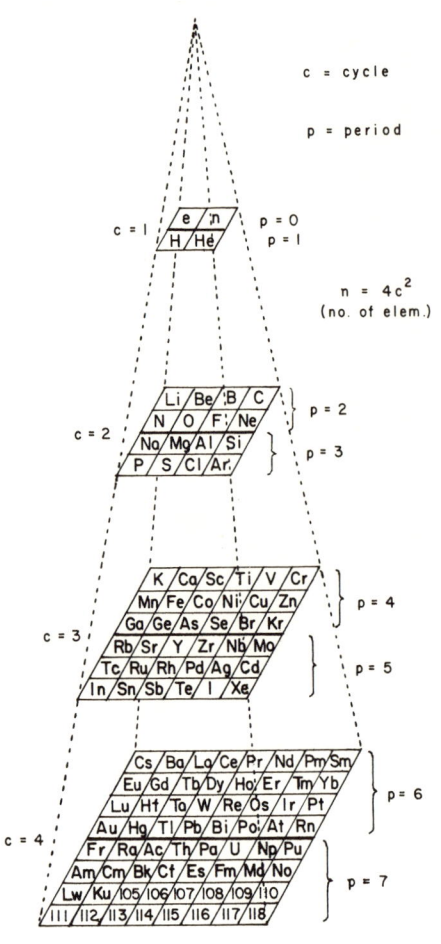

FIG. 78. Kapustinskii 1953. Type IIIA2-1.

Type IIIB2-1

Spiral of four revolutions with 4, 16, 36, and 64 elements on each revolution (Fig. 79). As shown on the drawing, the spiral can be divided by lines into zones of representative, transition, and inner transition elements. Rydberg was the originator in 1913 (1). This is a very interesting table, although not very convenient to read.

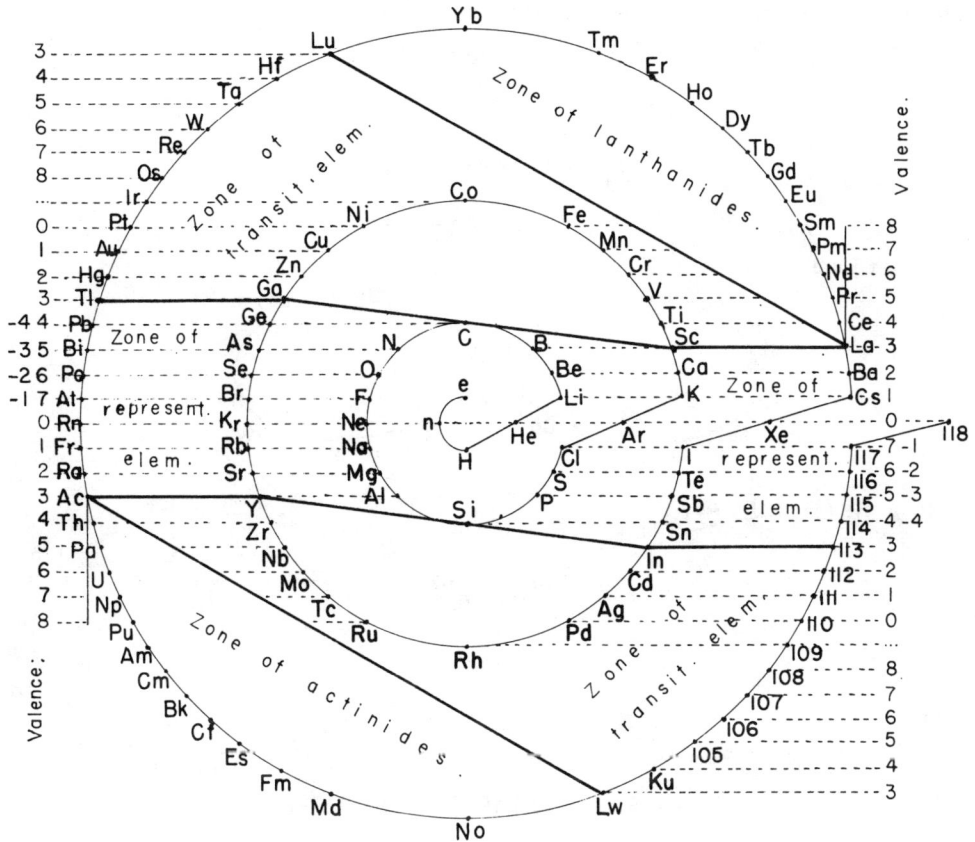

FIG. 79.   Rydberg 1913.   Type IIIB2-1.

Type IIIC2-1

Series table in cycles (Fig. 80). This table is adjusted to the new division into periods (see Introduction to the Electronic Configuration Tables). This table is not an important one, but is included to present the two previous types as a series table.

Cycles	Periods	Periods with odd numeration				Periods	Periods with even numeration			
		f = 3	d = 2	p = 1	s = 0		f = 3	d = 2	p = 1	s = 0
1.	1				H He	2				Li Be
2.	3			B C N O F Ne	Na Mg	4			Al Si P S Cl Ar	K Ca
3.	5		Sc Ti V Cr Mn Fe Co Ni Cu Zn	Ga Ge As Se Br Kr	Rb Sr	6		Y Zr Nb Mo Tc Ru Rh Pd Ag Cd	In Sn Sb Te I Xe	Cs Ba
4.	7	La Ce Pr Nd Pm Sm Eu Gd Tb Dy Ho Er Tm Yb	Lu Hf Ta W Re Os Ir Pt Au Hg	Tl Pb Bi Po At Rn	Fr Ra	8	Ac Th Pa U Np Pu Am Cm Bk Cf Es Fm Md No	Lw Ku 105 106 107 108 109 110 111	112 113 114 115 116 117 118	119 120

FIG. 80.   Mazurs 1965.   Type IIIC2-1.

Type 2: Table of Four Lemniscates

Type IIIA2-2

Four different size space lemniscates (Fig. 81). Bilecki originated this table in 1915 (1). He wrote schemes from which the lemniscates could be drawn. The drawing gives the impression that these lemniscates could result from the increasing swing of a pendulum. This type is not a successful one.

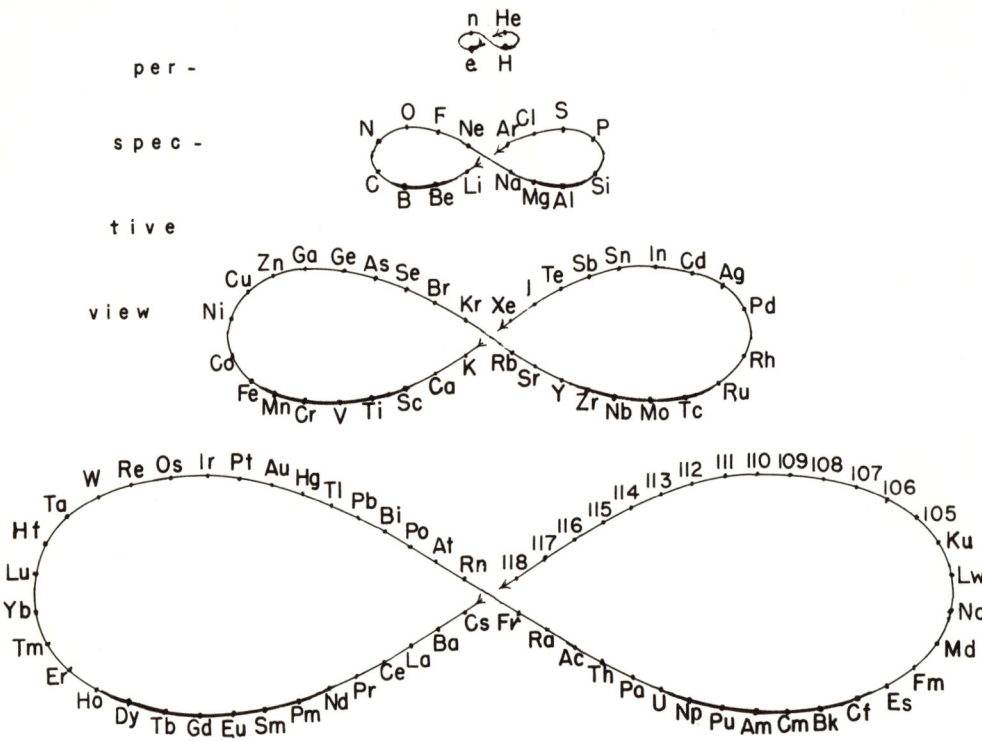

FIG. 81. Bilecki 1915. Type IIIA2-2.

Class 3. Tables with Disposition of Elements: 2, 8, 18, and 32 (Periods).

Type 1: Step Tables with Group Zero on One Side of the Table

Type IIIA3-1

Helix wound on three stepwise cylinders placed laterally with a common generating line at group 0 (Fig. 82). This helix is the large-size analogue of the medium-size helix, type IIA2–1 (Fig. 54), originated by Emerson in 1911. The originator of this helix was Stintzing in 1916 (1), who drew it on a rotational body similar to a cone. This table is not convenient to read.

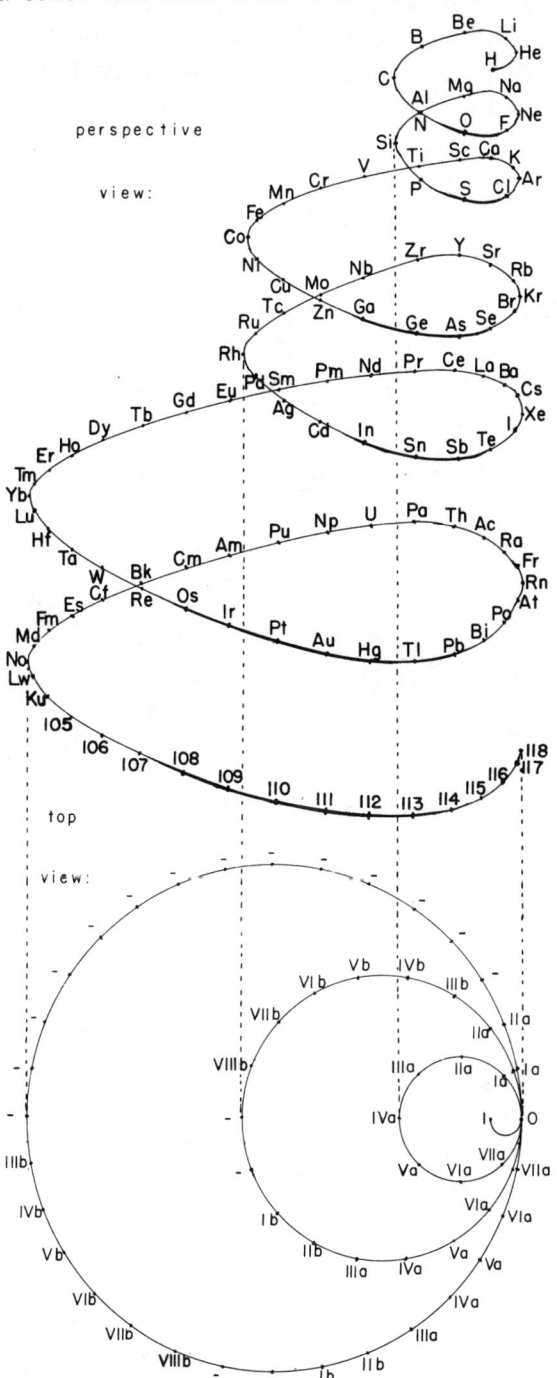

FIG. 82. Stintzing 1916. Type IIIA3–1

Type IIIB3-1

Spiral corresponding to the helix of basic type IIIA3-1 with three sizes of revolutions placed eccentrically (Fig. 83). The spiral is the top view of the helix. Originated by Hackh in 1914 (1), this spiral is an analogue of the medium-size one, type IIB2-1 (Fig. 55), that was originated earlier, in 1911, by Emerson. Hackh refers to him. The spiral as a table is not perfect. Havens (3) reduced the size of the table by drawing the largest revolutions of the long periods and the revolutions of the medium periods almost the same size and by inserting the lanthanides and actinides in group IIIb. In the construction of his table, Clark (4) used the "a-b-c" transition group which had been introduced by Bury in 1921 and used by many authors (see type IC3-2). The components of this transition group are: the noble gases in subgroup "a," the iron-platinum metals in subgroup "b," and seven inner transition elements in subgroups "c." Clark improved the drawing of the spiral by making the revolutions more quadrangular.

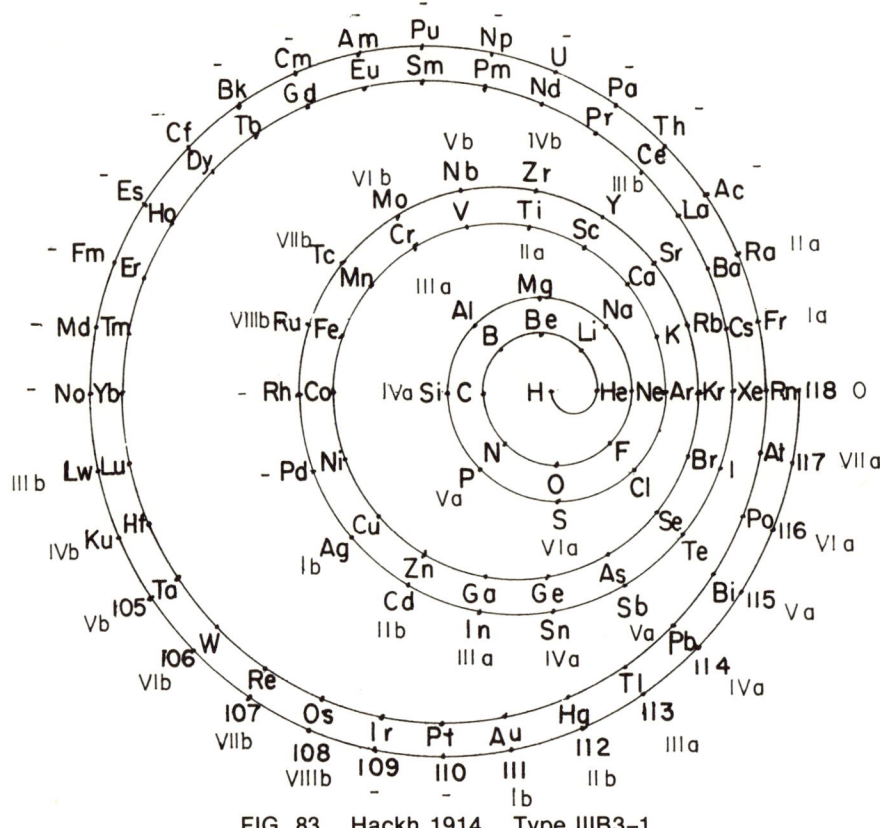

FIG. 83. Hackh 1914. Type IIIB3-1

Type IIIC3-1

Left-step table with group 0 on the right side of the table (Fig. 84). The table represents in series the space table, type IIIA3-1, and the plane table, type IIIB3-1. This table is the analogue of the medium-size table IIC2-1 originated earlier, in 1871, by Mendeleev (Fig. 56). The table was originated by Bassett in 1892 (1). Steele (2) suggested the table but did not draw it. The table is not successful because group I contains unrelated elements in two places. It would be acceptable if the s elements (groups Ia and IIa) were connected by lines. This was done by Klemenc (6), Tikhomirov (7), Ternström (8), and Chaikhorskii (9). Ternström arranged the inner transition elements in two series. In his table, Chaikhorskii (9) arranged 8th and 9th periods with 50 elements in each by including 5g and 6g elements; therefore, the total number of elements was 218.

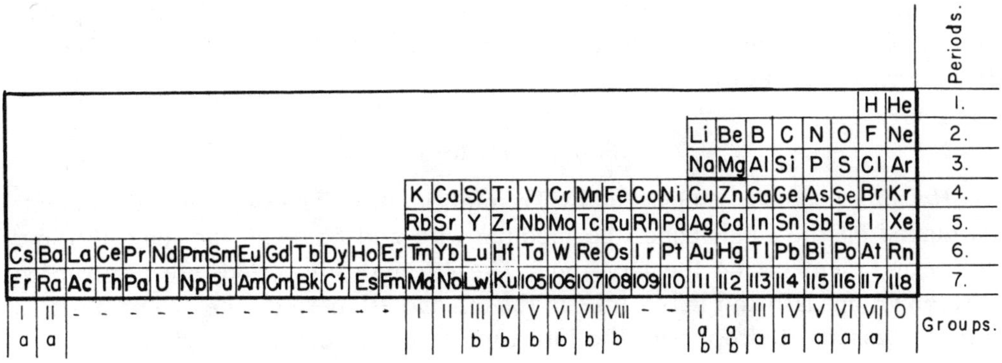

FIG. 84. Bassett 1892. Type IIIC3–1.

Subtype IIIA3–1a

Helix on a modified cone. The transition and inner transition elements have special revolutions in the form of loops. This table, originated by Stedman in 1947 (1), is not a successful one.

Subtype IIIB3–1a

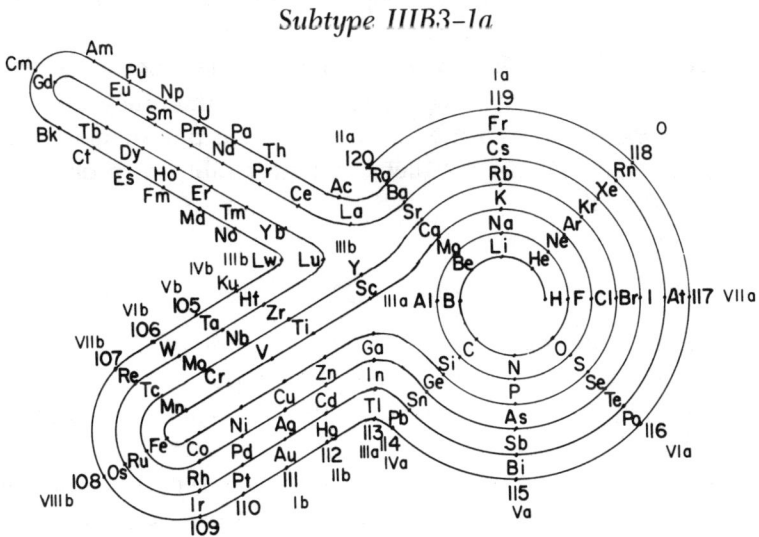

FIG. 85. Janet 1928. Subtype IIIB3–1a.

Spiral (Fig. 85) *as a top view of the helix, subtype IIIA3–1a.* The representative elements are placed on the main spiral, but the transition and inner transition elements are placed on loops of the same spiral. The spiral was originated by Janet in 1928 (1). The corresponding helix was originated later, in 1947, by Stedman. This spiral is not a successful table. Napol'skii (3) constructed his table a little differently.

Subtype IIIA3–1b

Helix similar to Emerson's medium size helix (type IIA2–1) *with a special sharp loop for the inner transition elements.* This helix, first published by Gamov (1) in 1940, has no advantages.

Subtype IIIA3–1c

Helix on two cones, suggested by Tremlelt in 1963. The short period elements are placed on the small cone, the medium (4th and 5th) and the long (6th and 7th) period elements on the large one. The 4th and 5th periods are interrupted between groups IIIb and IVb, while, in the 6th and 7th periods, the inner transition elements are inserted opposite the interruption in the previous periods.

Subtype IIIC3–1d

Left-step series table with group 0 on the right side of the table and with oblique series. Stareck originated this table in 1932 (1). Since the series are oblique, this subtype has the advantage that the related elements in groups can be compared not just vertically (the elements of the same block with each other) but also horizontally (the transition elements with the representative). This was done in the short tables by comparing the subgroups, but in the medium and long tables this property was lost. Stareck used the special transition group, consisting of a, b, and c parts (see Type IC3–2). The structure of this table is not successful because the elements of group I are arranged in such a way that they are not related to each other. Besides, many transition and inner transition elements are placed in groups which are not suitable for them. A better table with oblique series is subtype IIIC3–6b (Fig. 105).

Subtype IIIC3–1e

Right-step series table with group 0 on the left side of the table. This table is an analogue of the medium-size table subtype IIC2–1a, and it also is not acceptable. This one was published by Verschoyle in 1908 (1). This is the so-called wedge type of table. The table would be acceptable if connecting lines were drawn between the representative and transition elements as in type IIIC3–3. Langmuir's table (2) was slightly different; however, if the electrons of his table are taken into account, then it is similar.

DIVISION III. LONG TABLES

Type 2: Lemniscates and Zigzags with Group Zero on One Side of the Table

Type IIIA3–2

Complex of helix and space lemniscate with an additional loop for the inner transition elements (Fig. 86). There are three kinds of curves—one each for the representative, transition, and inner transition elements. This helix-lemniscate is analogous to the complex helix and space lemniscate of the medium-size type IIA2–2 (Fig. 57). Schirmeisen originated this long helix-lemniscate in 1900 (1), earlier than Soddy originated the medium-size lemniscate (in 1914). Actually Schirmeisen drew only the top views and the drawings were not very clear. This is not a convenient table.

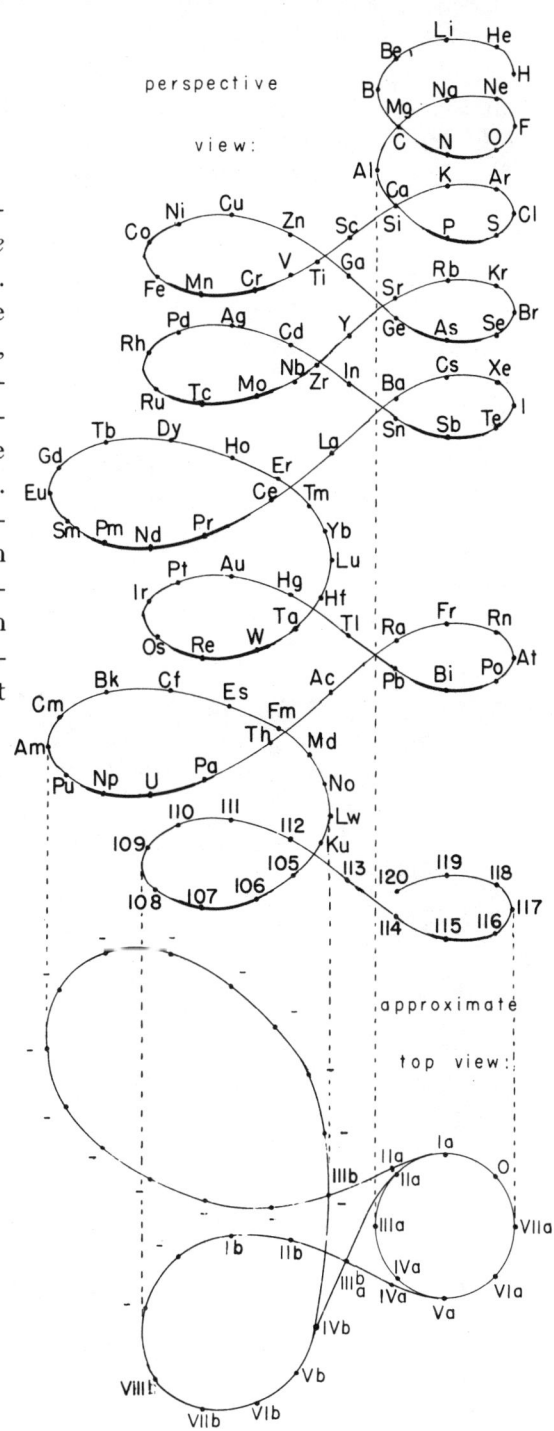

FIG. 86. Schirmeisen 1900. Type IIIA3–2.

Type IIIB3–2

Complex of a plane spiral-lemniscate for the representative and transition elements and an additional loop for the inner transition elements (Fig. 87). Spiral-lemniscate, originated by Janet in 1928 (1), corresponds to the helix-lemniscate of the same type, IIIA3–2 (Fig. 86), and is an analogue of the medium-size spiral-lemniscate, type IIB2–2 (Fig. 58), which was originated by Kipp in 1942.

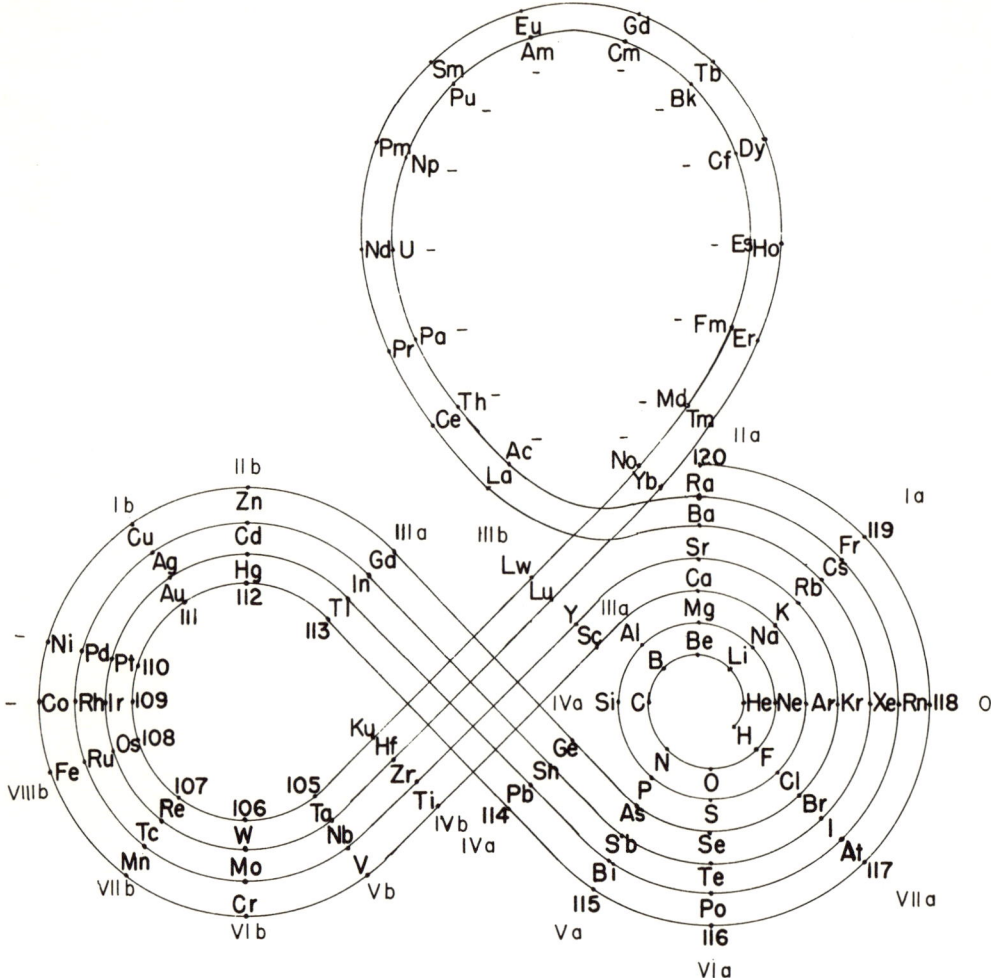

FIG. 87. Janet 1928. Type IIIB3–2.

Type IIIC3–2

Zigzag table (Fig. 88). This zigzag represents in lines the helix-lemniscate of Schirmeisen, type IIIA3–2 (Fig. 86). The zigzag table was originated by Saz in 1922 (1). His first table was a little different, but in 1931 he changed it to that shown in Fig. 88. This is not a successful table.

DIVISION III. LONG TABLES   81

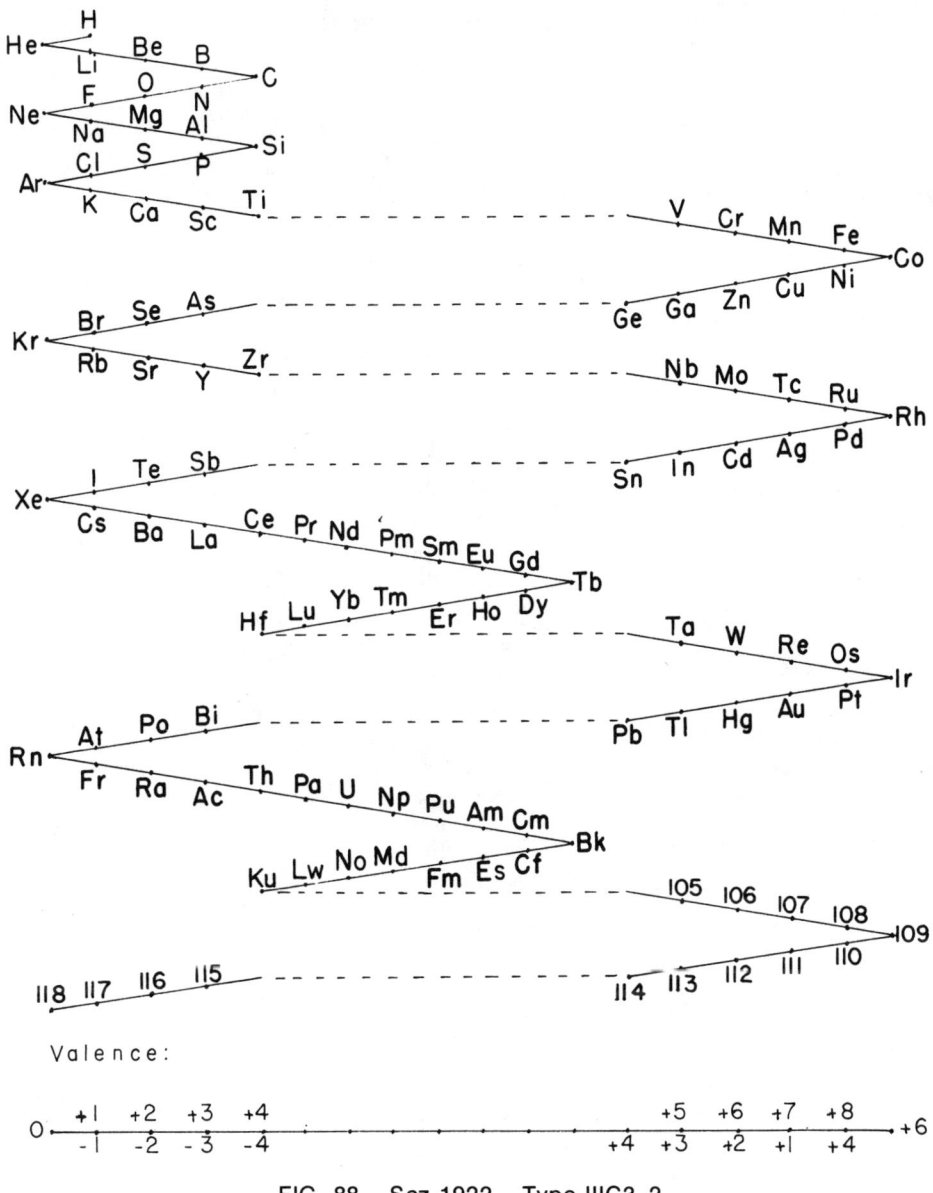

FIG. 88. Saz 1922. Type IIIC3–2.

Subtype IIIA3–2a

Complex of helix and space lemniscate with two additional loops for the inner transition elements with 7 elements on each loop (Fig. 89). This table, originated by Gooch and Walker in 1905 (1), is not a very convenient one.

82 GRAPHIC REPRESENTATIONS OF THE PERIODIC SYSTEM

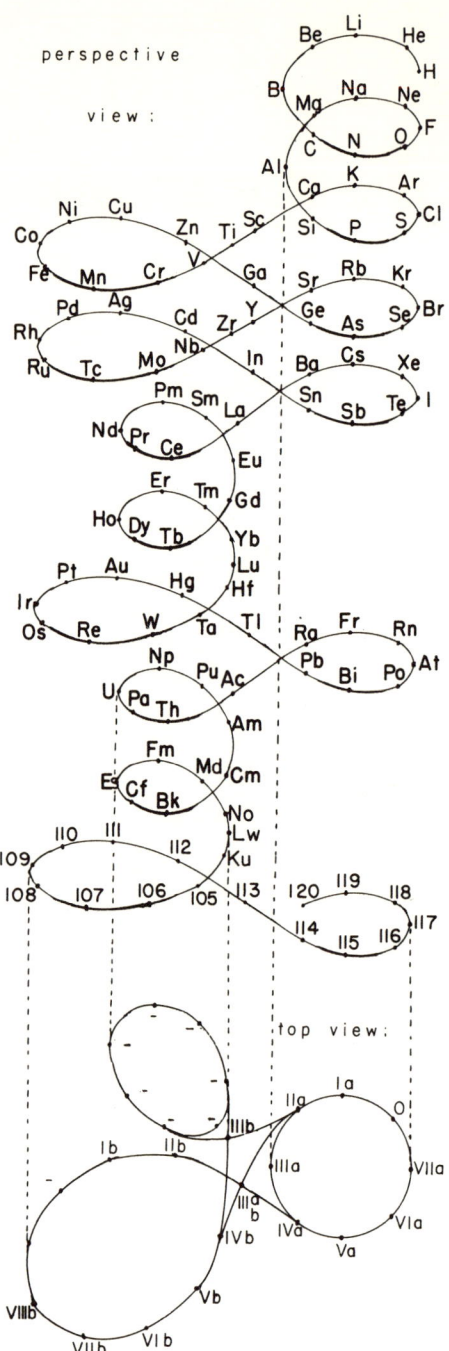

FIG. 89. Gooch a. Walker 1905. Subtype IIIA3–2a.

Subtype IIIB3–2a

Complex of sinusoids and partial lemniscates (Fig. 90). This is an attempt to draw the Gooch and Walker space lemniscate (subtype IIIA3–2a) in a single plane. This table is analogous to the medium size table type IIB2-2c, except that the lanthanides and actinides are included to form a long size table. As was mentioned there, the valences of the elements in this table also increase from the bottom to the top (from 0 to 8), with the exception of the second half of the transition elements (Co, Ni, Cu, Zn, etc.). The lanthanides also fit by valences; however, the actinides do not. This table was suggested by Rinck and Feschotte in 1962 (1).

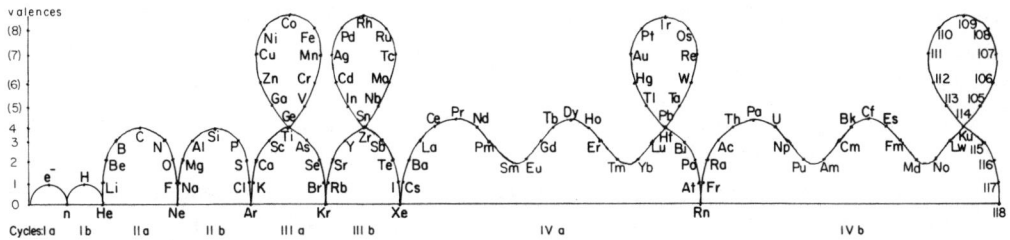

FIG. 90. Rinck a. Feschotte 1962. Subtype IIIB3-2a.

Type 3: Tables Symmetrical about a Vertical Line and with Group 0 on the Right Side of the Table

Type IIIA3-3

Helix on three main cylinders placed stepwise and centered one on top of the other, the so-called "Pagoda" (Fig. 91). This table, originated by Aucken in 1951 (1), cannot be conveniently read because the related elements are not on one vertical line.

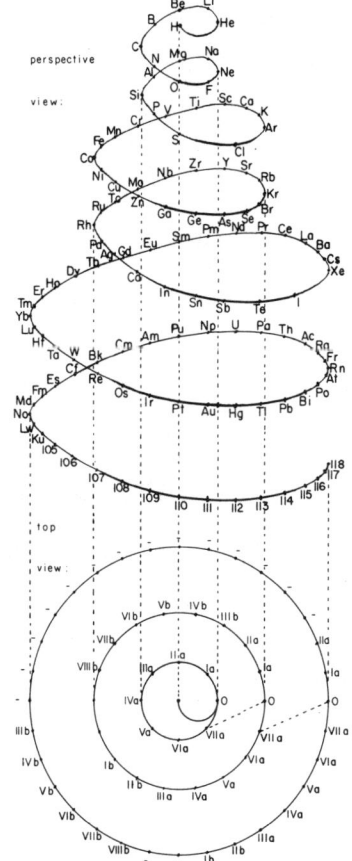

FIG. 91. Aucken 1951. Type IIIA3-3.

Type IIIC3-3

Pyramidal symmetrical series table with centered steps (Fig. 92). This table—one of the oldest, the most popular among the long tables, and a good one—is the well-known, so-called Thomsen-Bohr table. Originated neither by Thomsen in 1895 (2) nor by Bohr in 1922 (5) but by Bayley in 1882 (1), it should be called the Bayley table. Of course, Bayley's table was not complete because all elements were not known at that time. The related elements in this table are connected by lines. In the table published by Adams (3) only eight groups were foreseen for the "rare earth" elements and eight for the "radioactive" elements. Schaltenbrand (4) and Akhumov (10) called the non-connected elements the extensions ("Dehnungen" in German) of the periods. Extension 2 is identical to the transition elements and extension 3 to the inner transition elements (Ce-Lu and Th-Lw). Instead of actinides, Bohr (5) predicted a series starting from element No. 94 now known as Pu. Spedding (12) and Marson and Zucchi (17) tried to complete the table by introducing marks for shells and subshells. Seaborg (18), in his Russian articles in 1969 and 1971 (pp. 147 and 28), included the 5g elements in the 8th period as superactinides; his total number of elements equaled 168. Taube (22) added 8th and 9th periods with 5g and 6g transition elements, obtaining a total of 218 elements. Nebel (24) also used the superactinides, and his table had 168 elements. The medium-size analogue type IIC2-3 (Fig. 63) was originated by Carnelley four years later, in 1886.

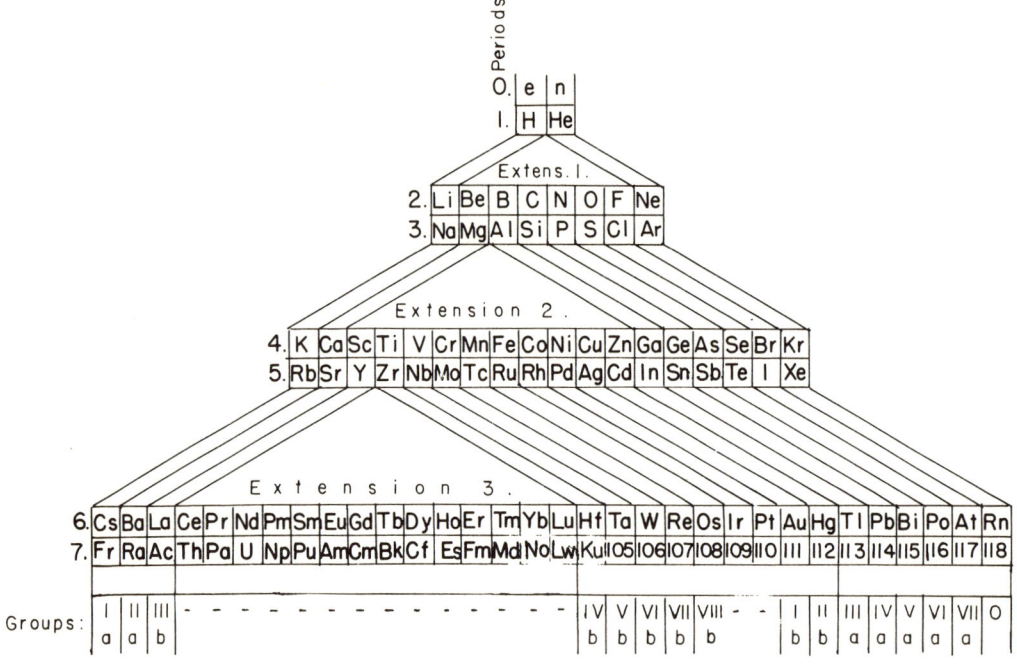

FIG. 92. Bayley 1882. Type IIIC3-3.

Subtype IIIA3–3a

Squares in space. The squares are arranged in planes on different levels with each plane denoting one period. The 1st period is presented by only one square with 2 elements in it. The 2nd and 3rd periods are each given in 4 squares: each square contains two elements which have the last electrons with opposite spin. The 4th and 5th periods require 9 squares each; 6th and 7th periods—16 squares (Fig. 93). The originators of this interesting table were Sugathan and Menon in 1956 (1).

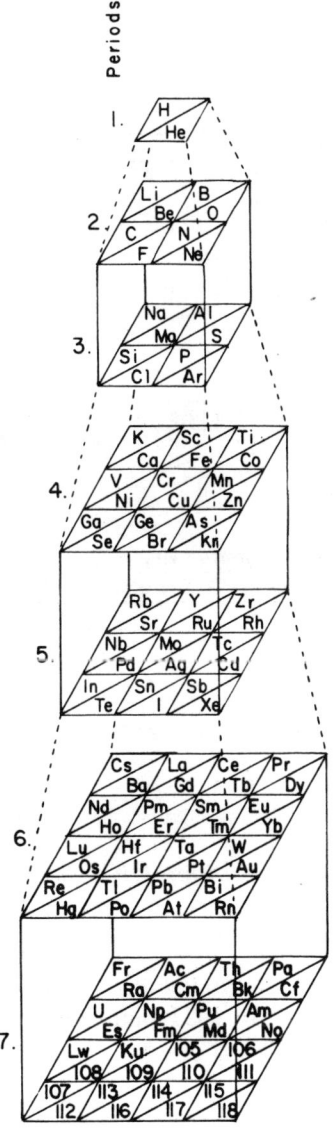

FIG. 93.   Sugathan a. Menon 1956.   Subtype IIIA3–3a.

Subtype IIIC3–3b

Symmetrical series table similar to the basic type, but constructed as a quadrangle (Fig. 94). For this purpose the short and medium periods are "stretched" in a horizontal direction. This table, originated by Hackh in 1914 (1), is a good one. He called his table a "synthetic system." Scheele (4) placed three of the inner transition elements into group IV and five into group VIII in approximately the same order as in his short table, subtype IC3–2b.

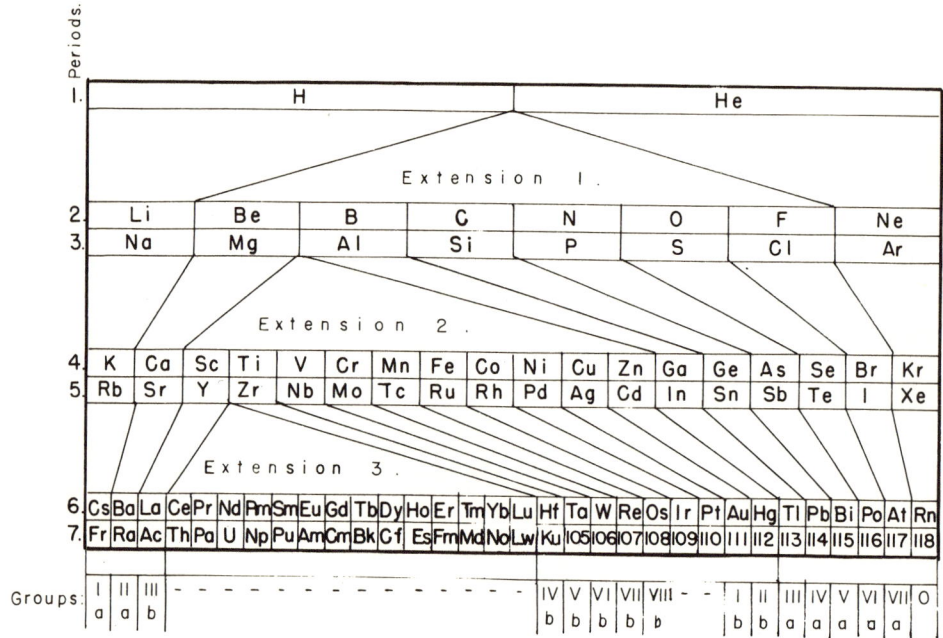

FIG. 94. Hackh 1914. Subtype IIIC3–3b.

Subtype IIIC3–3c

Symmetrical series table constructed as a triangle (Fig. 95). Wagner and Booth originated this table in 1945 (1). In 1937 Zmaczynski made an incomplete drawing of this type (0). The table is a good one.

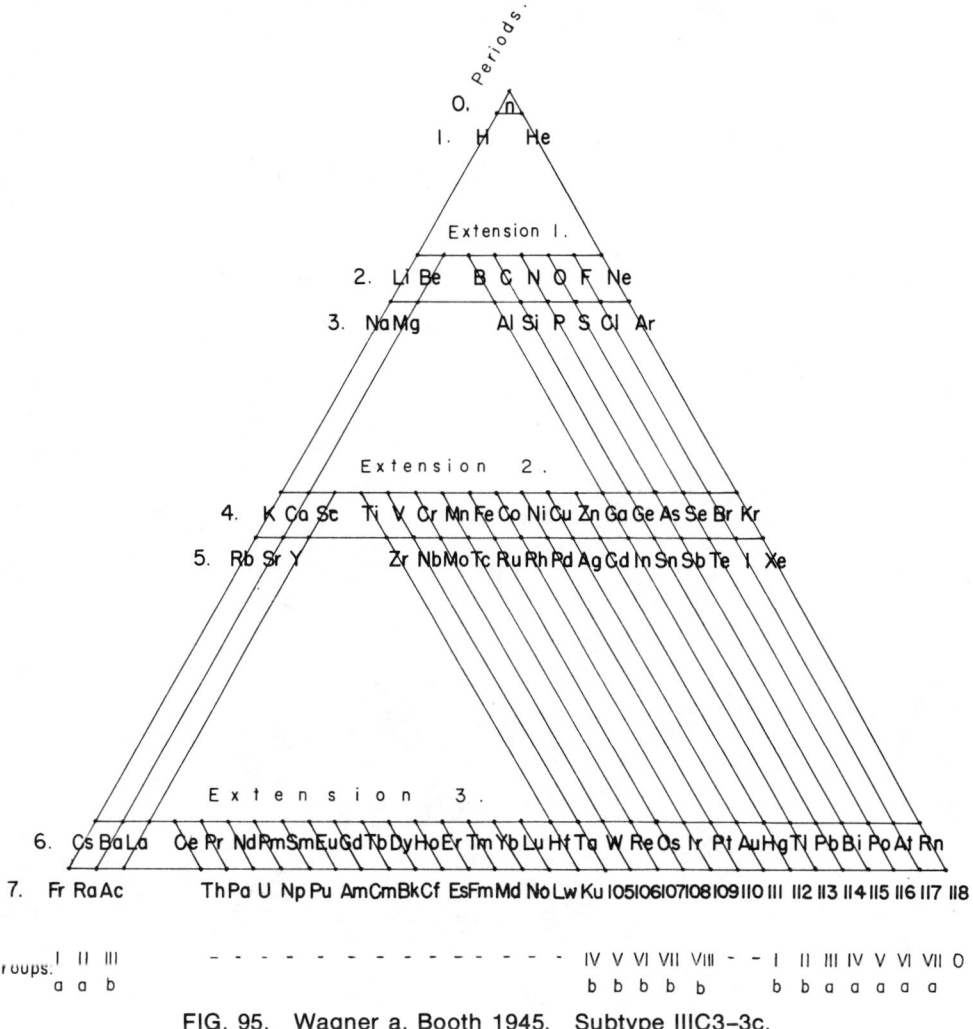

FIG. 95. Wagner a. Booth 1945. Subtype IIIC3–3c.

Type 4: Tables with Interrupted Short and Medium Periods

Type IIIA3–4

Helix wound on a cone or cylinder with interrupted short and medium periods (Fig. 96). This helix is analogous to the medium-size helix type IIA2–4 (Fig. 65) originated by Stackelberg much earlier, in 1911. This long-size helix was originated in 1937 by Zmaczynski (1), who drew it on a cone. Scherer and Shireby (3) drew it on a cylinder as it is shown in Fig. 96. This table could be judged a successful one.

88 GRAPHIC REPRESENTATIONS OF THE PERIODIC SYSTEM

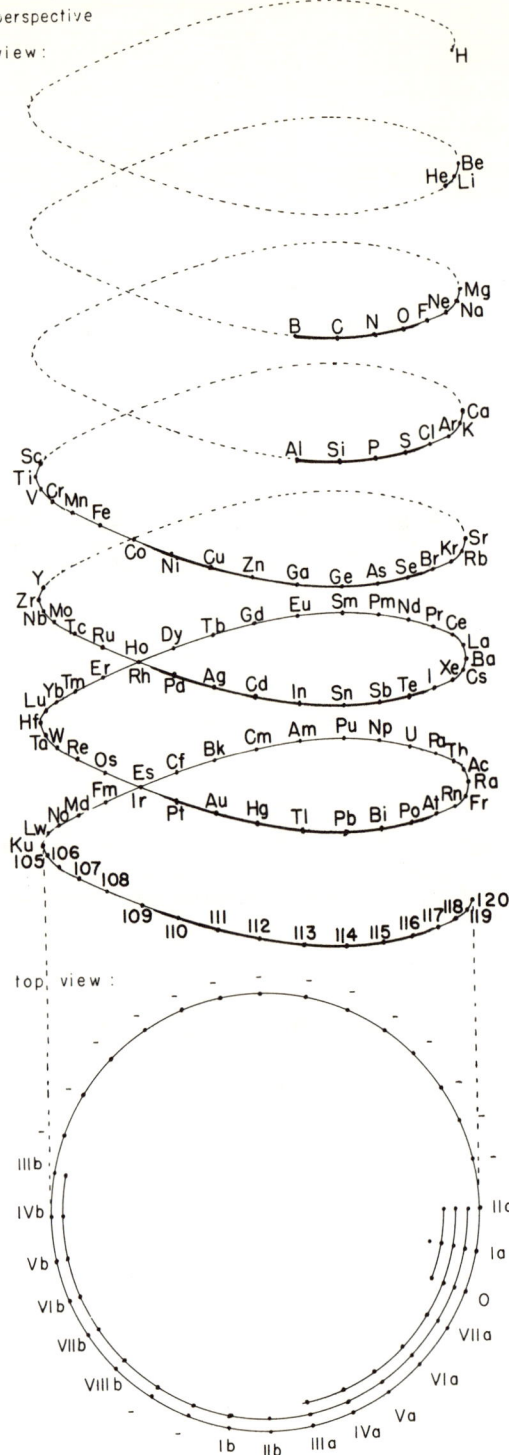

FIG. 96. Zmaczynski 1937. Type IIIA3–4.

Type IIIB3–4

Spiral with interrupted short and medium periods (Fig. 97). This spiral is analogous to the medium-size spiral, type IIB2–4 (Fig. 66), originated by Tocher in 1910. This long-size spiral, originated by Janet in 1928 (1), is a successful table. Dwight (5) changed this table by placing the lanthanides between groups IIIB and IVb and the actinides between groups IVb and Vb. Brown (2), Strack (3), Strong (4), Dwight (5), Griff (6), and Frassares (7) drew their tables in a clockwise direction as opposed to the Janet table.

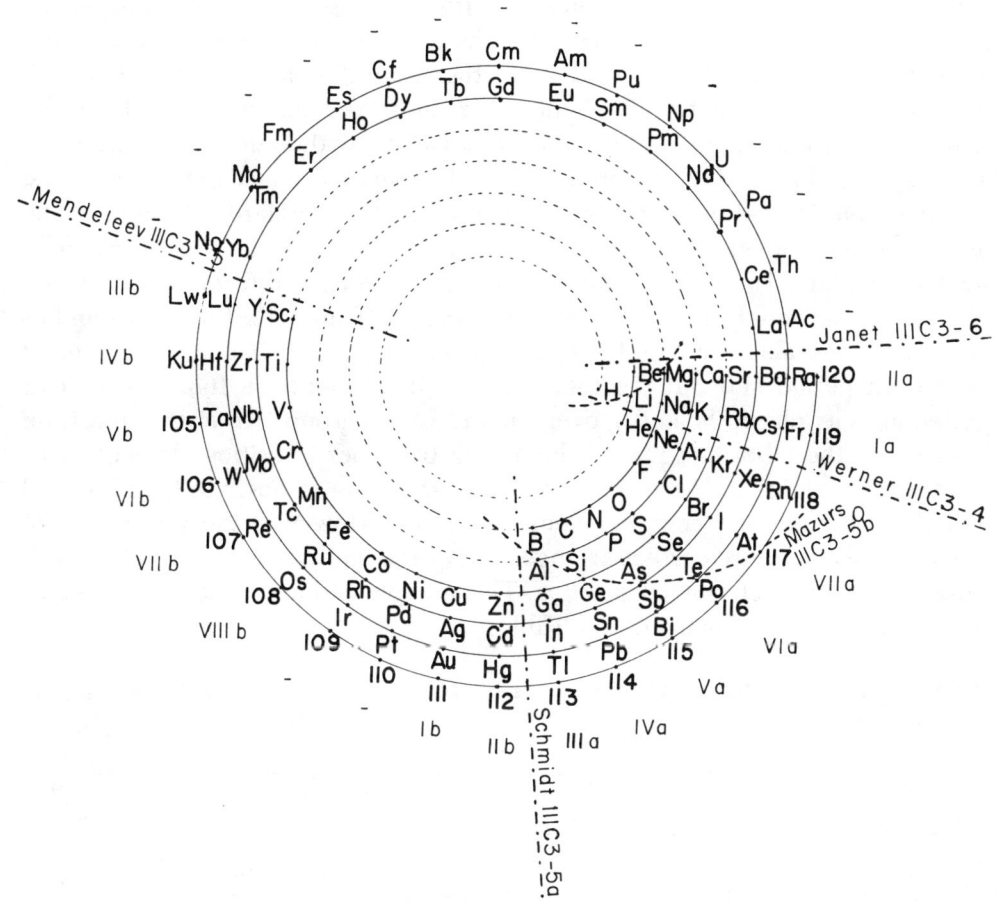

FIG. 97. Janet 1928. Type IIIB3–4.

Like the medium-size analogues (Fig. 65 and 66) of the preceding helix and spiral, the long-size helix and spiral are prototypes for the next types IIIC3–4, –5, –5a, –5b, and –6. Fig. 97 shows that by cutting the spiral the different types of series tables can be obtained.

Type IIIC3–4

Series table with interrupted short and medium periods (Fig. 98). This series table can be obtained from the spiral (Fig. 97) by cutting between groups 0 and Ia. The medium-size analogue, type IIC2–4, was originated by Mendeleev in 1869 (Fig. 67). This is the well-known Werner table, originated by him in 1905 (1). This table must be considered a successful one. As was mentioned in the discussion of type IIC2–4, this table also adheres to the phenomenon of changing metallic and nonmetallic properties in vertical and horizontal directions. As a result of this, the left lower corner of the table contains the elements with metallic properties and the right upper corner, the elements with the most nonmetallic properties. Wrigley, Mast, and McCutcheon (8) printed the symbols of the elements on different levels of a laminated model, closer to or further from the base in agreement with their belonging to different shells. Reuber (12) improved the table by adding the electronic configuration of the atoms. Wheeler (13) drew the same table vertically. Chissick's and Baldwin's (22) and Stokes' (26) tables were arranged as folding tables; by folding table IIC2–4 could be obtained. Loung (23) and Hardt (24) placed the inner transition elements in two rows, thus arranging them as a little periodic table. Gol'danskii (27) enlarged his table including the 5g and 6g elements, which he called octadecanoids in the 8th and 9th periods. Therefore, the total number of elements in his table was 218. In his book (coauthored by Polikanov) he included only the 8th period and put the total number of elements at 168. One half of the cited authors placed Sc and Y over Lu and Lw in group IIIb of the transition elements, and the other half placed them with the inner transition elements over La and Ac. According to the second approach, the table is closer to type IIIC3–3 where the 3rd extension contains elements Ce-Lu. However, the first group of authors accepts the more correct arrangement of the lanthanides which are La-Yb according to their electron configuration. This is explained further in the Introduction to the Electronic Configuration Tables.

Blocks:	"Alkali metals"		Inner transition elements													Transition elem.										"Nonmetals"						
Groups Periods	I a	II a														III b	IV b	V b	VI b	VII b	VIII b			I b	II b	III a	IV a	V a	VI a	VII a	0
1.	H																														He
2.	Li	Be																								B	C	N	O	F	Ne
3.	Na	Mg																								Al	Si	P	S	Cl	Ar
4.	K	Ca														Sc	Ti	V	Cr	Mn	Fe	Co	Ni	Cu	Zn	Ga	Ge	As	Se	Br	Kr
5.	Rb	Sr														Y	Zr	Nb	Mo	Tc	Ru	Rh	Pd	Ag	Cd	In	Sn	Sb	Te	I	Xe
6.	Cs	Ba	La	Ce	Pr	Nd	Pm	Sm	Eu	Gd	Tb	Dy	Ho	Er	Tm	Yb	Lu	Hf	Ta	W	Re	Os	Ir	Pt	Au	Hg	Tl	Pb	Bi	Po	At	Rn
7.	Fr	Ra	Ac	Th	Pa	U	Np	Pu	Am	Cm	Bk	Cf	Es	Fm	Md	No	Lw	Ku	105	106	107	108	109	110	111	112	113	114	115	116	117	118
Electrons	s^1	s^2	f^1	f^2	f^3	f^4	f^5	f^6	f^7	f^8	f^9	f^{10}	f^{11}	f^{12}	f^{13}	f^{14}	d^1	d^2	d^3	d^4	d^5	d^6	d^7	d^8	d^9	d^{10}	p^1	p^2	p^3	p^4	p^5	p^6

FIG. 98. Werner 1905. Type IIIC3–4.

Type 5: Tables with Group Zero in the Middle of the Table

Type IIIC3-5

Series table with group 0 in the middle of the table, the transition elements on the left side, and the inner transition elements on the right (Fig. 99). To obtain this table, the spiral (Fig. 97) is cut between the inner transition elements and group IIIb. The originator of the table was Mendeleev in 1869 (1). Its analogue, the medium-size table type IIC2-6 (Fig. 71), was also originated by Mendeleev (see Fig. 12). The inner transition elements on the right side of Mendeleev's table were not complete because they were not known at that time. This table is not a popular one.

FIG. 99. Mendeleev 1869. Type IIIC3-5.

Subtype IIIB3-5a

Two spirals connected with an oval. One spiral contains the representative elements, the other the transition elements. The inner transition elements are located on one side of the large oval. This spiral, originated by Nodder in 1920 (1), is not a very convenient one.

Subtype IIIC3-5a

Series table with group 0 in the middle of the table and the inner transition and transition elements on the right side of the table (Fig. 100). The order of the blocks is: the representative elements, the inner transition elements, and the transition elements. The table can be obtained by cutting the spiral (Fig. 97) between groups IIb and IIIa or by cutting the two spirals of subtype IIIB3-5a between groups II and III. This table, originated by Schmidt in 1911 (1), is not a popular one.

FIG. 100. Schmidt 1911. Subtype IIIC3-5a.

Subtype IIIC3–5b

Series table similar to the previous table with the difference that each series starts with a nonmetal (Fig. 101). The table is obtained by cutting the spiral (Fig. 97) with a curved line. This table is not an important one and is of interest only in comparison with its medium size analogue, type IIC2–5a. I suggested this table in 1957.

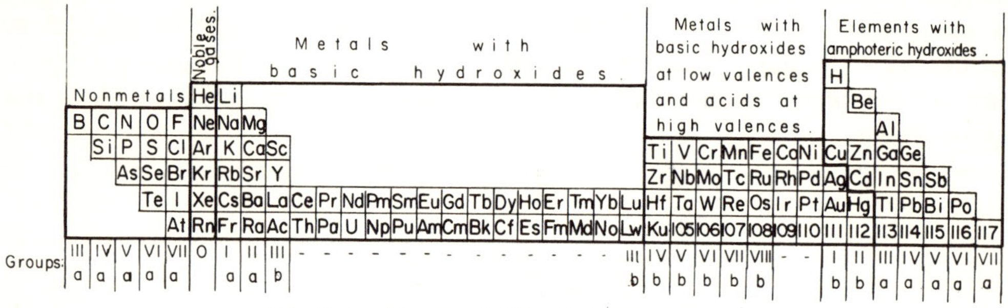

FIG. 101.   Mazurs 1957.   Subtype IIIC3–5b.

Subtype IIIC3–5c

Series table with group 0 in the middle of the table, the inner transition elements on the left side, and the transition elements on the right (Fig. 102). This table cannot be obtained by cutting the spiral IIIB3–4 (Fig. 97). The originator of this table was Sheehan in 1961 (1). The positions of the transition and inner transition elements are reversed in comparison with the basic type IIIC3–5. This is not a popular table.

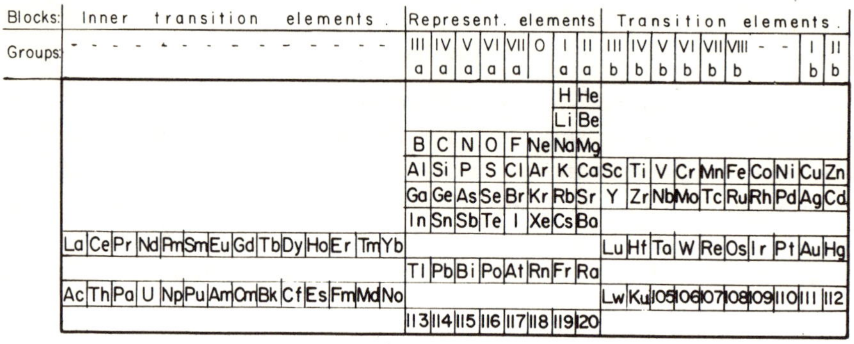

FIG. 102.   Sheehan 1961.   Subtype IIIC3–5c.

Type 6: Step Tables with Group IIa on the Right Side of the Table

Type IIIC3–6

Left-step series table with group IIa on the right side of the table (Fig. 103). The table is obtained by cutting the spiral (Fig. 97) between group IIa and the inner transition elements. It is analogous to Mendeleev's medium-size table type

IIC2–6 (Fig. 71) originated in 1869. This very good, modern table was originated by Janet in 1927 (1). Acceptance of this table requires the revision of the numeration of the periods. This matter is discussed in the Introduction to the Electronic Configuration Tables. The new division of periods was used by Janet (1), Simmons (8), von Auwers (9), Hakala (10), Klechkovskii (as a sum of $n+l$) (11), Lepsius and Asunmaa (12), Neville Smith (13), Farré-Torá (15), Tokarev (16), Neubert (17), and Mills (18). The subshells and electrons can be shown on this table in proper order; therefore, this table can be used as an electronic configuration table, as was done by Simmons (8), Hakala (10), Klechkovskii (11), Farré-Torá (15), and Tokarev (16). LeRoy's table (2) differed slightly from Janet's table in that he started the series with group IV instead of III. Sadikov refers to Lautié (6). Klechkovskii in 1961 (11) and Tokarev in 1966 (16) drew the periods in vertical columns. In his scheme, Klechkovskii in 1969 (11, p. 8) drew the 9th period with 50 elements by including eighteen 5g elements. Mills (18) in 1972 used special arrows to designate the symbols of elements which do not agree with the theoretical electron configuration. This approach is similar to my suggestion in subtype IIIC4–3c in 1969.

Blocks:	Inner transition elements.														Transition elements.										Representative elem.								New period numeration.
Groups:	-	-	-	-	-	-	-	-	-	-	-	-	-	-	III b	IV b	V b	VI b	VII b	VIII b	-	-	I b	II b	III a	IV a	V a	VI a	VII a	VIII a	I a	II a	
																															H	He	1.
																															Li	Be	2.
																									B	C	N	O	F	Ne	Na	Mg	3.
																									Al	Si	P	S	Cl	Ar	K	Ca	4.
															Sc	Ti	V	Cr	Mn	Fe	Co	Ni	Cu	Zn	Ga	Ge	As	Se	Br	Kr	Rb	Sr	5.
															Y	Zr	Nb	Mo	Tc	Ru	Rh	Pd	Ag	Cd	In	Sn	Sb	Te	I	Xe	Cs	Ba	6.
	La	Ce	Pr	Nd	Pm	Sm	Eu	Gd	Tb	Dy	Ho	Er	Tm	Yb	Lu	Hf	Ta	W	Re	Os	Ir	Pt	Au	Hg	Tl	Pb	Bi	Po	At	Rn	Fr	Ra	7.
	Ac	Th	Pa	U	Np	Pu	Am	Cm	Bk	Cf	Es	Fm	Md	No	Lw	Ku	105	106	107	108	109	110	111	112	113	114	115	116	117	118	119	120	8.
Subshells and electrons:	f^1	f^2	f^3	f^4	f^5	f^6	f^7	f^8	f^9	f^{10}	f^{11}	f^{12}	f^{13}	f^{14}	d^1	d^2	d^3	d^4	d^5	d^6	d^7	d^8	d^9	d^{10}	p^1	p^2	p^3	p^4	p^5	p^6	s^1	s^2	

FIG. 103. Janet 1927. Type IIIC3–6.

Subtype IIIC3–6a

Left-step series table constructed like the preceding series table, but with the symbols of the elements arranged according to their valences (Fig. 104). The valences are shown in one vertical column from -4 at the bottom to $+8$ at the top of each period. The symbols are repeated on different horizontal series if the elements have multiple valences. This subtype is reminiscent of the periodic chain, subtype IIIC1–1a, but is differently constructed. The table could be used to aid in the memorization of the valences of the elements. I originated this subtype in 1955 (1). Earlier, in 1924, Balarev (00) had tried to use the idea of multiple valences and suggested a table made of 8 glass plates, on each of which were placed elements with the same valence. Janek in 1944 (0) also gave a statistical table of valences and placed the elements of different periods in the group with the proper valence.

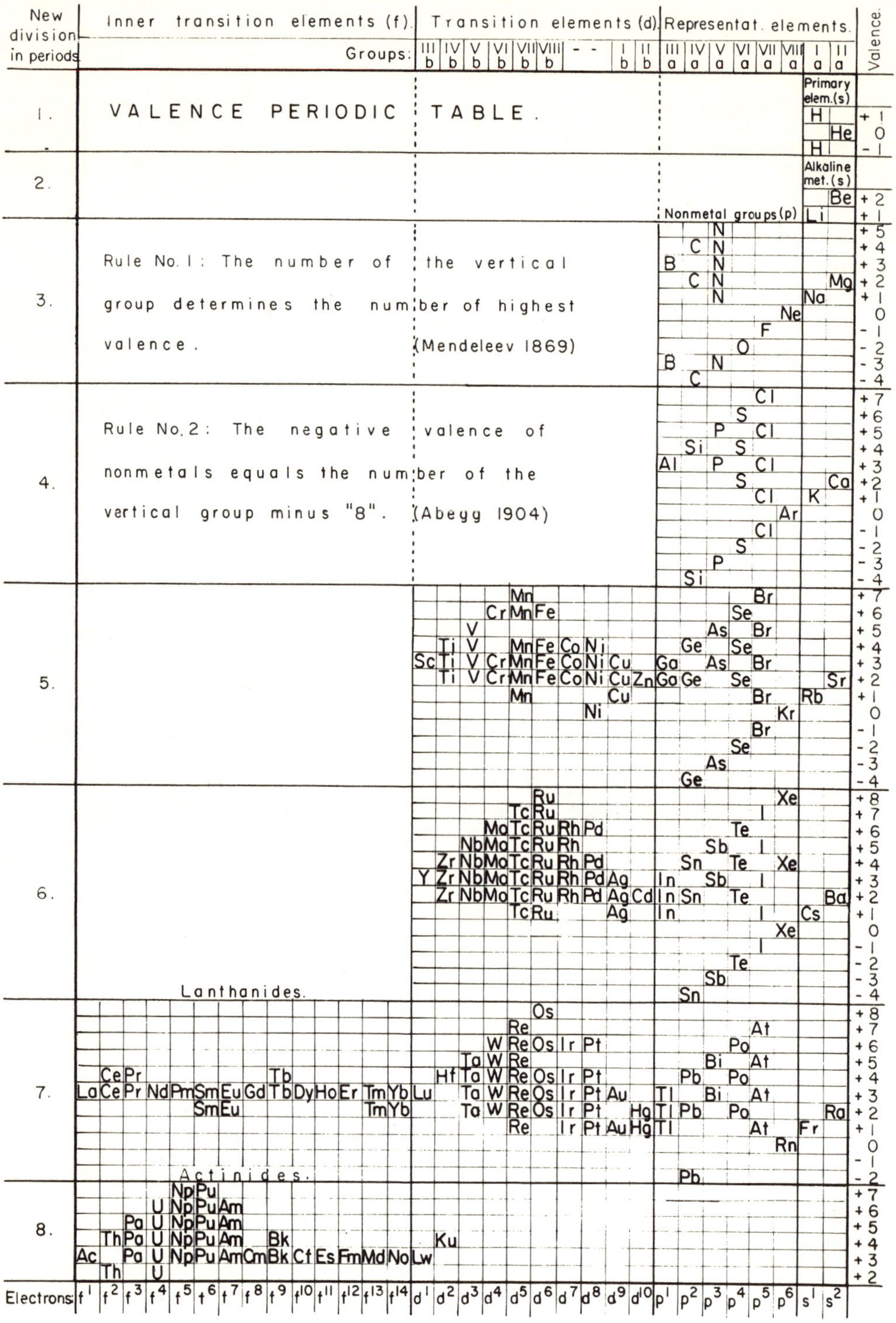
New division in periods
Inner transition elements (f).
Transition elements (d).
Representat. elements.
Groups:
Valence.
VALENCE PERIODIC TABLE.
Primary elem.(s)
Alkaline met.(s)
Nonmetal groups (p)
Rule No. 1: The number of the vertical group determines the number of highest valence.
(Mendeleev 1869)
Rule No. 2: The negative valence of nonmetals equals the number of the vertical group minus "8".
(Abegg 1904)
Lanthanides.
Actinides.
Electrons

FIG. 104. Mazurs 1955. Subtype IIIC3-6a.

Subtype IIIC3–6b.

Left-step series table with oblique series and with group IIa on the right side of the table (Fig. 105). This table is similar to the table with oblique series by Stareck (1932), but his table belongs to a different type (IIIC3–1d) because group 0 is on the right side of the table. For the construction of the present table, the "a-b-c" transition group, proposed by various authors, was used (see type IC3-2). To this transition group belong the noble gases from subgroups "a," the iron and platinum metals from subgroups "b," and the seven inner transition metals from subgroups "c." The transition group elements are in the same horizontal rows and are denoted as group "0." This table has the advantage that not only the

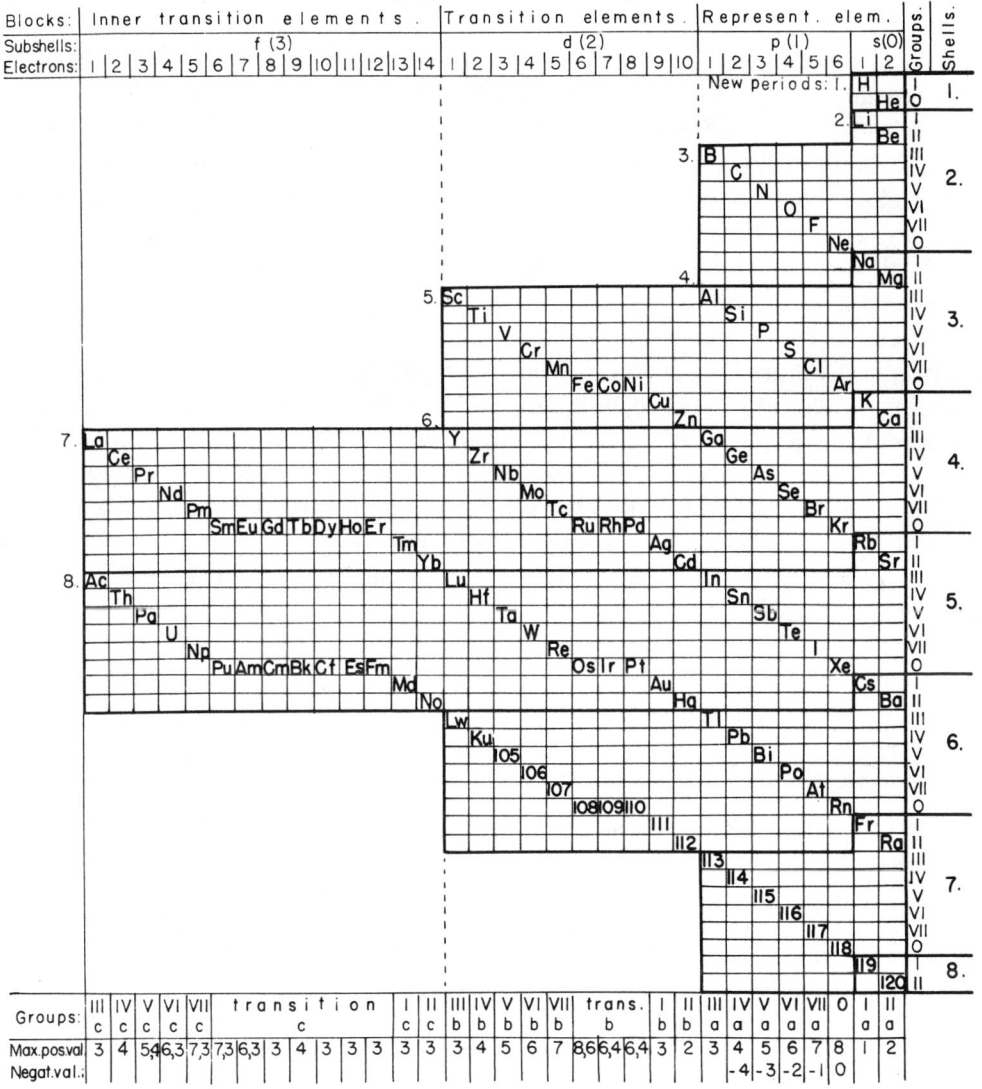

FIG. 105. Mazurs 1955. Subtype IIIC3–6b.

elements of the same block can be compared but also the elements of different blocks, because the series are oblique. The table could easily be used as an electronic configuration table because the elements having the last electron on similar subshells are located in the same vertical columns. In this aspect, the table is similar to the Gibson table, type IIIC4-3 (Fig. 119). I originated the table in 1955 (1). Earlier, in 1934, Romanoff (0) had drawn a table with oblique series, but he did not place the symbols in definite vertical columns or horizontal rows.

INTRODUCTION TO ELECTRONIC CONFIGURATION TABLES

BASIS IN ATOMIC STRUCTURE

So far, all the tables have been arranged by placing the elements in order of increasing atomic numbers and by taking into consideration their chemical and physical properties. These tables are the so-called chemical tables. The electronic configuration tables were introduced in the nineteen-twenties. In these tables the extranuclear electron configuration of the atoms of the elements is also taken into account.

If the chemical elements are placed in ascending order by their atomic numbers, then the configuration of the electrons changes periodically—a similar construction of electrons is repeated from time to time. In agreement with this electronic structure of the atom, the chemical and physical properties of the elements also change and repeat periodically. The elements whose atoms have similar electronic structure are chemically related and these related elements have the same number of valence electrons. The valence electrons are the two outside electrons on subshell s plus the electrons on any incomplete subshell. Hence, the Periodic System reflects the extranuclear electron configuration of atoms and this electronic structure is the sole reason for the existence of the Periodic System. As Tomkeieff said:

In the present state of our knowledge, it is clear that the periodicity of the Periodic System is due to the fundamental periodicity of the electronic structure of the elements, in other words, the Periodic System is a projection of the atomic structure (28).

Even earlier, in 1938, Emeléus and Anderson had written:

Two distinct systems of classification for the chemical elements have been developed during the last hundred years. The first has been evolved primarily from variations in chemical and physical properties in passing from one element and its compounds to the next in order of increasing atomic weight. The second is based on differences in the atomic structure. Chemistry is now approaching a stage in its evolution at which these two subjects, which may be termed the chemical and physical classification of the elements, may be correlated. We may seek to explain the chemical periodicity in terms of atomic structure and to place the pictorial representations of valency on a surer physical basis (9).

More than thirty years have passed since this was published. The time has come to accept the electronic structure tables.

Heavier atoms are constructed from lighter ones by adding electron by electron to the extranuclear part of the atom and proton by proton, with some neutrons, to the nucleus itself. The electron added last to obtain a certain element is called the *differentiating* electron. The distribution of the electrons around the atomic nucleus on discrete energy levels in the extranuclear structure of the atom is governed, according to quantum theory, by four quantum numbers: the principal quantum number n, the subsidiary or azimuthal quantum number l, the magnetic quantum number m, and the spin quantum number s.

The *principal quantum number* n has the values 1, 2, 3, 4, 5, 6, 7, and 8. These values enumerate the energy levels of the electrons around the atomic nucleus; in other words, the distance from the nucleus. These energy levels are also called shells.

The *subsidiary (azimuthal) quantum number* l determines complexes of orbitals on the shells as subdivisions of these shells, the so-called subshells. The notations of this quantum number are s, p, d and f and the values—0, 1, 2, and 3 respectively. The values of l are restricted according to the following inequality:

$$0 \leqslant l \leqslant n - 1$$

According to this inequality:
shell 1 ($n=1$) can have only one subshell, $s(0)$;
shell 2 ($n=2$) has 2 subshells, $s(0)$ and $p(1)$;
shell 3 ($n=3$) has 3 subshells, $s(0)$, $p(1)$, $d(2)$; and
shell 4 ($n=4$) has 4 subshells, $s(0)$, $p(1)$, $d(2)$, and $f(3)$.

For the fifth shell and those following, the number of subshells should increase. However, for the elements known at the present time, subshells with an l value higher than 3 are not known (at the ground state of the elements).

The *magnetic quantum number* m determines the orbitals in the subshells. The number of orbitals (L_m) in each subshell is determined by the following equation:

$$L_m = 2l + 1$$

According to this equation:
subshell s ($l=0$) can have only 1 orbital,
subshell p ($l=1$) has 3 orbitals,
subshell d ($l=2$) has 5 orbitals, and
subshell f ($l=3$) has 7 orbitals.

According to the *spin quantum number* s, each orbital can possess a maximum of 2 electrons with opposite spin, in agreement with the Pauli exclusion principle. The equation for the number of electrons (L_s) in each subshell is as follows:

$$L_s = 2(2l + 1)$$

Therefore, the number of electrons in subshells s, p, d, and f is 2, 6, 10, and 14 respectively.

As a result of this, the electronic configuration periodic tables are divided into blocks s, p, d, and f with 2, 6, 10, and 14 elements in each.

Table 3 shows the complete picture of shells, subshells, orbitals, and electrons around the nucleus of a hypothetical element No. 120.

Table 3. Shells and Subshells in the Extranuclear Atomic Structure of a Hypothetical Element No. 120

n (shells)	l (subshells)	Number of orbitals in subshells	Number of electrons in subshells	Number of electrons in shells = magic numbers.
1	s(0)	1	2	$2 = 2 \cdot 1^2$
2	s(0)	1	2	$8 = 2 \cdot 2^2$
	p(1)	3	6	
3	s(0)	1	2	
	p(1)	3	6	$18 = 2 \cdot 3^2$
	d(2)	5	10	
4	s(0)	1	2	
	p(1)	3	6	$32 = 2 \cdot 4^2$
	d(2)	5	10	
	f(3)	7	14	
5	s(0)	1	2	
	p(1)	3	6	$32 = 2 \cdot 4^2$
	d(2)	5	10	
	f(3)	7	14	
6	s(0)	1	2	
	p(1)	3	6	$18 = 2 \cdot 3^2$
	d(2)	5	10	
7	s(0)	1	2	$8 = 2 \cdot 2^2$
	p(1)	3	6	
8	s(0)	1	2	$2 = 2 \cdot 1^2$

The true picture of electron locations was demonstrated by DeVault in his diagram of electron energy levels in 1944 (13). The heaviest element in his diagram was uranium (U). A more complete diagram with Lr (Lw) as the last element, based on the orbital energy levels published by Herman and Skillman in 1963 (37), appears as a foldout at the back of this book. Their figures were calculated "approximately" (as they say) with the help of a computer. The energy levels of the last electrons on orbitals of certain subshells of the extranuclear electron configuration are shown on the diagram. The electrons on subshells s are shown as •, on subshells p as ^{0}, on subshells d as $|^+|$, and on subshells f as $'^\circ$. In the diagram energy is expressed on a logarithmic scale in rydberg units. The energies in DeVault's articles were expressed in ev (electron volts)—1 rydberg = 14.3 ev. In general, Herman and Skillman's diagram is related to the Stoye table, subtype IIIB1–1b (Fig. 76)—but Stoye's table is theoretical, while this diagram is more or less experimental.

By comparing the diagram (enclosure 2) and Table 3, we can observe some disagreements. For example, Sc, belonging to the subshell 3d, has exactly the same arrangement of electrons in all shells and subshells as it should according to the Table 3. However, its homologue Y, belonging to subshell 4d, has a different arrangement starting with shell 4. For the lanthanides (4f), the arrangement is true only for shells 1 to 3 and the construction of shells 4, 5, and 6 is different. Likewise for the actinides, the arrangement agrees only in shells from 1 to 4; in shells 5, 6, and 7 there are disagreements.

The filling of subshell orbitals with electrons does not occur in the order in which the subshells are listed in Table 3. The order is determined by the mathematical expression of the Periodic Law. This equation (1) is presented below. The law was formulated indistinctly by Madelung in 1936 (8). He constructed, as he called it, a "catalog" of electrons for atomic structure (Table 4) and accepted it as an empirical rule that had no theoretical basis. Others have also used this "catalog."[*]

The examination of the third column of Table 4 shows the following: If the completion of the shells with electrons occurs on subshells s and p, then these subshells belong to the outermost shell with the highest n value for this element. The elements obtained in this case are the *representative* elements. If the differentiating electron is placed in the subshell s, then the elements obtained are primary (H and He) or alkali and alkaline earth elements. If this electron goes to the subshell p, then the obtained elements are nonmetals (including noble gases) or metals belonging to the nonmetal groups. If, however, the electron completes subshell d, then this subshell belongs to an inner shell, the principal quantum value of which is $n-1$, meaning the shell next to the highest that this element already possesses. The elements obtained in this case are the usual *transition* elements (also called related elements). Furthermore, if the completed subshell is f, then this subshell belongs to a shell with a lower principal quantum number $n-2$ than the highest n value for this element. The elements obtained by this completion are the *inner transition* elements (also called similar elements), the lanthanides and actinides.

Using common language, we can say that the electrons located on the outermost subshells s and p serve to "swell" the atom—they increase its volume or decrease its density. The electrons on subshells d and f penetrate these outermost subshells and, therefore, are located on inner subshells. These electrons increase the density of the atom. This fact has been proved experimentally and can be seen by observing the curves of density of elements belonging to different blocks s, p, d, and f on Fig. 139.

The above mentioned mathematical expression of the Periodic Law is as follows:

$$t = n + l \qquad (1)$$

---

[*] Klechkovskii 1951–54 (23), Rubinowicz 1959 (34), Platonov 1961 (36), Balaban 1963 (38), Chepelevetskii 1966 (39), Tokarev 1966 (41), Farré-Torá 1966/67 (42), Madras 1967 (43), and Neubert 1969 (45).

t—ordinal number of periods, n—principal quantum number, l—subsidiary (azimuthal) quantum number.

As was also shown above, l is related to n by the following inequality:

$$0 \leqslant l \leqslant n - 1 \qquad (2)$$

As a law the equation (1) was established by Yeou Ta in 1946 (15). Earlier, in 1945, it was used by Wiswesser (14) as a sum of the two quantum numbers. Later this equation was used by Simmons in 1947/48 (17) and others°. Yeou Ta called the sum a "mixed quantum number of transition"; Simmons called it a "period number"; and Klechkovskii (23), simply, "$(n+l)$-groups." Actually, the sum t represents the ordinal numbers of the periods in the periodic table.

Table 4. Construction of the Periodic Table According to the Equation:
$t = n + l$

Ordinal numbers of periods t	Combination possibilities of $n+l$ which give periods and corresponding subperiods	Subperiod notation	Number of elements in sub-periods	Number of elements in periods	Ordinal numbers of cycles x
1	1 + 0	1s(s=0)	2	2	1
2	2 + 0	2s	2	2	
3	2 + 1 3 + 0	2p(p=1) 3s	6 2	8	2
4	3 + 1 4 + 0	3p 4s	6 2	8	
5	3 + 2 4 + 1 5 + 0	3d(d=2) 4p 5s	10 6 2	18	3
6	4 + 2 5 + 1 6 + 0	4d 5p 6s	10 6 2	18	
7	4 + 3 5 + 2 6 + 1 7 + 0	4f(f=3) 5d 6p 7s	14 10 6 2	32	4
8	5 + 3 6 + 2 7 + 1 8 + 0	5f 6d 7p 8s	14 10 6 2	32	

° Hakala 1948/52 (18), Carroll and Lehrman 1948 (19), von Auwers 1948 (20), Klechkovskii 1951–1971 (23), Keller 1956 (31), Nissen 1956 (32), Rubinowicz 1959 (34), Platonov 1961–69 (36), Tokarev 1966 (41), Farré-Torá 1966/67 (42), Madras 1967 (43), Neubert 1970 (45), and Otake 1971 (49).

Table 4 shows the use of the mathematical equation (1) of the Periodic Law, which gives the actual construction of the periodic table. By substituting n and l into this equation with their different values, different period numbers are obtained: 1, 2, 3, 4, 5, 6, 7, and 8. Not only the periods, but also *subperiods* are obtained. Since the addition of one electron to the extranuclear atomic structure produces a new element, the number of elements in a subperiod coincides with the number of electrons in a subshell.

There are always two periods with an equal number of elements. Each odd numbered period is duplicated by an even numbered one. A pair of periods forms one cycle (diad or binod). The term "cycles" was introduced by Bilecki in 1913 (2), "diads" by Janet in 1927 (4), and "binods" by Baca-Mendoza in 1953 (26). The number of elements in a cycle is indicated by a very simple equation:

$$L_c = 4x^2$$

where x is the cycle number. Among other things it should be noted that:

$$x = l+1$$

where l is the value of the new subshell that appears in this cycle.

Additional rules derived from the Periodic Law for the construction of the periodic table are as follows:

1) to form chemical elements starting from lighter to heavier ones, the filling of electronic levels (shells 1 to 8 and subshells s to f) with electrons occurs gradually from a smaller value of t to a larger value of t;

2) in the limits of each period t, the filling proceeds from a subperiod (subshell) with a smaller value of n and larger value of l to a subperiod with a larger n and smaller l.

For example, for the period $t=3$, there are two possible sums of $n+l$: $2+1$ and $3+0$. The sum combination $2+1$ precedes and $3+0$ follows.

Both rules were published by Yeou Ta in 1946 (15). Various other authors have also discussed these rules.°

The construction of the periodic table can also be shown by a scheme (Fig. 106) that was drawn a little differently by various authors°°. Sommerfeld seems to have been the originator of this scheme in 1925/26 (3).

° Simmons 1948 (17), Carroll and Lehrman 1948 (19), Klechkovskii 1951, 54, 57, and 65 (23), Nissen 1956 (32), Balaban 1963 (38), and Madras 1967 (43).

°° Janet 1930 (4), Rabinowitsch and Thilo 1930 (5), Pauling 1939 (10), Hazlehurst 1941 (11), Pao-Fang Yi 1947 (16), Campbell 1949 (21), Hähnel 1949 (22), Klechkovskii 1952 (23), Moeller 1952 (24), Neville Smith 1955 (29), Ganesan 1955 (30), Keller 1956 (31), Petrovici 1956 (33), Rubinowicz 1959 (34), Frassares 1966 (40), Madras 1967 (43), Leyh 1969 (46), and Layzer 1971 (48).

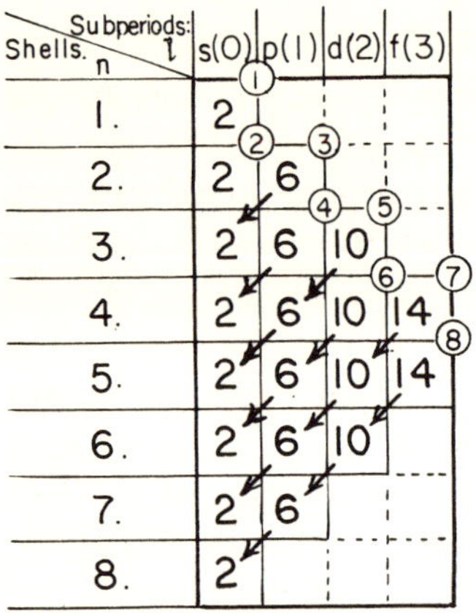

FIG. 106. Scheme of periodic table by Sommerfeld.

The numerals in squares (2, 6, 10, and 14) denote the numbers of elements in the subperiods. The periods ($n + l$) are located on diagonals of the scheme from the upper right corner to the lower left corner (shown by small numerals in circles).

Daudel in 1943 (12) and others suggested a different scheme (Fig. 107)°

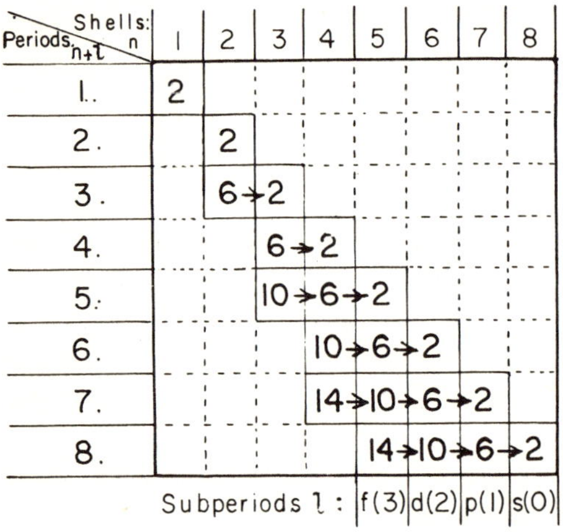

FIG. 107. Scheme of periodic table by Daudel.

The numerals (2, 6, 10, and 14) mean numbers of elements in subperiods. The subperiods (l) are located on diagonals by reading from the lower right corner to the upper left corner.

Until recently, the Rydberg method of division into periods, developed in 1914, was used: $2 \cdot 1^2 + 2 \cdot 2^2 + 2 \cdot 2^2 + 2 \cdot 3^2 + 2 \cdot 3^2 + 2 \cdot 4^2 + 2 \cdot 4^2 =$
$= 2 + 8 + 8 + 18 + 18 + 32 + 32$ (Table 5).

Table 5. Old Distribution of Subshells in Periods

Number of period.	Subshells included into the periods.	Number of elements in periods.
1.	$1s^2$	2
2.	$2s^2, 2p^6$	8
3.	$3s^2, 3p^6$	8
4.	$4s^2, 3d^{10}, 4p^6$	18
5.	$5s^2, 4d^{10}, 5p^6$	18
6.	$6s^2, 4f^{14}, 5d^{10}, 6p^6$	32
7.	$7s^2, 5f^{14}, 6d^{10}, 7p^6$	32

° Hakala 1948 (18), Klechkovskii 1953 and 1954 (23), Dockx 1959 (35), Gol'danskii 1970 (47), and Otake 1971 (49).

The periods started with alkali metals and ended with noble gases; in other words, the periods started with subshell s and ended with subshell p (except the first period). The order of subshells in long periods was s–f–d–p, which is actually illogical considering their values 0–3–2–1.

Table 4 shows that the arrangement of the periods should be as follows: $2+2+8+8+18+18+32+32 = 2(1^2+1^2+2^2+2^2+3^2+3^2+4^2+4^2)$. Table 6 is a rearrangement of table 5, and it shows the distribution of subperiods in periods according to the Periodic Law and construction rules. This table represents a scheme of the electron configuration periodic table.

Table 6. Distribution of Subperiods in Periods According to the New Arrangement

Ordinal number of a period.	Blocks or subperiods of periods.				Number of elements in one period.
	Inner transition metals.	Transition metals.	Nonmetal group elements.	Primary elements and "alkaline" metals.	
1.				$1s^2$	2
2.				$2s^2$	2
3.			$2p^6$	$3s^2$	8
4.			$3p^6$	$4s^2$	8
5.		$3d^{10}$	$4p^6$	$5s^2$	18
6.		$4d^{10}$	$5p^6$	$6s^2$	18
7.	$4f^{14}$	$5d^{10}$	$6p^6$	$7s^2$	32
8.	$5f^{14}$	$6d^{10}$	7^6	$8s^2$	32

The contents of this table were expressed by Dockx' rules in 1959 (35):
1) the periods terminate with vertical column s^2, similar to the first period;
2) the periods are divided into subperiods corresponding to subshells in reversed order of their values: $f(=3)$, $d(=2)$, $p(=1)$, and $s(=0)$;
3) the periods do not contain subshells belonging to the same shell; the subshells belong to different shells, starting from a shell with a lower value of the principal quantum number to a larger value of it: for example, for the period No. 7, $n = 4, 5, 6, 7$.

The order of subshells in the long periods is f–d–p–s, which is logical considering their values 3–2–1–0. As can be seen from Tables 4 and 6 and Figs. 106 and 107, the numeration of periods that should be accepted is different from what is used now. The latter numeration was not established by the discoverer, Mendeleev. In the first edition of the second volume of his *Principles of Chemistry* (1871), he suggested a numeration of five periods (Fig. 17). However, in the fourth edition (1881), he had six periods, using a table similar to type IIC1-1. Finally in the 5th edition (1889), he had two short periods and five long (actually medium) periods—seven periods for the same table, type IIC1-1. Today's numeration was suggested by Palmer in the United States in 1890 and by Rang in Sweden (published in England) in 1893. These authors used type IIC2-4 as a sample, the most popular type even now. This numeration was finally accepted around 1914 although some authors designated H and He as period zero and started the real numeration with element Li.

As early as 1927, Janet (4) showed the new numeration of periods in his drawings. In 1931, Remy (6) explained the necessity of changing the periods so that they would end with alkaline earth metals and the noble gases would be between the halogens and alkali metals. His reason for this was the possibility of having in one row, i.e., in one new period, the isoelectronic ions of the elements. In 1934 Nikol'skii (7) was also close to this idea of new periods, and Daudel (12) in his scheme (Fig. 107) in 1943 showed the new period division. The new period numeration was suggested definitely by Simmons in 1947 (17) and was supported by many authors.° In 1955 Neville Smith (29) said:

> Period is mistakenly thought to end with an inert gas, whereas in fact the inert gases are, of course, intermediate in chemical character and electronic structure between the halogens and the alkali metals.

Although the electronic configuration tables should be accepted as fitting into the present status of scientific knowledge, they are not accepted everywhere in the world. For example, in Russia the old eight group periodic table is still in favor. In 1969, Kedrov, an advocate of the classical eight group periodic table, said:

> This is why, in our opinion, those scientists who suggest renouncing Mendeleev's short (classical) form of the table of elements and wish to return to the primeval long form of it make an inexcusable and unjustifiable mistake. (?!)°°

Because the Russian scientists try to keep the division of the table into subgroups, they do not know where to put the lanthanides. In the United States, the medium-size table with 18 groups is preferred over other tables; and, here too, the lanthanides and actinides are outside the element classification.

Mendeleev discovered the Periodic System by comparing the chemical and physical properties of the elements. However, the representative elements, placed in the main subgroups "a," and the transition elements, in the subordinate subgroups "b," can be compared as related elements only in groups III, IV, V, and VI. The elements in the subgroups of groups I, II, VII, and VIII are not related by their properties in any way: for example, K and Cu, Ca and Zn, Mn and Br, Fe and Kr, etc. The arrangement of the elements into a short table with main subgroups "a" and subordinate subgroups "b" actually retarded the development of the Periodic System itself. The historical fact should be remembered that, after the discovery of the noble gases, the scientists did not know where to place them. In addition, many difficulties have arisen in placing the rare earth (inner transition) elements into the regular groups of an eight column table.

It should be repeated that the periodic table contains many discrepancies

---

° Hakala 1948 and 52 (18), von Auwers 1948 (20), Hähnel 1949 (22), Klechkovskii 1951, 52, 61, and 65 (23), Lepsius and Asunmaa 1952 and 56 (25), Nevill Smith 1955 (29), Nissen 1956 (32), Dockx 1959 (35), Chepelevetskii 1966 (39), Tokarev 1966 (41), Farré-Torá 1966/67 and 1967/68 (42), Platonov 1969 (36), and Neubert 1970 (45).

°° Vot pochemu, na nash vzgliad, neprostitel'nuiu i nichem ne opravdannuiu oshibku sovershaiut te uchenye, kotorye predlagaiut segodnia otkazat'sia ot korotkoi mendeleevskoi (klasicheskoi) formy tablitsy elementov i vernut'sia k pervonachal'noi dlinnoi ee formy (44, p. 51).

when constructed according to the properties of the elements. Therefore, it is more convenient to adjust the periodic table to one physical phenomenon; that is, to the electronic structure of the atoms, which is the basis for the Periodic System. The periodicity of atomic structure must be accepted as a Natural Law. Therefore, scientists have to change their minds, get away from the conservatism that accepts only Mendeleev's chemical table as right, and adjust the other phenomena to this phenomenon; that is, derive the chemical and physical properties of the elements from the electronic structure of the atoms. Even Mendeleev himself had the feeling that the table created by him was not perfect and needed improvement. This is evident in the following quotation from his Faraday lecture in London in 1889:

> . . . I unhesitatingly say that although greatly enlarging our vision, even now the Periodic Law needs further improvements in order that it may become a trustworthy instrument in further discoveries (1).

Also, Shchukarev, who presently occupies the chair of Mendeleev at Leningrad University, wrote in 1954:

> The knowledge of spectral terms and ionization energies of the great majority of the elements presents, in connection with the electronic configuration of neutral and ionized atoms, such a significant picture of periodicity of properties of electronic shells that this picture must be preferred to the comparison of types of chemical compounds in understanding the primary characteristics of the Periodic Law (27).

PROBLEM OF INNER TRANSITION ELEMENTS

Although the basis of electronic configuration tables is scientifically confirmed, some difficulties remain in the arrangement of the tables themselves. It is observable as an empirical rule that the properties of the top series in each block differs from the other series.

In block s, H and He are the so-called primary elements, and they differ from the other elements of this block, which are alkali and alkaline earth metals.

In block p, O and F do not reach their highest oxidation state (6 and 7 respectively) in chemical combinations with other elements as do the other elements in their groups. N and O exist in a different physical state (gaseous at room temperature) than the other elements in their groups.

In block d, Fe, Co, and Ni do not reach their highest oxidation state as do the other elements of this group.

In block f, concerning the lanthanides, only La and Ce have the proper oxidation state, while the other elements of this row predominantly possess oxidation state "3"; the properties of the other elements of this row are similar to each other. The second row of inner transition elements, the actinides, should be accepted as a regular one because they have increasing oxidation states from 3 to 7 (Ac to Np) and a change in the chemical properties corresponding to the changes in the rows of representative (p) and transition elements (d).

Special difficulties in the arrangement of the electronic configuration tables concern the placing of the first elements of the lanthanide and actinide rows in the right block. Table 7 shows the discrepancies in the experimental electronic configuration of the lanthanide series compared with the theoretical configuration.

Table 7. The Lanthanide Completion with Electrons

Symbol of the element.	Experimental location.	Theoretical location.
La	$5d^1$	$4f^1$
Ce	$4f^2$	$4f^2$
Pr	$4f^3$	$4f^3$
Nd	$4f^4$	$4f^4$
Pm	$4f^5$	$4f^5$
Sm	$4f^6$	$4f^6$
Eu	$4f^7$	$4f^7$
Gd	$5d^1\ 4f^7$	$4f^8$
Tb	$4f^9$	$4f^9$
Dy	$4f^{10}$	$4f^{10}$
Ho	$4f^{11}$	$4f^{11}$
Er	$4f^{12}$	$4f^{12}$
Tm	$4f^{13}$	$4f^{13}$
Yb	$4f^{14}$	$4f^{14}$
Lu	$5d^1\ 4f^{14}$	$5d^1\ 4f^{14}$

Lu, with the differentiating electron on subshell $5d$, and subshell $4f$—completed with 14 electrons, must belong to the transition elements. Hamilton, in 1965 (1), and Merz and Ulmer in 1967 (2) came to the same conclusion after comparing the trend of the following physical properties of La and Lu with Sc and Y: the melting points, the crystal structure, the atomic spectra, and the superconductivity. Also, Chistiakov in 1968 shared this opinion after comparing the trends of the atomic radii and ionization potentials (3).

Since 14 elements should be included in the inner transition elements, La must belong to them although the differentiating electron enters irregularly into subshell d instead of f. This discrepancy should be accepted because the discrepancies involve only the 1st (La) and 8th element (Gd), i.e., the first elements of each half of a subperiod.

What was said about the lanthanides can be said about the actinides. Table 8 shows that here the discrepancies are not only for the 1st (Ac) and 8th elements (Cm), but also for the 2nd (Th), 3rd (Pa), 4th (U), and 5th elements (Np).

Table 8. The Actinide Completion with Electrons

Symbol of the element.	Experimental location.	Theoretical location.
Ac	$6d^1$	$5f^1$
Th	$6d^2$	$5f^2$
Pa	$6d^1\ 5f^2$	$5f^3$
U	$6d^1\ 5f^3$	$5f^4$
Np	$6d^1\ 5f^4$	$5f^5$
Pu	$5f^6$	$5f^6$
Am	$5f^7$	$5f^7$
Cm	$6d^1\ 5f^7$	$5f^8$
Bk	$5f^9$	$5f^9$
Cf	$5f^{10}$	$5f^{10}$
Es	$5f^{11}$	$5f^{11}$
Fm	$5f^{12}$	$5f^{12}$
Md	$5f^{13}$	$5f^{13}$
No	$5f^{14}$	$5f^{14}$
Lw	$6d^1\ 5f^{14}$	$6d^1\ 5f^{14}$

Because of these many discrepancies, some scientists do not accept the actinide series, but, instead, assume that Ac, Th, and Pa are transition elements; U, Np, Pu, and Am form a new uranide series; and a new curide series starts with Cm. In 1922, Bohr (I. 1) predicted a plutonide series. List I of literature references in the Bibliography section contains the "Supporters of the uranide and curide series." Among them, Bedreag (I. 10), Haissinsky (I. 14), and Paneth (I. 15) are especially noticeable.

List II includes the "Supporters of the actinide series." The actinide series was predicted by Bassett in 1892 (II. 1), first used by Werner in 1905 (II. 2), and especially propagated by Villar (II. 11), beginning in 1938. This series was suggested also by Seaborg in 1942 (II. 12) and authoritatively established by him in 1949. This latter opinion is employed in this book.

If the new electronic configuration tables are accepted, the placement of the second half of the transition elements into definite valence groups, as has been the practice until now, should be suspended because they do not fit at all into these groups by their valences, as is shown in Table 9.

Table 9. Maximum Valence of Old VIIIb, Ib, and IIb Groups

Former group numbers:	VIIIb			Ib	IIb
	Fe	Co	Ni	Cu	Zn
Maximum valence:	6	4	4	3	2
	Ru	Rh	Pd	Ag	Cd
Maximum valence:	8	6	6	3	2
	Os	Ir	Pt	Au	Hg
Maximum valence:	8	6	6	3	2

SUBDIVISION IIIB: ELECTRONIC CONFIGURATION TABLES

Electronic configuration tables are of two classes, numbered 4 and 5.

4) tables with electronic configuration disposition of elements: 2, 6, 10, and 14 (blocks or subperiods).
5) shell and subshell tables with the same disposition of elements: 2, 6, 10, and 14.

Class 4 of the division of long tables contains three types:

Type 1 of this class contains symmetrical electronic configuration tables, which cannot be conveniently read.

To type 2 belong the right-side electronic configuration tables. The very popular Gardner table (type IIIC4–2) belongs here.

Type 3 contains the left-side electronic configuration tables, which are as important as the right-side tables. The Gibson table (type IIIC4–3) and the Mazurs table (subtype IIIC4–3c) are of considerable importance to this type.

Class 5 of the shell and subshell tables has three types:

Type 1 tables are symmetrical tables of concentric circles and parallel lines. These tables are constructed according to the electronic shells and subshells with the periods not so clearly visible as in class 4. Therefore, these tables cannot be used for teaching purposes since it is hard to read the element symbols in the order of increasing atomic numbers. In this connection, attempts have been made to improve some subtypes of this type.

Type 2 presents the right-side shell and subshell tables with the periods more visible than in type 1. The Mazzucchelli table, subtype IIIC5–2a, is important here, and the Mazurs table, subtype IIIC5–2d, is somewhat interesting.

Type 3 contains the left-side shell and subshell tables. Among them two are notable: the Scheele spiral, type IIIB5–3; and a series table, type IIIC5–3.

CLASS 4. TABLES WITH ELECTRONIC CONFIGURATION DISPOSITION OF ELEMENTS 2, 6, 10, AND 14 (BLOCKS OR SUBPERIODS)

Type 1: Symmetrical Helices, Spirals, and Series Electronic Configuration Tables

Type IIIA4–1

Helix with different sizes of revolutions for 2, 6, 10, and 14 elements (Fig. 108). This helix must be considered as the best table for presenting the Periodic Law in all its details. However, the space curves are not conveniently constructed nor easily read. Therefore, the right- and left-side electronic configuration tables (types 2 and 3) are the best tables for common use. This helix was originated

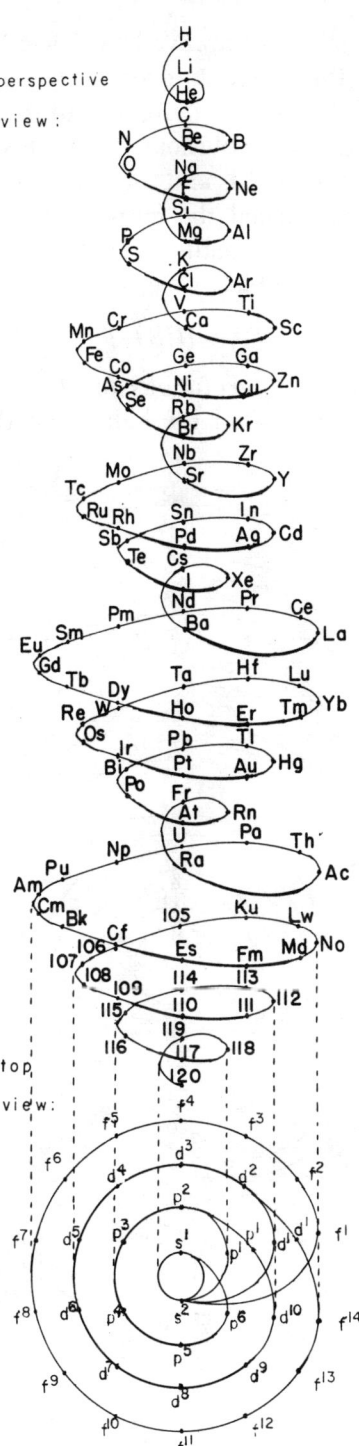

FIG. 108. Schaltenbrand 1920. Type IIIA4–1.

110 GRAPHIC REPRESENTATIONS OF THE PERIODIC SYSTEM

by Schaltenbrand in 1920 (1) and improved by Janet in 1928 (4). Schaltenbrand drew only group zero on the smallest revolutions; therefore, the disposition of the elements was a little different—1, 7, 10, and 14. Monroe and Turner (2) drew groups 0 and Ia on the smallest revolutions. In the first paper, Janet (4) had the disposition of the elements as 8, 10, and 14; but later he changed to the right order—2, 6, 10, and 14. He placed the correct elements of groups Ia and IIa belonging to subperiods *s* on the smallest revolutions. Efremov's (5) suggestion only approximated this kind of table.

Type IIIB4–1

Spiral with different sizes of revolutions for 2, 6, 10, and 14 elements (Fig. 109). The spiral is a true top view of the helix type IIIA4–1. It was originated

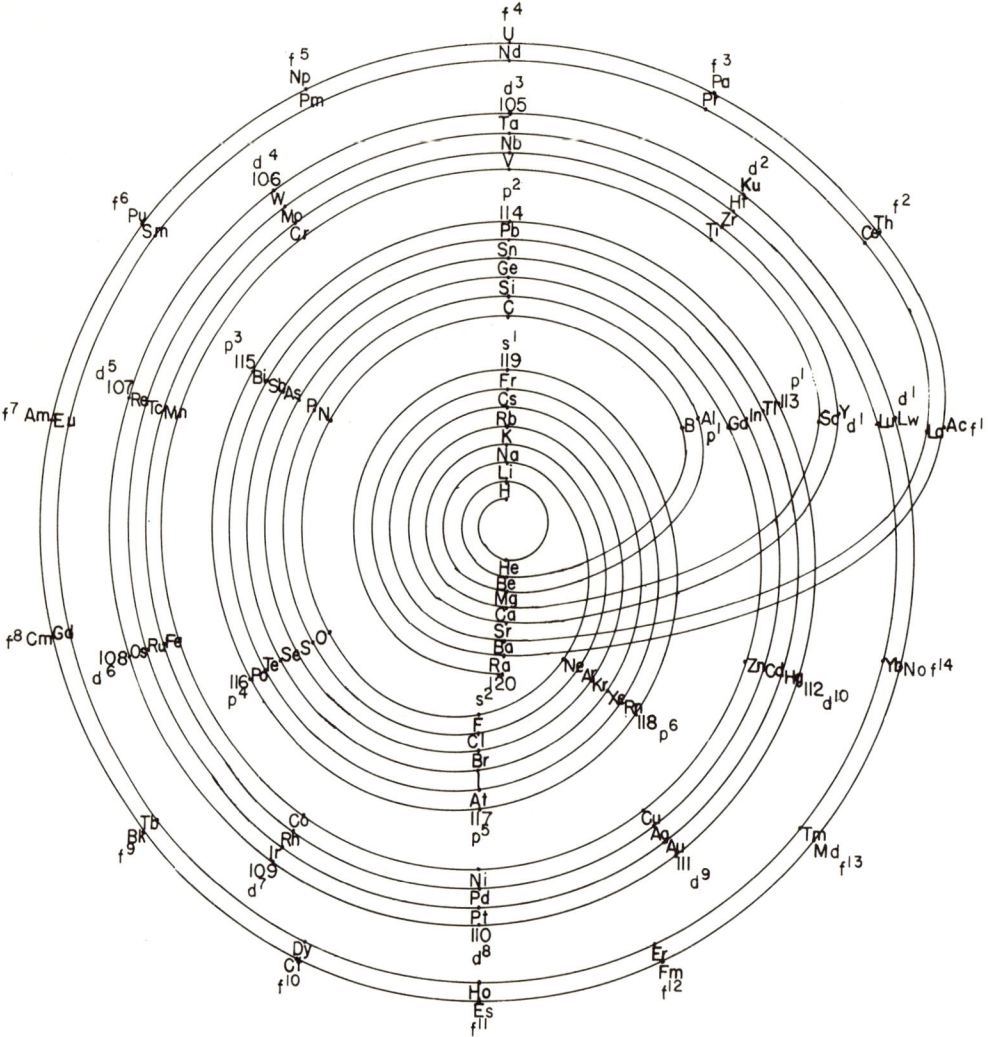

FIG. 109. Monroe a. Turner 1926. Type IIIB4–1.

DIVISION III. LONG TABLES   111

by Monroe and Turner in 1926 (1) and improved by Janet in 1928 (3). Monroe and Turner drew the elements of groups 0 and Ia on the smallest inside circles. In his first paper, Janet stated the disposition of elements as 8, 10, and 14. Later, he gave the correct disposition of elements—2, 6, 10, and 14—and on the smallest circles he placed the elements of groups Ia and IIa. Actually, the spiral is hard to read.

Subtype IIIA4–1a

Helix with four sizes of revolutions on four separate axes. Lepsius and Asunmaa originated this table in 1954 (1). However, they did not draw helices on four separate axes, but rather used only two axes, combining two sizes of revolutions under one axis. In this kind of helix, the subperiods are more easily compared than is possible with the helix type IIIA4–1. In 1966 Giguère (2) succeeded in constructing a very good space model of this subtype. He put s, p, d, and f elements on four plastic sheets, half of them on each side of the sheets. The top view of this model resembles subtype IIIB4–1a except that the loops are replaced by sheets. If this kind of table is read in the order of atomic numbers, it is apparent that this table is actually the same as subtype IIIC4–1a, but without interruptions. Therefore,

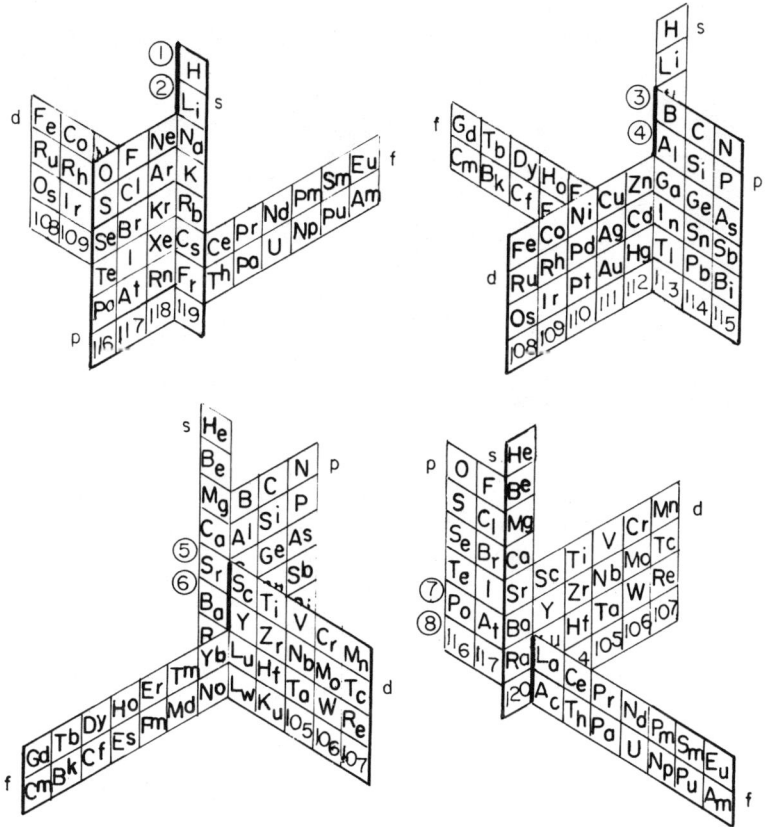

FIG. 110. Giguère 1966. Subtype IIIA4–1a.

Subtype IIIB4-1a

Spiral similar to the basic spiral except that the revolutions are separated to form a figure similar to a triple lemniscate (Fig. 111). This two-dimensional spiral represents the helix subtype IIIA4-1a. Janet originated this spiral in 1928 (1). Lepsius and Asunmaa (2) drew two lemniscates, one inside the other. One lemniscate had s and d elements on the two different loops and the other bore p and f elements; the p circles were outside the s circles and the f circles outside the d circles. However, in principle it is little different from Janet's triple lemniscate. This spiral is more easily read than the basic type IIIB4-1.

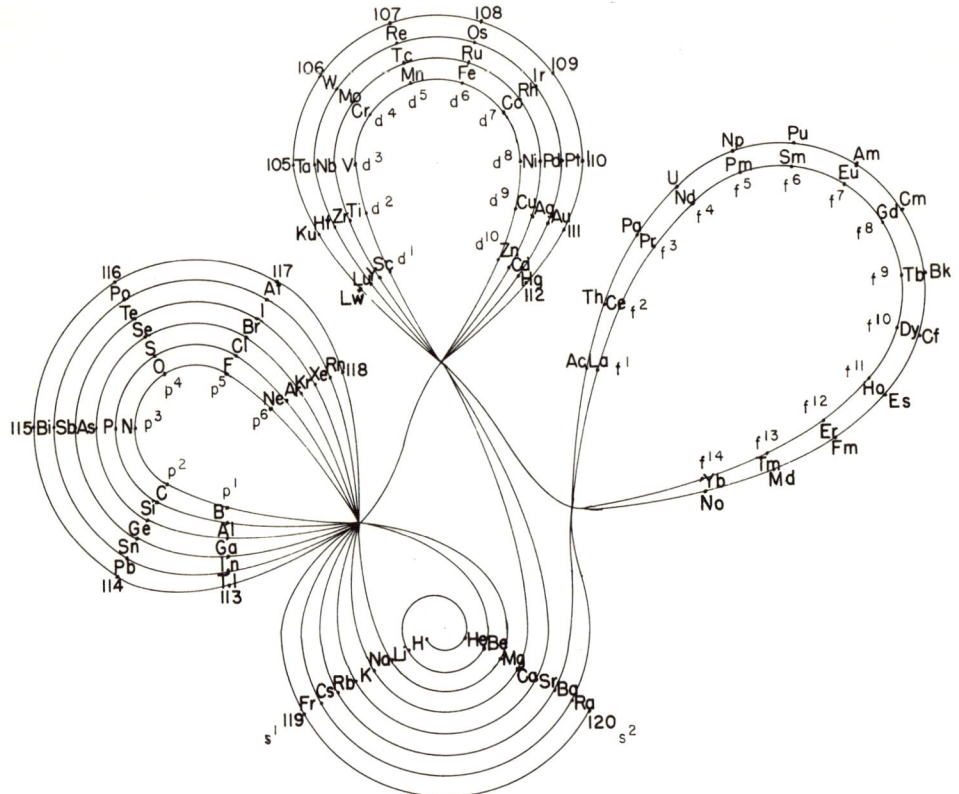

FIG. 111. Janet 1928. Subtype IIIB4-1a.

Subtype IIIC4-1a

Symmetrical series electronic configuration table (Fig. 112). An approximation of this subtype was suggested by Efremov (1). The table is not very convenient because there are too many interruptions. The beginning of each period (according to the old divisions) is shown by a numeral in the vertical column. On one side of the table are the elements with one electron spin and on the other side those with opposite spin.

Subshells:	f							d					p			Periods.	s		p			d					f						
Electrons:	8	9	10	11	12	13	14	6	7	8	9	10	4	5	6		1	2	1	2	3	1	2	3	4	5	1	2	3	4	5	6	7
																1.	H	He															
																2.	Li	Be	B	C	N												
												O	F	Ne	3.	Na	Mg	Al	Si	P												
												S	Cl	Ar	4.	K	Ca				Sc	Ti	V	Cr	Mn							
							Fe	Co	Ni	Cu	Zn							Ga	Ge	As												
												Se	Br	Kr	5.	Rb	Sr				Y	Zr	Nb	Mo	Tc							
							Ru	Rh	Pd	Ag	Cd							In	Sn	Sb												
												Te	I	Xe	6.	Cs	Ba				Lu	Hf	Ta	W	Re	La	Ce	Pr	Nd	Pm	Sm	Eu
Gd	Tb	Dy	Ho	Er	Tm	Yb	Os	Ir	Pt	Au	Hg							Tl	Pb	Bi												
												Po	At	Rn	7.	Fr	Ra				Lw	Ku	105	106	107	Ac	Th	Pa	U	Np	Pu	Am
Cm	Bk	Cf	Es	Fm	Md	No	108	109	110	111	112							113	114	115												
												116	117	118																		
Groups:													VI a	VII a	VIII a		I a	II a	III a	IV a	V a	III b	IV b	V b	VI b	VII b							

FIG. 112. Efremov 1951. Subtype IIIC4-1a.

Subtype IIIB4-1b

Spiral without differentiation in the sizes of the revolutions, but on which the disposition of the elements is shown by different lengths of the radii. The

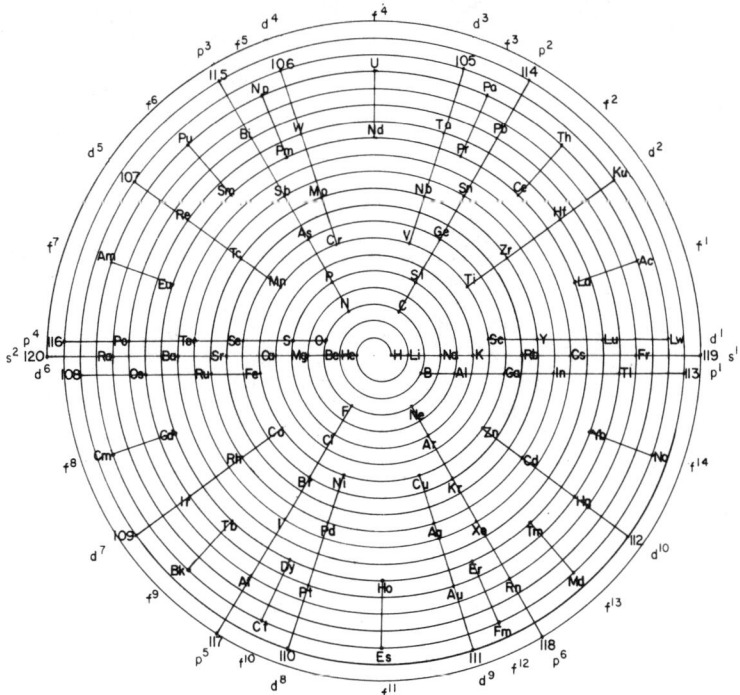

FIG. 113. Cueilleron 1946. Subtype IIIB4-1b.

114 GRAPHIC REPRESENTATIONS OF THE PERIODIC SYSTEM

spiral (Fig. 113) resembles the short spiral type IB3-1 (Fig. 42), which was originated by Green and Jackson in 1950. Cueilleron originated this one in 1946 (1). It is a projection of helix IIIA4-1 on a plane. The spiral is very difficult to read.

Type 2: Right-Side Electronic Configuration Tables

Type IIIB4-2

Spiral with interruptions between the subperiods (Fig. 114). The subperiods (or blocks) of elements s, p, d, and f are separated by placing them on different sections of the spiral. On the drawing, the starting points of periods (according to the old numeration) are shown by circled numerals. The spiral was originated by Zapffe (1), but the order of blocks in his table was s, p, f, d instead of s, p, d, f. Plichta (2) split block p into two parts; Therefore, his drawing was slightly different from that shown in Fig. 114. From this spiral, the next series table, type IIIC4-2, is easily derived by cutting between s^1 and f^{14}. This is a successful table. It is reminiscent of the spiral type IIIB3-4 (Fig. 97), but is not the same since it is an electronic configuration table.

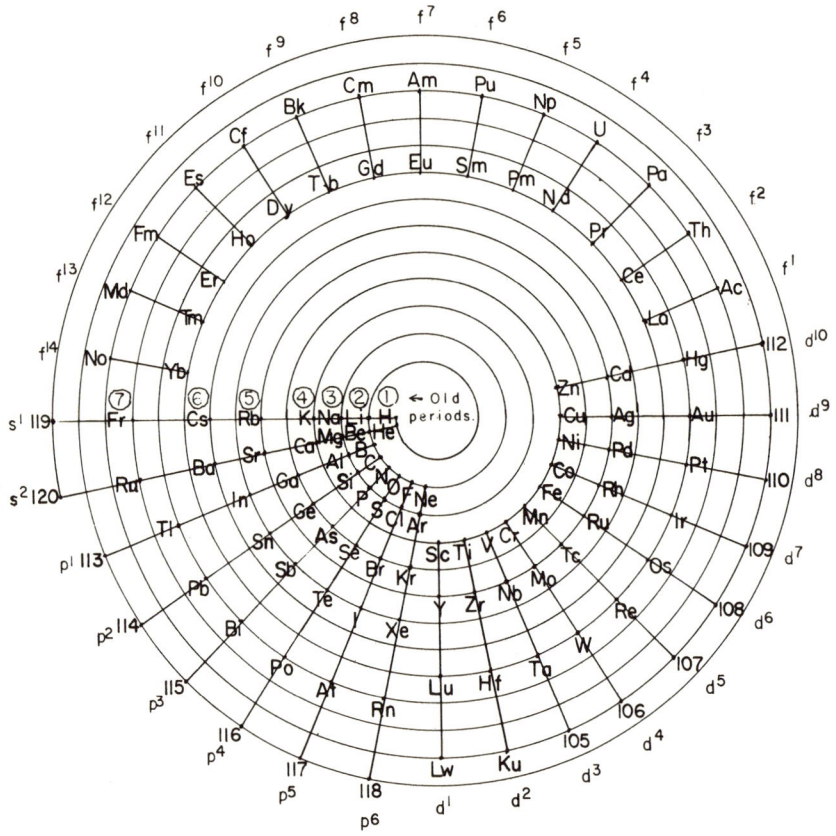

FIG. 114. Zapffe 1947. Type IIIB4-2.

Type IIIC4–2

Right-side electronic configuration series table (Fig. 115). This series table can be derived from the previous spiral table, type IIIB4–2, by cutting the spiral between s^1 and f^{14} and flattening it. Originated by Gardner in 1930 (1), this table is one of the best tables and is now a popular one. Earlier, in 1905, Gooch and Walker (0000) had drawn a similar table, although not for use as an electronic configuration table. They divided the table into primary (now the representative elements), secondary (transition elements), and tertiary series (inner transition elements). Kirchhof (000) had also drawn a similar table, except that the lanthanides were incorrectly separated from the transition elements. The two forerunners of the discovery of this table were Rolla and Piccardi in 1926 (00) and Maghnad Saha in 1927 (0) who drew a scheme for this table and fragments of the table itself. Znoiko (8) drew a table very similar to that of Gooch and Walker (0000). The tables of Finke (4) and Grjébine (7) had oblique series. The tables of Finke (4), Scheibe, Baumgärtner, and Genzer (12), and Walker and Curthois (13) were constructed in double series of subshells. Some authors, Finke (4), Scheibe, Baumgärtner, and Genzer (12), Walker and Curthois (13), Longuet-Higgins (16), Lu Shao-chi (18), and Smith (21), constructed the table similar to the scheme of subshell energy levels so that the table seems upside down in comparison with Fig. 115 and has to be read from bottom to top.

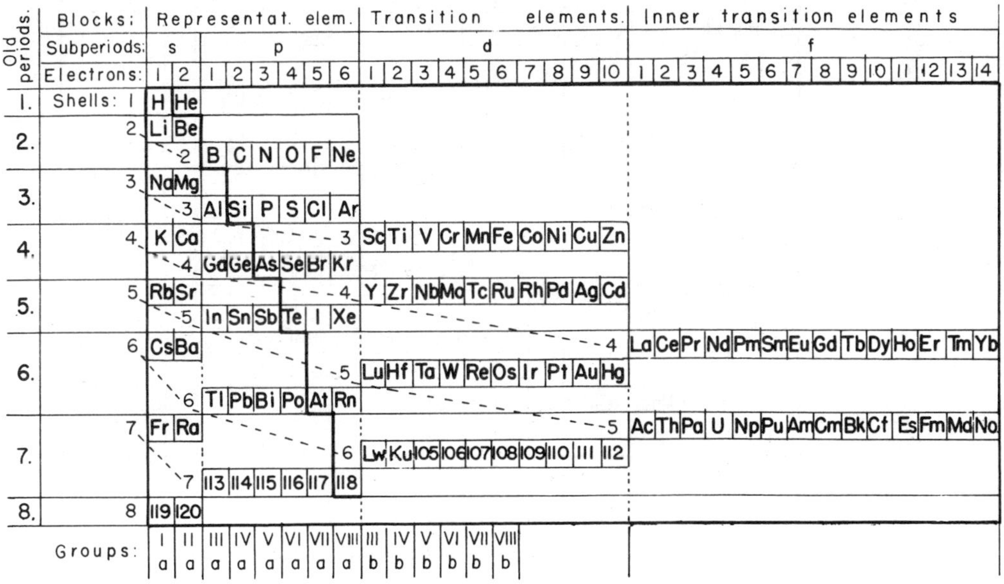

FIG. 115. Gardner 1930. Type IIIC4–2.

Subtype IIIC4–2a

Series table similar to the previous table, but with five blocks of elements instead of four. The fifth block contains only two elements: the primary elements

116   GRAPHIC REPRESENTATIONS OF THE PERIODIC SYSTEM

H and He. Ramírez-Torres originated this subtype in 1955 (1). There is no reason to separate the primary elements from the alkali and alkaline earth elements despite the fact that they are not chemically similar. This was explained in the Introduction to Subdivision IIIB.

Subtype IIIC4–2b

Series table similar to type IIIC4–2 with some groups of block d as subgroups of the block p groups. This subtype, suggested by Shtandel' in 1949, cannot be considered as successful as the preceding types IIIC4–2 and IIIC4–2a.

Subtype IIIC4–2c

Series table with blocks d and f combined. The table, originated by Rose Aynard (1) and based on magnetic susceptibility of elements, was slightly different from the one shown (Fig. 116). Grigorovich's table (3) was a little different in the placement of s^1 and s^2 elements. Combining the blocks of the table only shortens it without significant improvement.

Old periods	Blocks:	Representative elements								Transition elements													
	Subperiods:	s		p						d and f														
	Electrons:	1	2	1	2	3	4	5	6	1	2	3	4	5	6	7	8	9	10	11	12	13	14
1.		H	He																				
2.		Li	Be	B	C	N	O	F	Ne														
3.		Na	Mg	Al	Si	P	S	Cl	Ar														
4.		K	Ca							Sc	Ti	V	Cr	Mn	Fe	Co	Ni	Cu	Zn				
				Ga	Ge	As	Se	Br	Kr														
5.		Rb	Sr							Y	Zr	Nb	Mo	Tc	Ru	Rh	Pd	Ag	Cd				
				In	Sn	Sb	Te	I	Xe														
6.		Cs	Ba							La	Ce	Pr	Nd	Pm	Sm	Eu	Gd	Tb	Dy	Ho	Er	Tm	Yb
										Lu	Hf	Ta	W	Re	Os	Ir	Pt	Au	Hg				
				Tl	Pb	Bi	Po	At	Rn														
7.		Fr	Ra							Ac	Th	Pa	U	Np	Pu	Am	Cm	Bk	Cf	Es	Fm	Md	No
										Lw	Ku	105	106	107	108	109	110	111	112				
				113	114	115	116	117	118														
8.		119	120																				
Groups:		I a	II a	III a	IV a	V a	VI a	VII a	VIII a	III b	IV b	V b	VI b	VII b	VIII b								

FIG. 116.   Aynard 1959.   Subtype IIIC4–2c.

Subtype IIIC4–2d

Table constructed on the basis of the main type IIIC4–2 (Fig. 117). The difference is that the symbols of the elements belonging to the same block are placed in different rows and columns if irregularities exist in the electron configurations of these elements. The changes in electron configuration occur in neighboring subshells. The notation of these subshells, together with the included number of electrons, is shown in front of the rows in which the changes occur (blocks d and f). Horizontally, the element symbols are placed in appropriate rows in agreement with these changes in neighboring subshells. Vertically, they are placed in columns according to the number of electrons in a subshell, the notation of which is in the heading of the table. This table was originated by Condon and Shortley in 1951 (1).

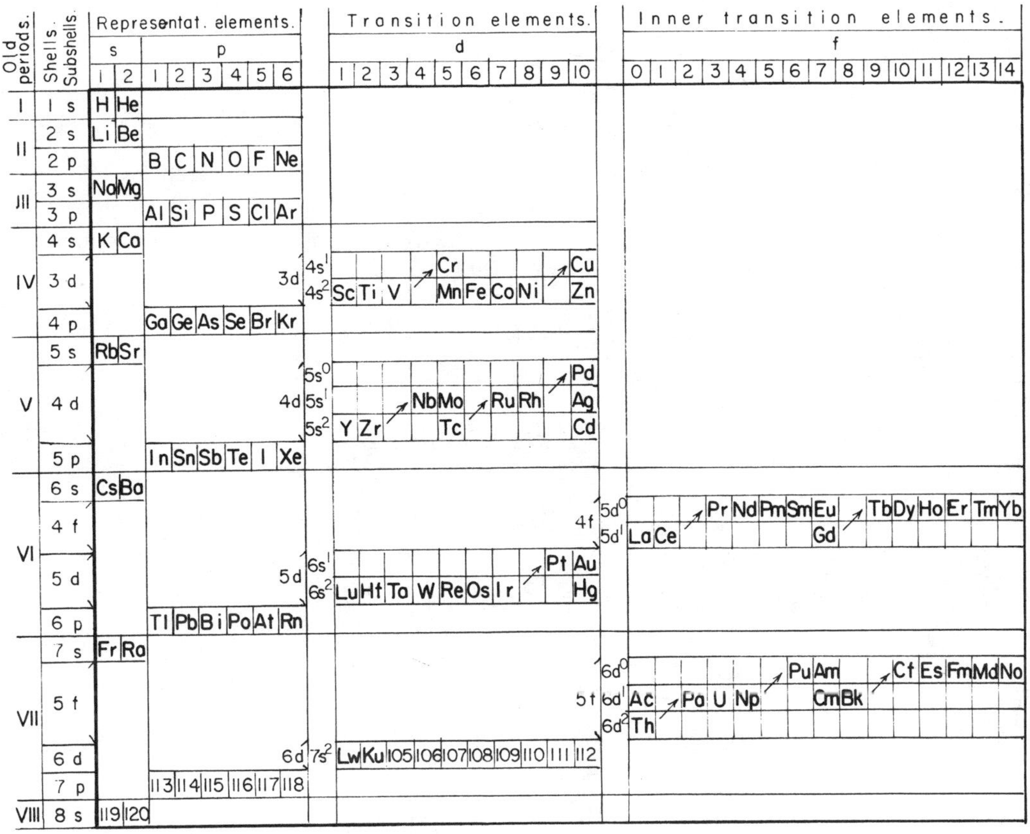

FIG. 117. Condon and Shortley 1951. Subtype IIIC4-2d.

Subtype IIIC4-2e

Right-side zigzag table (Fig. 118). This zigzag table, originated by Douglas Clark in 1931 (1), has a slight advantage over its equivalent helix, type IIIA4-1. The zigzags show that the orbitals of the subshells are half filled with electrons of one spin and then completed with electrons of the opposite spin. It is also similar in construction to the Giguère table, subtype IIIA4-1a.

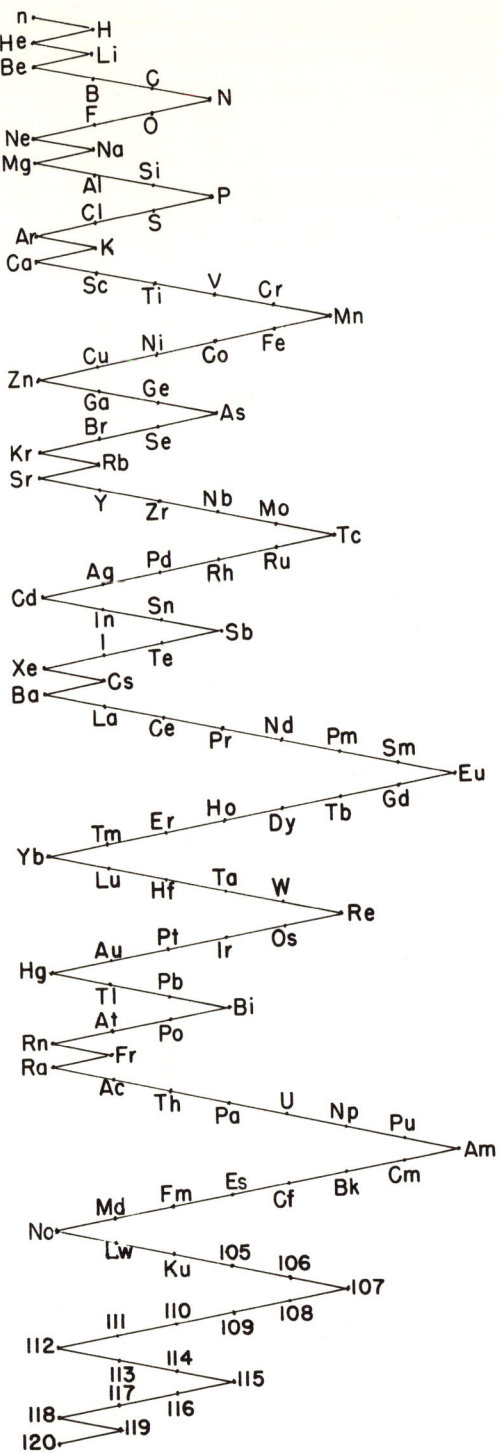

FIG. 118. Douglas Clark 1931.   Subtype IIIC4-2e.

Type 3: Left-Side Electronic Configuration Tables

Type IIIC4-3

Left-side electronic configuration series table (Fig. 119). This table resembles the left-step series table by Janet, type IIIC3-6 (Fig. 103), only "stretched" in a vertical direction. Gibson originated this table in 1948 (1). Earlier, in 1921, Bury (00) had devised the first electronic configuration table, which partially belongs to this type. In 1934 Romanoff (0) had drawn this kind of table with oblique series. He placed the elements in four blocks of shells but did not place them in definite vertical columns. Bolívar and others in 1957 and Keller in 1958 (2) drew the table on a paraboloid; however, the space figure did not play any role in the construction of the table. The periods were turned around and must be read from right to left. Mazurs (3) placed the inner transition elements in two series to shorten the table horizontally. Platonov's table in 1961 (4) was constructed to be read from the bottom to the top; the division into new periods was not used. This type is one of the best electronic configuration tables.

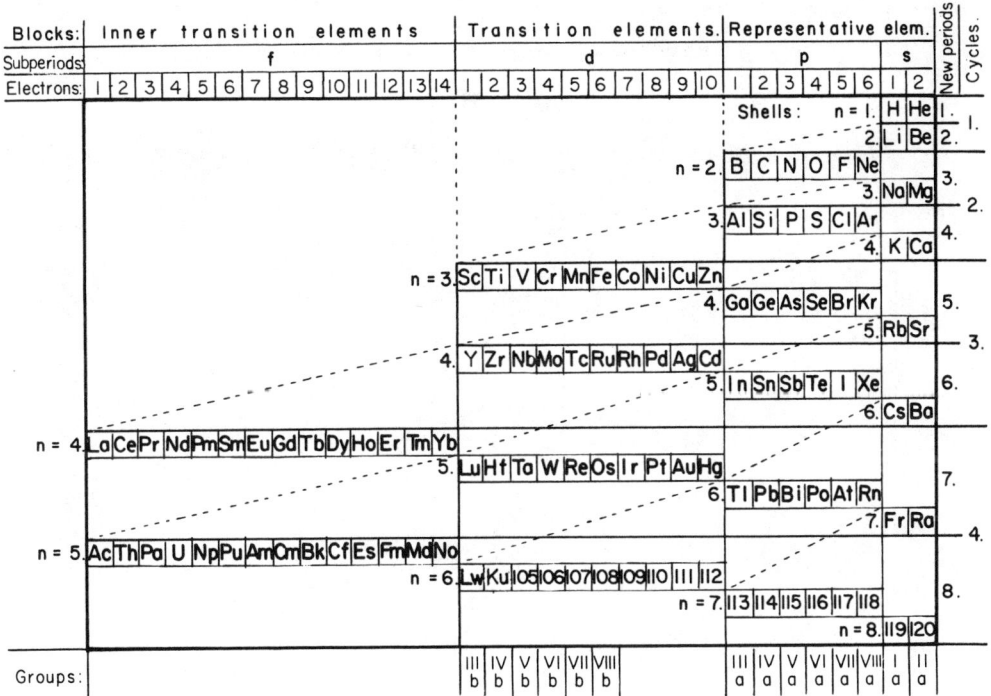

FIG. 119.   Gibson 1948.   Type IIIC4-3.

Subtype IIIC4-3a

Series table similar to the preceding table with the blocks of "alkali metals" and "nonmetals" placed in consecutive order and combined into one block of representative elements (Fig. 120). Chaverri-Rodríguez originated this subtype in 1953 (1). In some ways this table combines the best aspects of an electronic and a "chemical" table. Mazurs (2) placed the inner transition elements in two series.

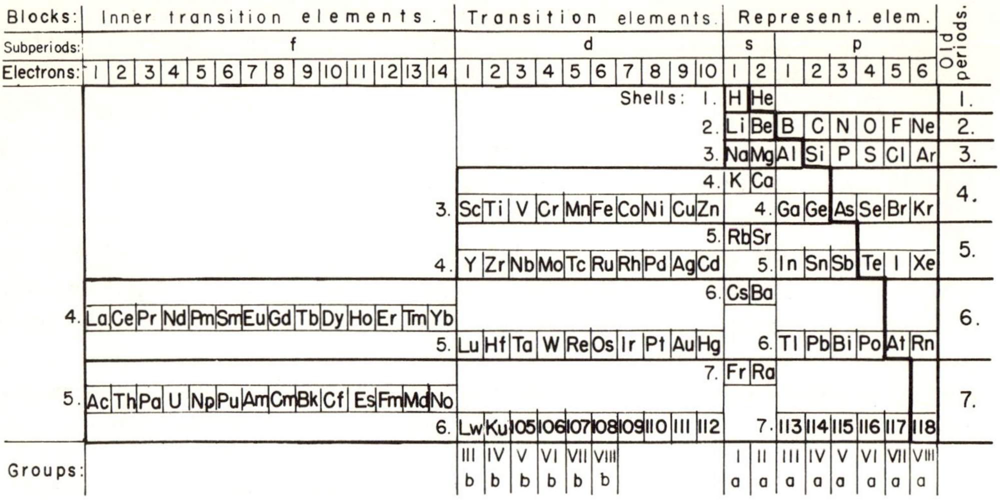

FIG. 120. Chaverri-Rodríguez 1953. Subtype IIIC4-3a.

Subtype IIIC4-3b

Table similar to the main type IIIC4-3 (Fig. 121), but changed by placing

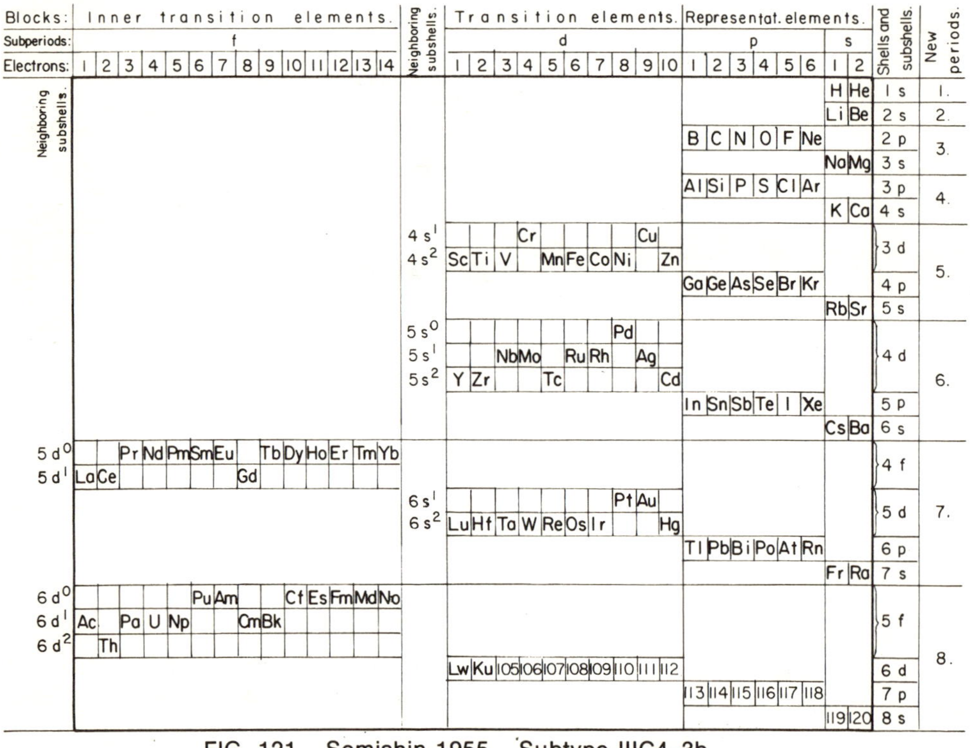

FIG. 121. Semishin 1955. Subtype IIIC4-3b.

DIVISION III. LONG TABLES   121

the elements in different rows if they do not have regular electronic configurations. The change in electronic configurations occurs in neighboring subshells. The notations of these subshells, together with included number of electrons, are shown in front of the horizontal rows of blocks (*d* and *f*) where the element symbols with the changes are located. Vertically they are placed in columns where they should theoretically belong according to the subshell notations and electron numbers shown in the heading of the table. This is different from the table of subtype IIIC4–2d where the element symbols are also placed vertically according to the irregularities in electron configurations. In 1955 Semishin (1) actually used type IIIC3–4 as the basis of his table.

Subtype IIIC4–3c

Left-side electronic configuration table based on the main type IIIC4–3 with placement of element symbols in various horizontal rows according to their differentiating electrons (Fig. 122). As is the case with the other electronic configuration

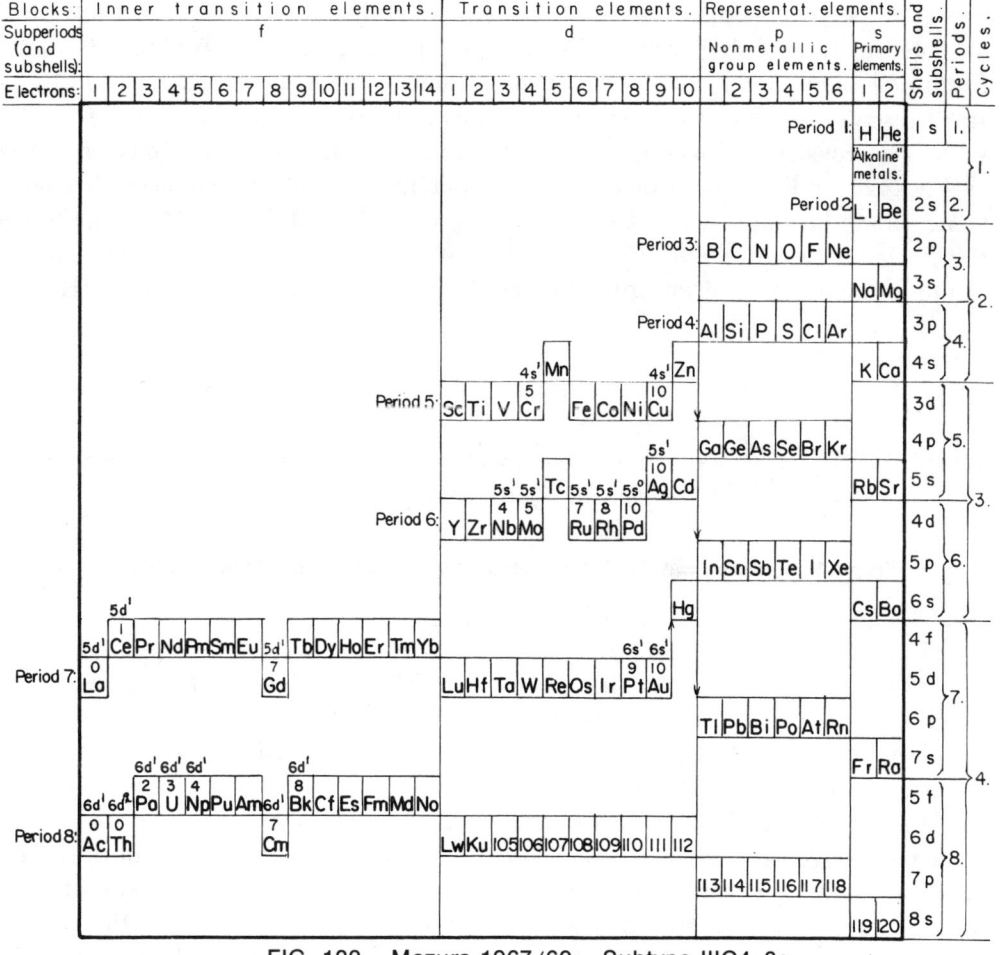

FIG. 122.   Mazurs 1967/69.   Subtype IIIC4–3c.

tables, the heading of the table contains the numbers of electrons in various subshells that would be expected if the subshells were filled according to the theoretical point of view. The element symbols are placed vertically according to this ideal number of electrons on subshells. However, if discrepancies occur between the actual position of electrons on the subshells and this theoretical ideal, then these changes in electron contents are shown as follows:

1) the number of electrons *for the same subshell* is placed on the top of the element symbol and

2) the change occurring *in a different subshell* is shown above the element square.

The horizontal rows of the table are denoted by their shells and subshells. This is shown in one of the last vertical columns of the table. The order of subshells in these horizontal rows is the same as in the construction scheme of the periodic table (Table 4, p. 100). The element symbols are placed in the horizontal rows according to their last completing electron, the so-called differentiating electron, as was suggested by Pohl in 1958 (see subtype IIIC5–2b) and by Kessler in 1961 (see subtype IIIC5–3b). As a result, 6 elements belonging to the d subshell block are placed in s subshell rows: Mn, Zn, Tc, Ag, Cd, and Hg; 5 elements of the f subshell block are located now in d subshell rows: La, Gd, Ac, Th, and Cm. In this way, the table was made more realistic than the main type IIIC4–3. This table was constructed by myself in 1967 and published in 1969. It is presented as Fig. 122 and it also appears as a foldout in color at the back of this book. Concerning the coloration of this table, see pages 140–141, below. Mills (2) showed the irregularities in electron configuration with arrows; however, he did not use the differentiating electron principle and, therefore, his type was actually IIIC3–6.

CLASS 5: SHELL AND SUBSHELL TABLES WITH ELECTRONIC CONFIGURATION DISPOSITION OF ELEMENTS 2, 6, 10, AND 14.

Type 1: Symmetrical Tables of Concentric Circles and Parallel Lines

Type IIIA5–1

Space concentric circles in eight planes (Fig. 123). The planes represent the shells of atomic electron structure. The circles represent the subshells with different numbers of elements—2, 6, 10, and 14. Haenzel originated this table in 1943 (1). He used polygons instead of circles, but that does not make any difference. Talpain (3) drew quadrangles. Since this table is a shell table and not a period table, it is not convenient to read according to periods. I attempted to improve this table by connecting the circles (subshells) belonging to the same period with lines. In this way, truncated and inverted cones are formed that represent the periods. The table should be read in order of the period numeration, which is given in small circles on the drawing. Periods begin at the largest circles and proceed to the smaller circles.

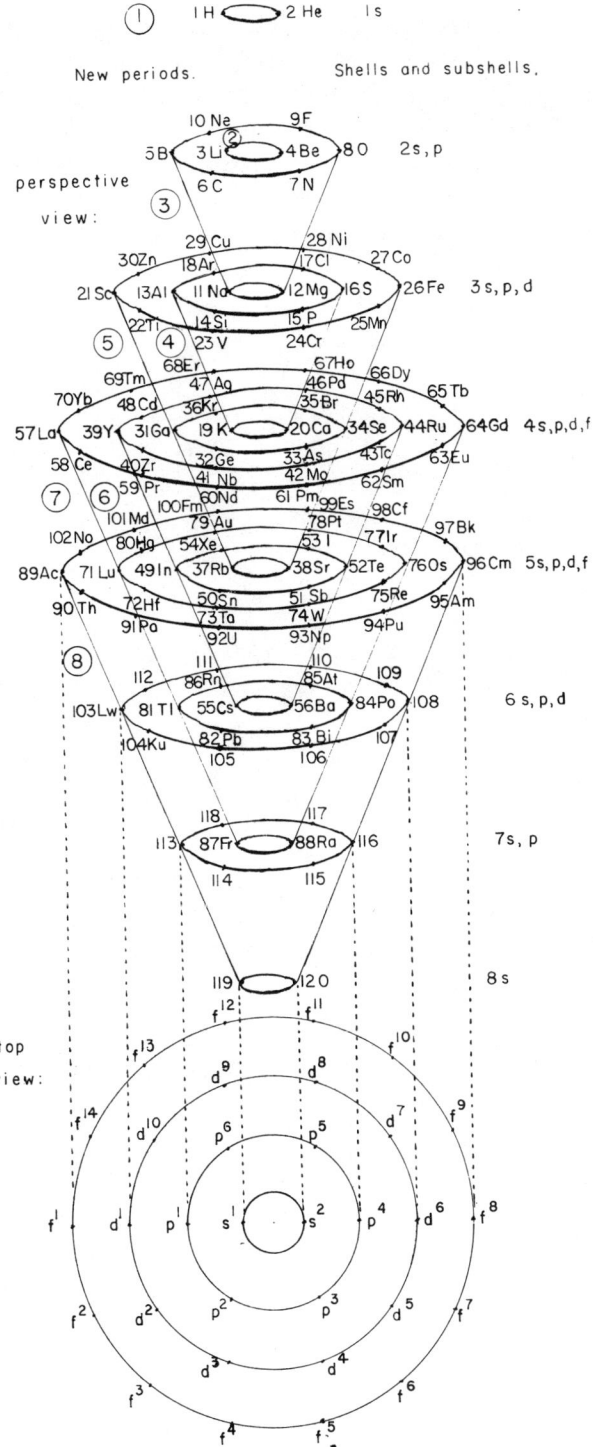

FIG. 123. Haenzel 1943. Type IIIA5-1.

124 GRAPHIC REPRESENTATIONS OF THE PERIODIC SYSTEM

Type IIIB5-1

Concentric circles on a plane (Fig. 124). The circles represent the subshells. The groups of circles (circles that lie closest together) belong to the same shell. A different number of elements is found on the circles—2, 6, 10, and 14. The elements belonging to the same block of elements (*s, p, d,* or *f*) lie on radii of the same length, and the length of radii of different blocks differs from each other. These concentric circles do not differ from the spiral type IB3-1 (Fig. 42); however, the principle involved in drawing the table is entirely different. Sibaiya originated these circles in 1941 (1). Schultze's spiral (2) was not drawn very clearly. The coloring of Flores-Cabral's table (3) was suggested by C. L. Alves da Fonseca. This table is not convenient to read.

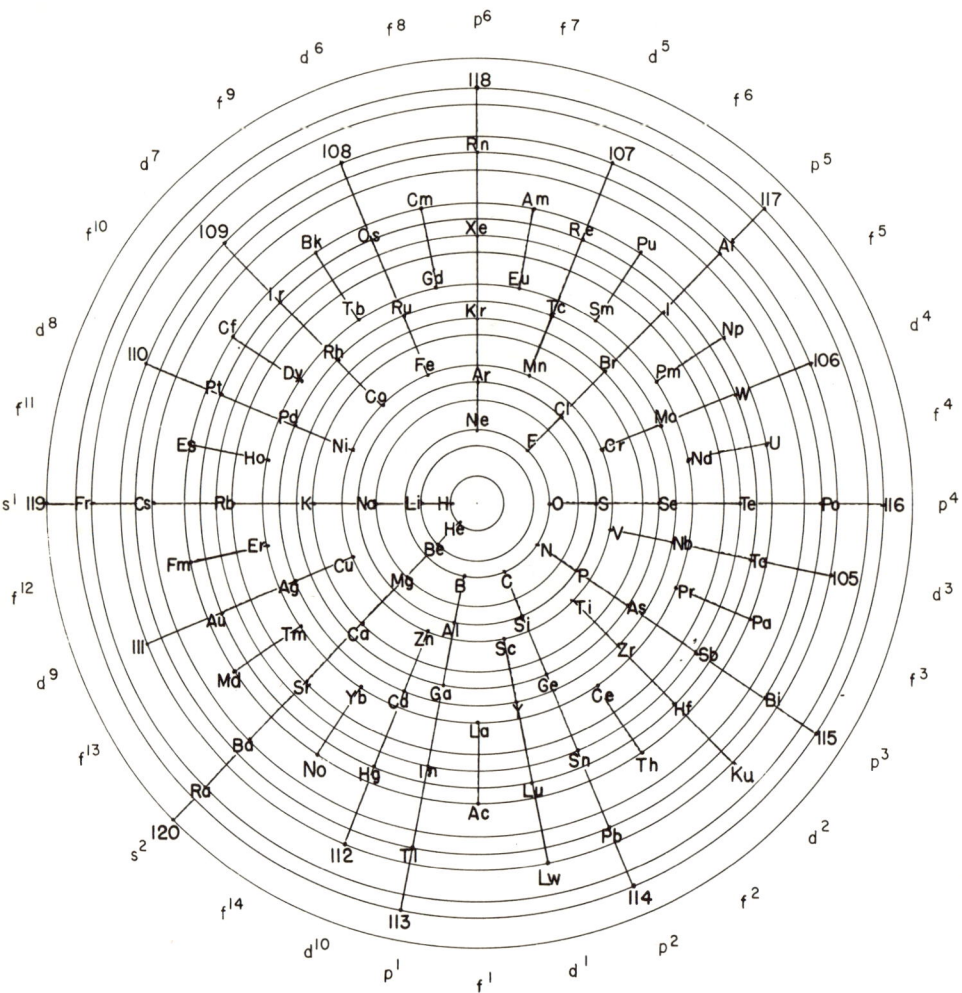

FIG. 124. Sibaiya 1941. Type IIIB5-1.

Type IIIC5-1

Parallel-line table (Fig. 125). Parallel lines designate subshells, the groups of parallel lines represent the different shells. Cerasoli originated this table in 1941 (1). The table is not convenient for common use since it is divided into shells instead of into periods, and it cannot be read in the order of increasing atomic numbers. Yeou Ta (3) had triangles in a single line with 2 elements in each triangle.

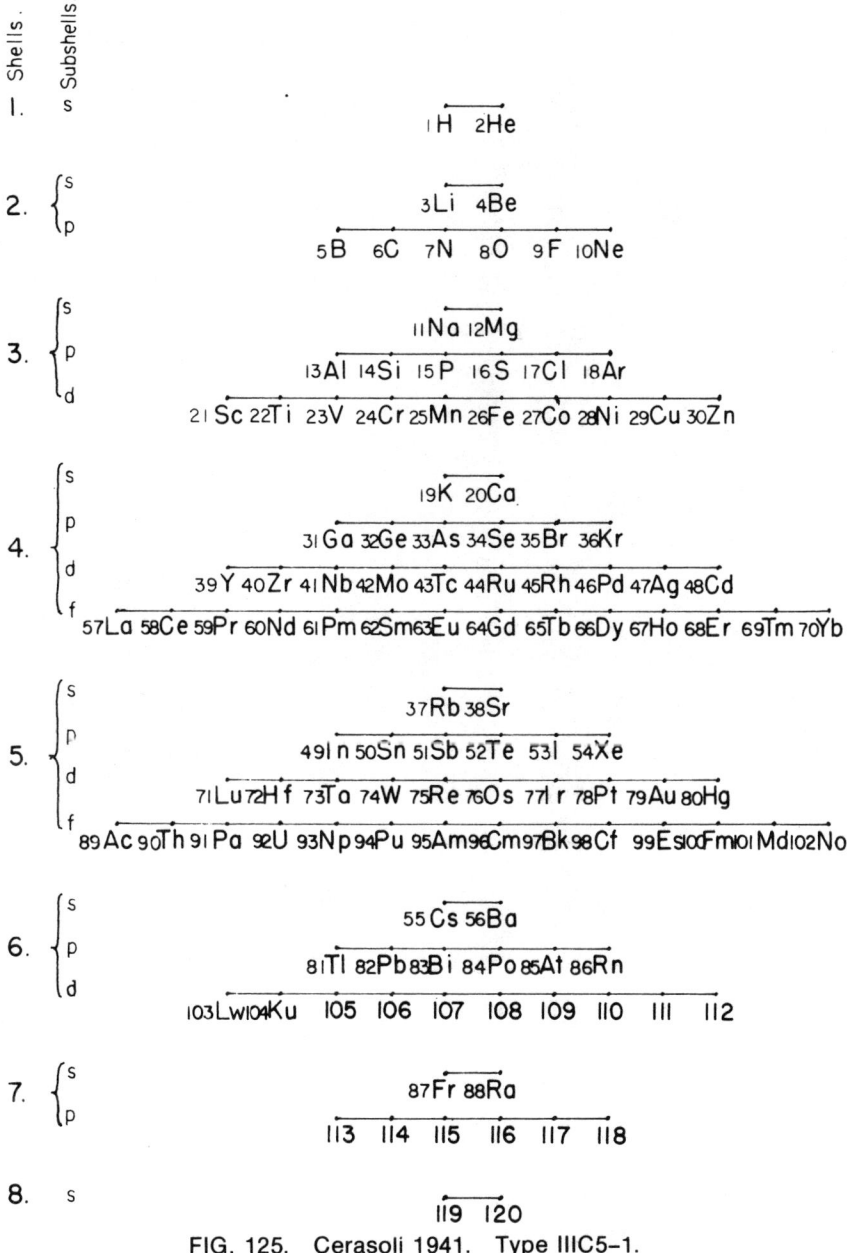

FIG. 125. Cerasoli 1941. Type IIIC5-1.

Subtype IIIA5–1a

Concentric circles in space representing the subshells 2, 6, 10, and 14 elements on them. This subtype differs from the previous main type, IIIA5–1, because the "period cones" are stretched vertically (Fig. 126). At the same time, the subshell circles are arranged approximately in order of decreasing energy from the top to the bottom. This subtype was suggested by myself in 1967 (1).

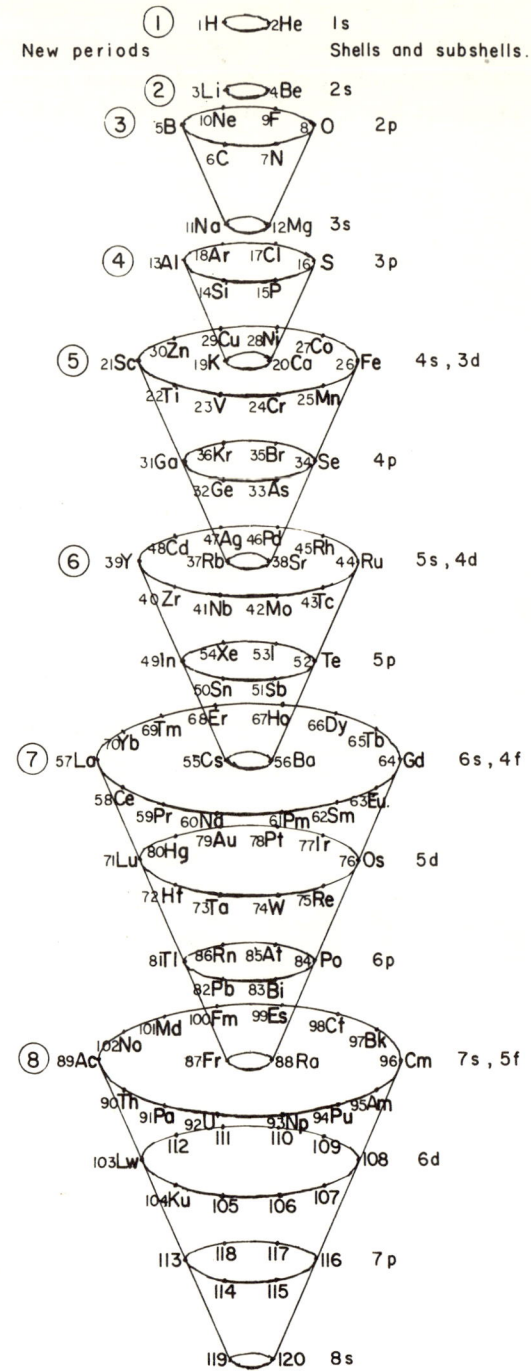

FIG. 126. Mazurs 1967. Subtype IIIA5–1a.

DIVISION III. LONG TABLES

Subtype IIIC5-1a

Parallel-line table (Fig. 127). The basic type, IIIC5-1, is changed as follows: Firstly, the subshell lines are arranged in the order in which they are filled by electrons according to Table 4 (p. 100). Secondly, the subshell lines belonging to the same period are connected to form inverted trapezoids. In this way, the table appears to be more like a periodic table and it can be read in order of increasing atomic numbers. This arrangement was done by myself in 1967 (1).

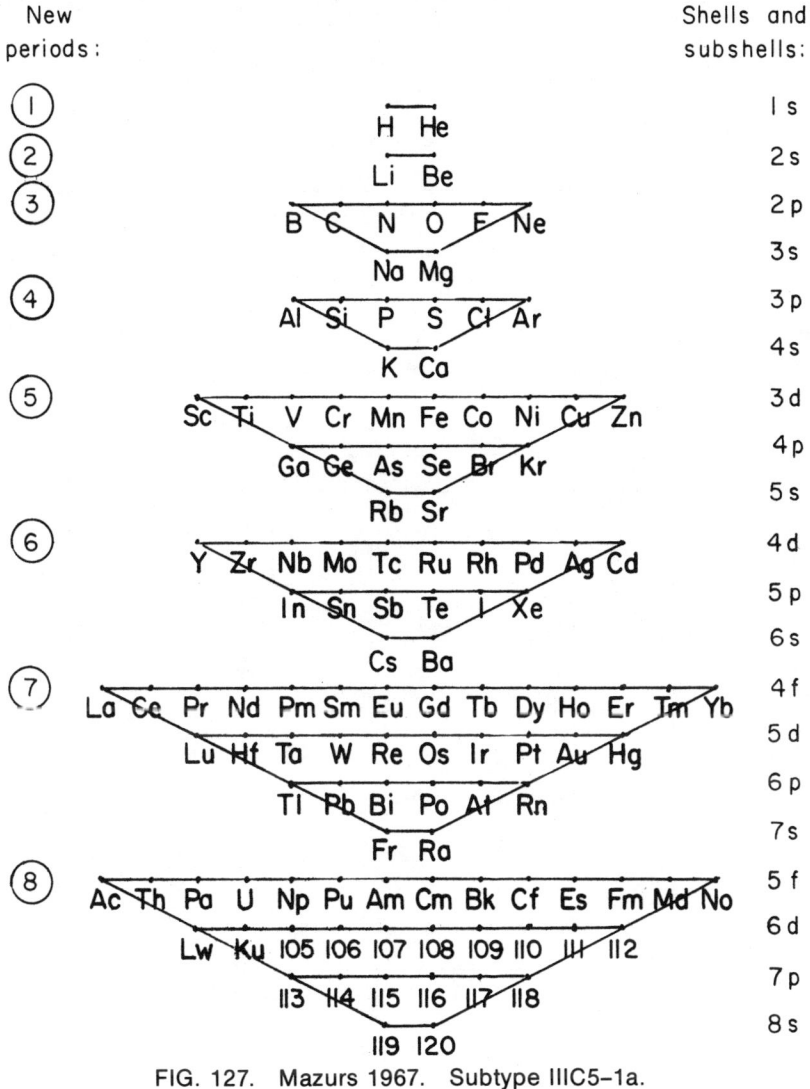

FIG. 127. Mazurs 1967. Subtype IIIC5-1a.

Subtype IIIC5-1b

Shell table (Fig. 128) where the elements of odd atomic numbers are placed to the left and those of even atomic numbers to the right. The originator was Kerker (1) in 1957. This is not a convenient table.

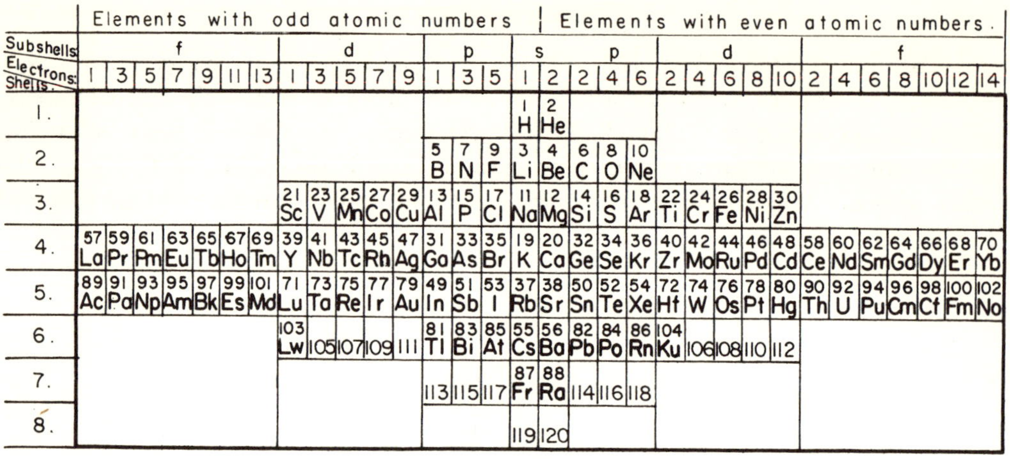

128 GRAPHIC REPRESENTATIONS OF THE PERIODIC SYSTEM

FIG. 128. Kerker 1957. Subtype IIIC5–1b.

Subtype IIIC5–1c

Shell and subshell series table originated by Mitra in 1931 (1). This table (Fig. 129) is actually adjusted to the construction of type IC2–4.

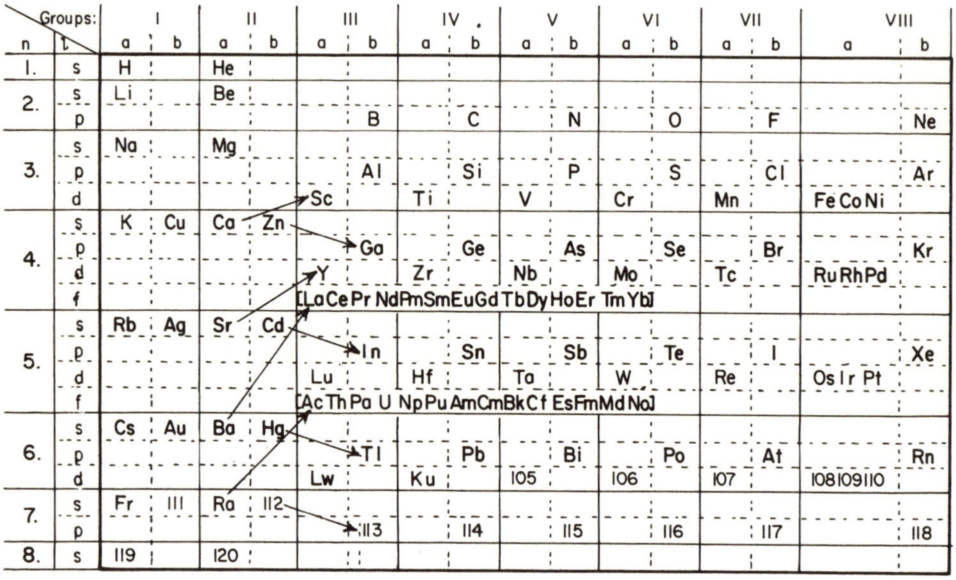

FIG. 129. Mitra 1931. Subtype IIIC5–1c.

Type 2: Right-Side Shell and Subshell Tables

Type IIIC5_2

Right-side electronic shell and subshell series table symmetrical about a horizontal line (Fig. 130). This table is more a "shell" table than a periodic table.

Therefore, it is not good for teaching purposes since reading the elements in order of increasing atomic numbers requires jumping from one series to another. However, this table originated by Corbino in 1928 (1) became popular. Van Rysselberghe (2) drew only fragments of this table. Gardiner's table (11) was a little different. This type was suggested by many authors, including Luder in 1939 (4).

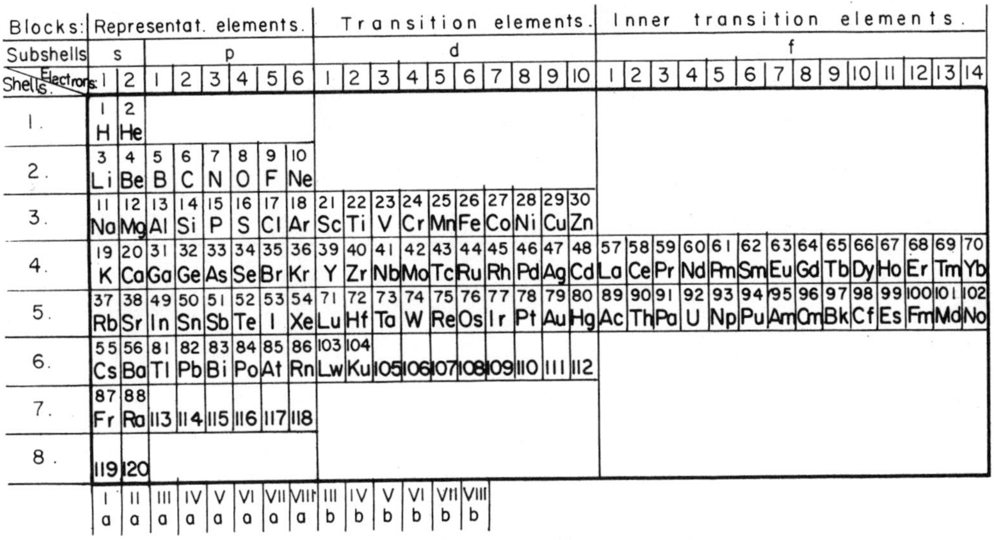

FIG. 130. Corbino 1928. Type IIIC5-2.

Subtype IIIC5-2a

Right-side electronic shell and subshell table stretched vertically so that the subshells are separated (Fig. 131). Originated by Mazzucchelli in 1930 (1), this is a shell and subshell table; the periods are shown by diagonal dotted lines. Taylor (3) placed the lanthanides and actinides together with the transition (*d*) elements.

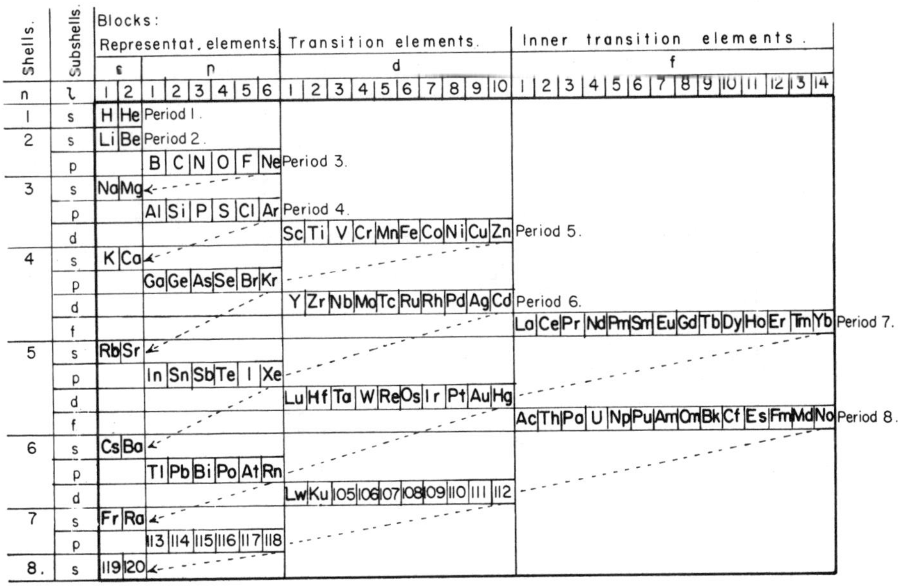

FIG. 131. Mazzucchelli 1930. Subtype IIIC5-2a.

Subtype IIIC5–2b

Right-side electronic shell and subshell table where the element symbols are placed in horizontal rows according to their differentiating electron (Fig. 132). This table was originated by Pohl (1) in 1958 and is based on the preceding subtype. The only difference between this table and the preceding one is in repositioning some element symbols to make the table more realistic. Actually, Pohl's table was not very clearly drawn and its relationship to subtype IIIC5–2a was difficult to see. This table is not as convenient to read as subtype IIIC4–3c because long jumps must be made from one block (or line) to another.

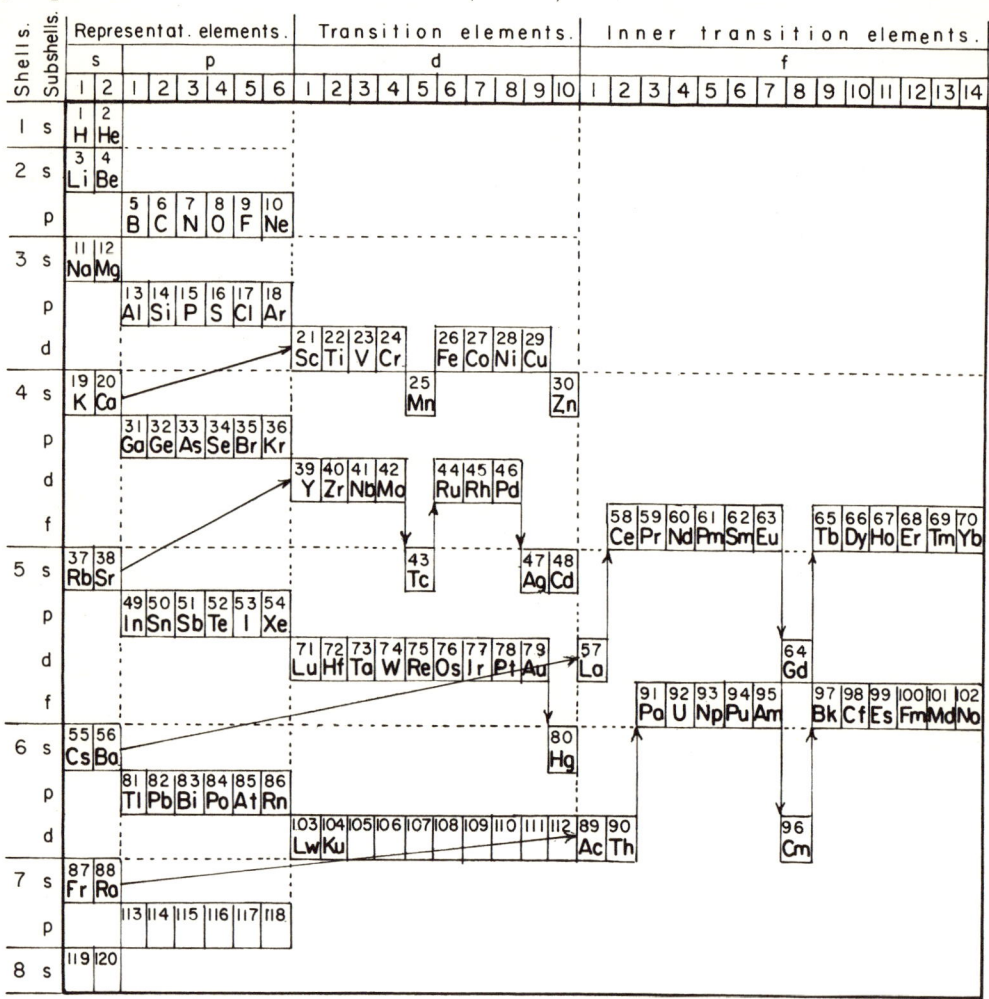

FIG. 132. Pohl 1958. Subtype IIIC5–2b.

Subtype IIIC5–2c

Right-side shell and subshell table (Fig. 133). This table contains 14 vertical columns in agreement with the maximum number of electrons found in block

DIVISION III. LONG TABLES 131

f. In these columns, the elements of other blocks are located according to their electron numbers. This table is similar to type IIIC5-1, except that this table is not symmetrical. Originated by Steinberg in 1938 (1), the table is difficult to read since big jumps are required to follow an order of increasing atomic numbers. This table shows that the difference in length of neighboring subperiods is four units according to the formula $L_s = 2+4l$ (see Appendix III-3).

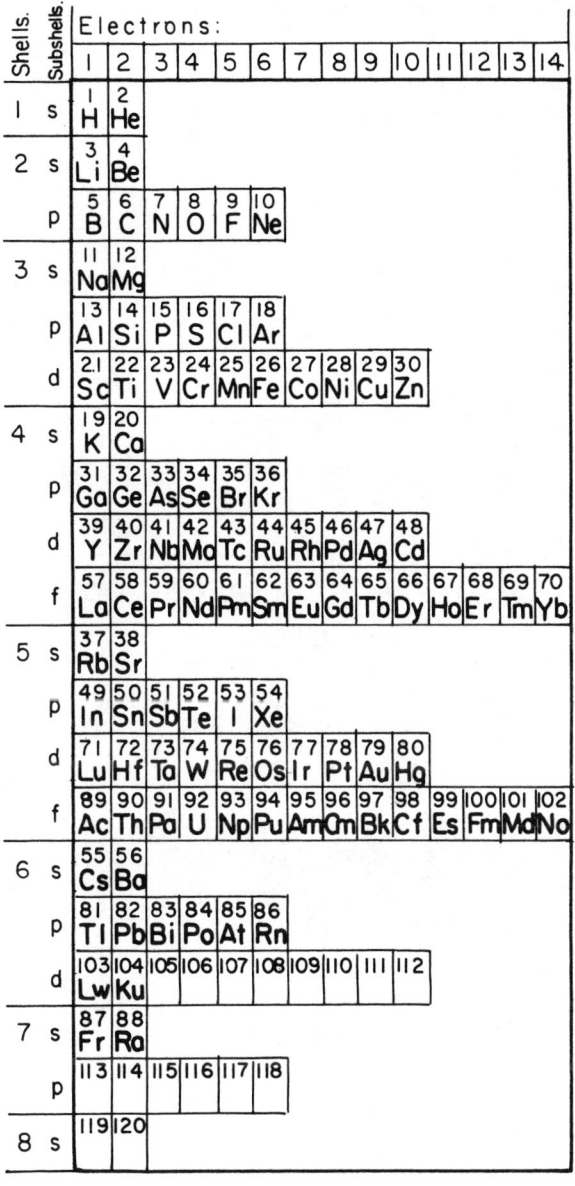

FIG. 133. Steinberg 1938. Subtype IIIC5-2c.

Subtype IIIC5-2d

Right-side combined electronic and shell and subshell table (Fig. 134). This table also incorporates some properties of the chemical tables since the element family groups are designated on it. The new period numeration is seen clearly. The order of subshell rows is arranged according to the mathematical expression of the Periodic Law, $t = n+l$, demonstrated by Table 4 (p. 100). Therefore, the element symbols are read naturally in order of increasing atomic numbers. The valence shown by group numbers is obtained by summing the electrons of block *s* and an appropriate block (*p*, *d*, or *f*). The coloring of the table shows the elements that are related to one another by their chemical and physical properties. The color reproduction appears at the back of this book. This table was published for the first time in 1973 (1).

New periods	Shells and subshells	Groups: I	II	III	IV	V	VI	VII	VIII						
		Electrons: 1	2	1	2	3	4	5	6	7	8	9	10	11	12	13	14
1.	1 s	1 H	2 He														
2.	2 s	3 Li	4 Be														
3.	2 p			5 B	6 C	7 N	8 O	9 F	10 Ne								
	3 s	11 Na	12 Mg														
4.	3 p			13 Al	14 Si	15 P	16 S	17 Cl	18 Ar								
	4 s	19 K	20 Ca														
5.	3 d			21 Sc	22 Ti	23 V	24 Cr	25 Mn	26 Fe	27 Co	28 Ni	29 Cu	30 Zn				
	4 p													31 Ga	32 Ge	33 As	34 Se	35 Br	36 Kr
	5 s	37 Rb	38 Sr														
6.	4 d			39 Y	40 Zr	41 Nb	42 Mo	43 Tc	44 Ru	45 Rh	46 Pd	47 Ag	48 Cd				
	5 p													49 In	50 Sn	51 Sb	52 Te	53 I	54 Xe
	6 s	55 Cs	56 Ba														
7.	4 f			57 La	58 Ce	59 Pr	60 Nd	61 Pm	62 Sm	63 Eu	64 Gd	65 Tb	66 Dy	67 Ho	68 Er	69 Tm	70 Yb
	5 d			71 Lu	72 Hf	73 Ta	74 W	75 Re	76 Os	77 Ir	78 Pt	79 Au	80 Hg				
	6 p													81 Tl	82 Pb	83 Bi	84 Po	85 At	86 Rn
	7 s	87 Fr	88 Ra														
8.	5 f			89 Ac	90 Th	91 Pa	92 U	93 Np	94 Pu	95 Am	96 Cm	97 Bk	98 Cf	99 Es	100 Fm	101 Md	102 No
	6 d			103 Lw	104 Ku	105	106	107	108	109	110	111	112				
	7 p													113	114	115	116	117	118
	8 s	119	120														

FIG. 134. Mazurs 1958/1973. Subtype IIIC5-2d.

Type 3: Left-Side Shell and Subshell Tables

Type IIIB5–3

Concentric circles representing the shells on which the symbols are located (Fig. 135). From this type, originated by Scheele in 1950 (1), the type IIIC5–3 can be derived. A very interesting table, it resembles the spiral type IIIB4–2 (Fig. 114), which is an electronic configuration table while this one is a shell table. In addition, the former was constructed on a spiral while this one is constructed on concentric circles. The small circled figures show the different shells.

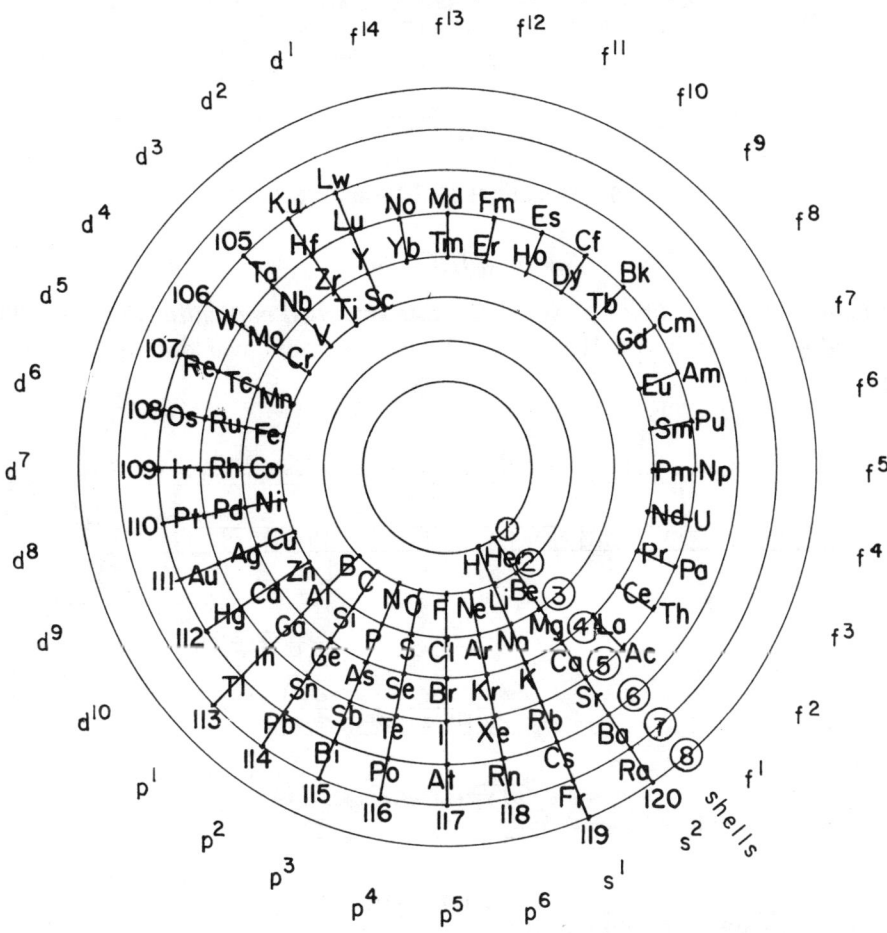

FIG. 135. Scheele 1950. Type IIIB5–3.

Type IIIC5–3

Left-side electronic shell and subshell table symmetrical about a horizontal line. This table (Fig. 136), originated by myself in 1965 (1), provides a better picture of shells, but can also be used as a periodic table. The beginning of each

134 GRAPHIC REPRESENTATIONS OF THE PERIODIC SYSTEM

period is shown by a small circled numeral. The curved lines designate the periods. Shells are read by ignoring these lines and reading straight across the table horizontally. This table can be obtained from the preceding spiral IIIB5–3 by cutting it between s^2 and f^1 and stretching the spiral into a series table.

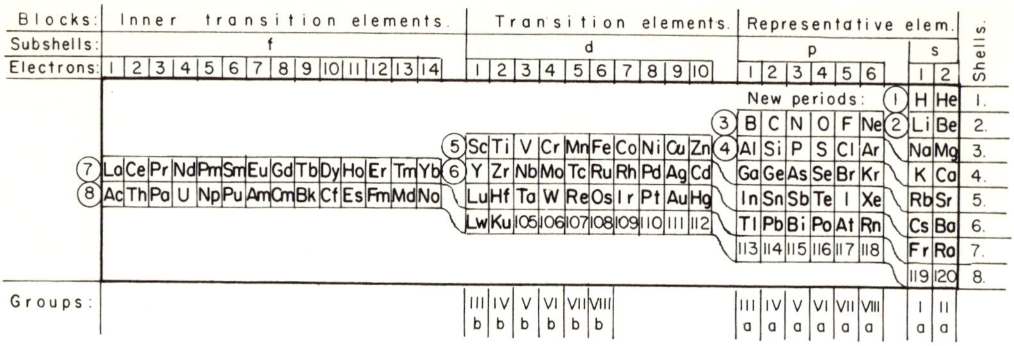

FIG. 136. Mazurs 1965. Type IIIC5–3.

Subtype IIIC5–3a

Left-side electronic shell and subshell table stretched vertically (Fig. 137). This table is analogous to subtype IIIC5–2a. The arrangement was done by myself in 1967 (1).

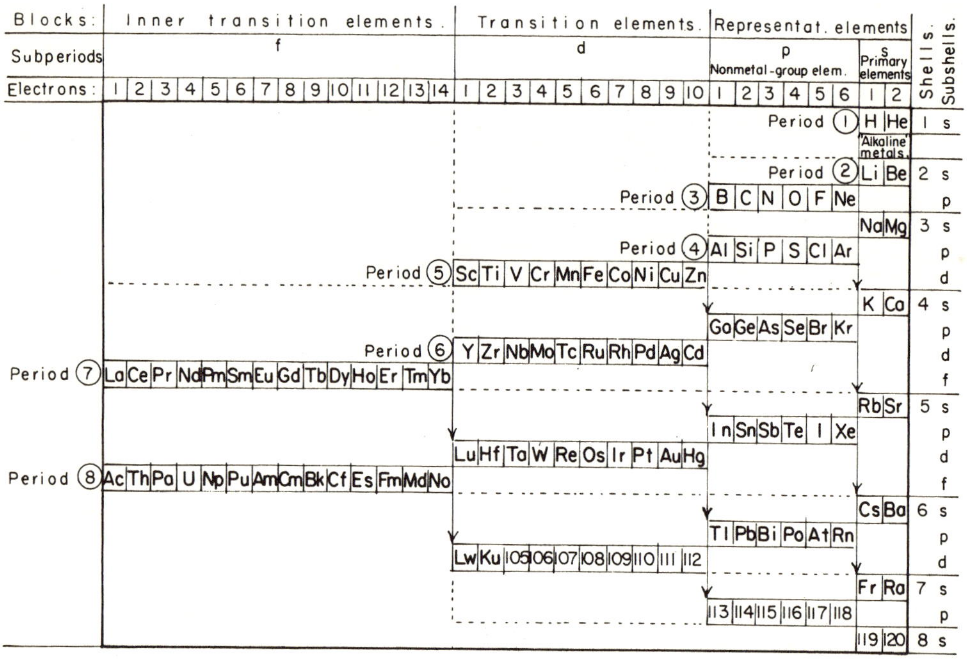

FIG. 137. Mazurs 1967. Subtype IIIC–3a.

Subtype IIIC5–3b

Left-side electronic shell and subshell table stretched vertically with subshells in different horizontal rows (Fig. 138). The subshells are arranged in the sequence of shells shown in Table 3 (p. 98). The element symbols are placed on those rows which belong to subshells where the differentiating electron of the element appears. This method was accepted by both Pohl (1) and Kessler (2) who used it to construct a table based on type IIC2–4. Dolgushin (3) constructed the same shell and subshell table on the basis of type IIIC3–4. Fig. 138 is based on the preceding subtype IIIC5–3a. This table is not as convenient as the table of subtype IIIC4–3c since the jumps from one row to another are longer.

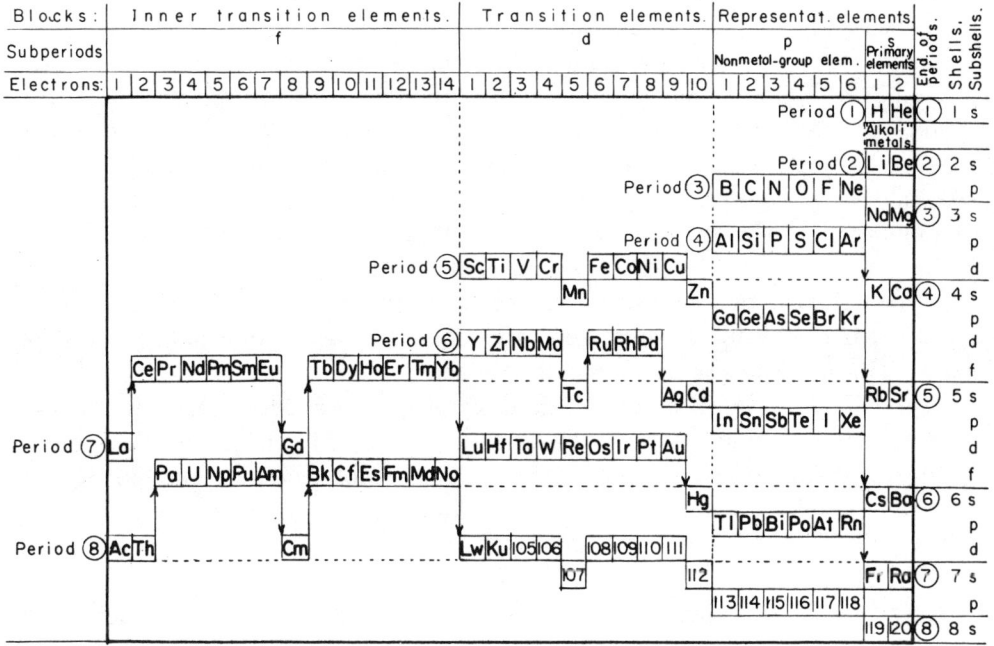

FIG. 138. Pohl 1958. Subtype IIIC5–3b.

4. Conclusion

A multitude of periodic tables has been presented in this book. Which one is the best table?

No definite answer can be given to this question. The answer depends on the purpose for which the tables are to be used. Three different uses can be distinguished: 1) as a "chemical" table in order to see clearly the valence of elements belonging to certain vertical families; 2) as an electronic configuration table in order to see the connection of the elements with their atomic structure, on which their properties depend; and 3) as shell and subshell tables in order to see the dependence of the periods on shell and subshell construction.

My opinion is that the only tables that can be taken into account are those in which all three kinds of elements—the representative, transition, and the inner transition—are arranged in the construction of the chart. Therefore, short tables with two subgroups as well as the medium size tables must be excluded.

As "chemical" tables, the following types can be suggested as good ones:

1) subtype IC3-2a by Murashov, 1949. This table also combines the best properties of the chemical and electronic configuration tables.

2) type IIIC3-3 by Bayley, 1882, known as the Thomsen (1895) and Bohr (1922) table, is preferable because the extensions (transition and inner transition elements) within each new cycle are clearly seen.

3) type IIIC3-4 by Werner, 1905, shows the sequence of element blocks. This table, which becomes type IIC2-4 (originated by Mendeleev in 1869) with the exclusion of the inner transition elements, is probably the most popular table.

4) Some teachers prefer a separation of blocks. For them, the best table is type IC1-2 by Blanshard, 1895, which was also recommended by Sanderson in 1964.

As good electronic configuration tables, the following ones are mentioned:

1) type IIIC3-6 by Janet, 1927, which was suggested as a "chemical" table, but can be used as an electronic configuration table even with the new division of periods.

2) type IIIC4-2 by Gardner, 1930, and

3) type IIIC4-3 by Gibson, 1948, can be equally suggested as good ones, although the first one is more popular.

As shell and subshell tables, where the connection between the periods and shells is seen, the following types are suggested:

1) subtype IIIC5-1a by Mazurs, 1967;
2) subtype IIIC5-2a by Mazzucchelli, 1930;
3) subtype IIIC5-2d by Mazurs, 1958/1973; and
4) type IIIC5-3 by Mazurs, 1965.

CONCLUSION

All the mentioned tables are idealistic; i.e., they are constructed according to certain rules. However, they do not reflect exactly what is found in nature. Many tables have these discrepancies, even the electronic configuration tables.

In 1969 I published a realistic electronic configuration table of subtype IIIC4–3c that agrees in its details with experimental data. This table is constructed on the basis of Gibson's table, type IIIC4–3. From all twelve tables previously mentioned, I prefer three:

1) Janet's table, type IIIC3–6.
2) Gibson's table, type IIIC4–3.
3) Mazurs' table subtype IIIC4–3c.

The reasons for these preferences can be summarized as follows:

1) *Hydrogen* and *helium* are arranged according to their electronic configuration into the *s* block of the periodic table in the same vertical block where the alkali and alkaline earth metals are located. It is no longer necessary to move them from one place to another as was done in earlier times. However, although they have the same electronic structure as the alkali and alkaline earth metals, their properties are different. The explanation for this is that the electrons of the H and He atoms are in close proximity to the positive nucleus and are not shielded from it by other electrons as is the case with the atoms of alkali and alkaline earth elements. Therefore, they are called the primary elements. This subject is discussed in more detail by Dash in 1964. The following quotation is taken from his article:

> The uniqueness of the first quantum shell resides in the peculiarity that while hydrogen and helium both belong to *s* orbitals, they exhibit chemical properties typical of *p* orbitals (6).

2) Mazurs' table, subtype IIIC4–3c, abolishes the dispute over which element should begin the *lanthanide* and *actinide* series since the problem of their arrangement in the table is solved by placing them in different rows according to their differentiating electrons.

3) Attempts have often been made to correlate the *position of Zn, Cd,* and *Hg* with that of Be and Mg in the table because of the similarity in properties. This problem is also solved by Mazurs' table. This table shows that the similarity in properties is caused by the differentiating electron, which appears on the same subshell, *s*, for all these elements.

4) Some authors placed the *electron* and *neutron* above H and He to form a cycle of two very short periods. With the new numeration of periods, this is not necessary because the first cycle consists of H–He and Li–Be. The first element is the element with atomic number 1, which means hydrogen. The electron or neutron represents only a part of a chemical element, and these element parts have never been considered metals or nonmetals with physical or chemical properties.

5) If different *physical property curves* of the elements are drawn in the frame of Janet's table, another proof is received that this table and the other two are the best constructed tables, because uninterrupted curves are obtained, which cannot be obtained in the frame of other types of periodic tables. The physical properties of elements in subperiods can be conveniently compared. The curvature of subperiods belonging to different periods are unexpectedly similar. As examples, curves of density, melting point, atomic radius, and first ionization potential are presented in Figs. 139, 140, 141, and 142. In these drawings, the small circled numerals indicate the beginning of periods according to the new numeration. As is shown by dotted lines, in some cases H and He fit better into p^5 and p^6 than into s^1 and s^2 (Fig. 140 and 141). This is compatible with the quotation from Dash's article (6).

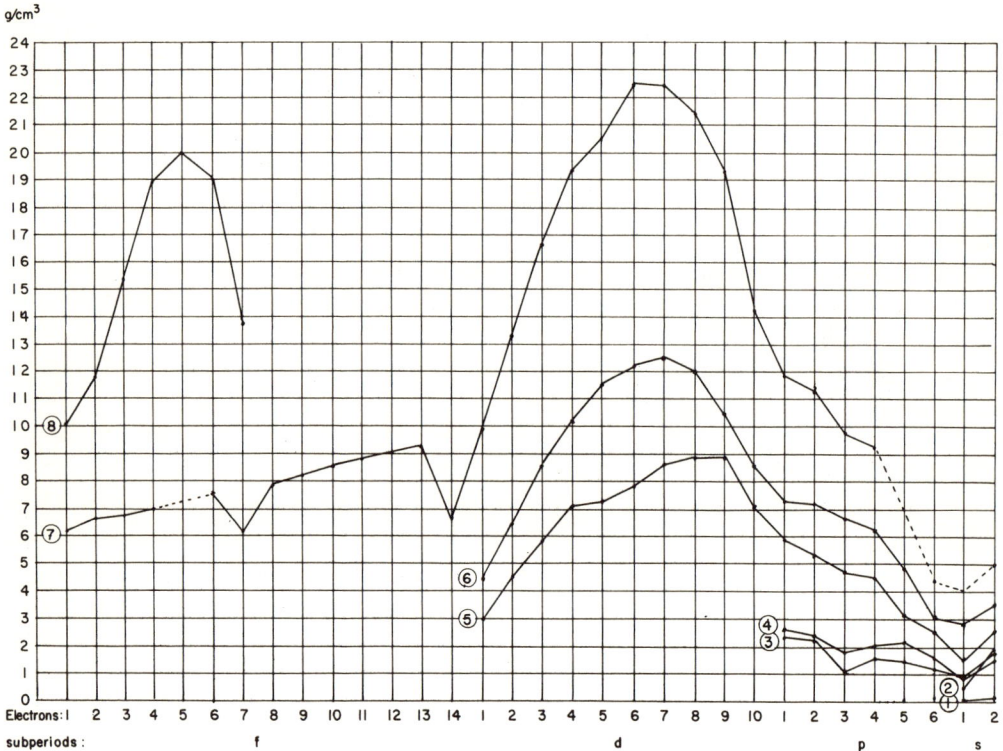

FIG. 139. Density of elements (g/cm³).

CONCLUSION 139

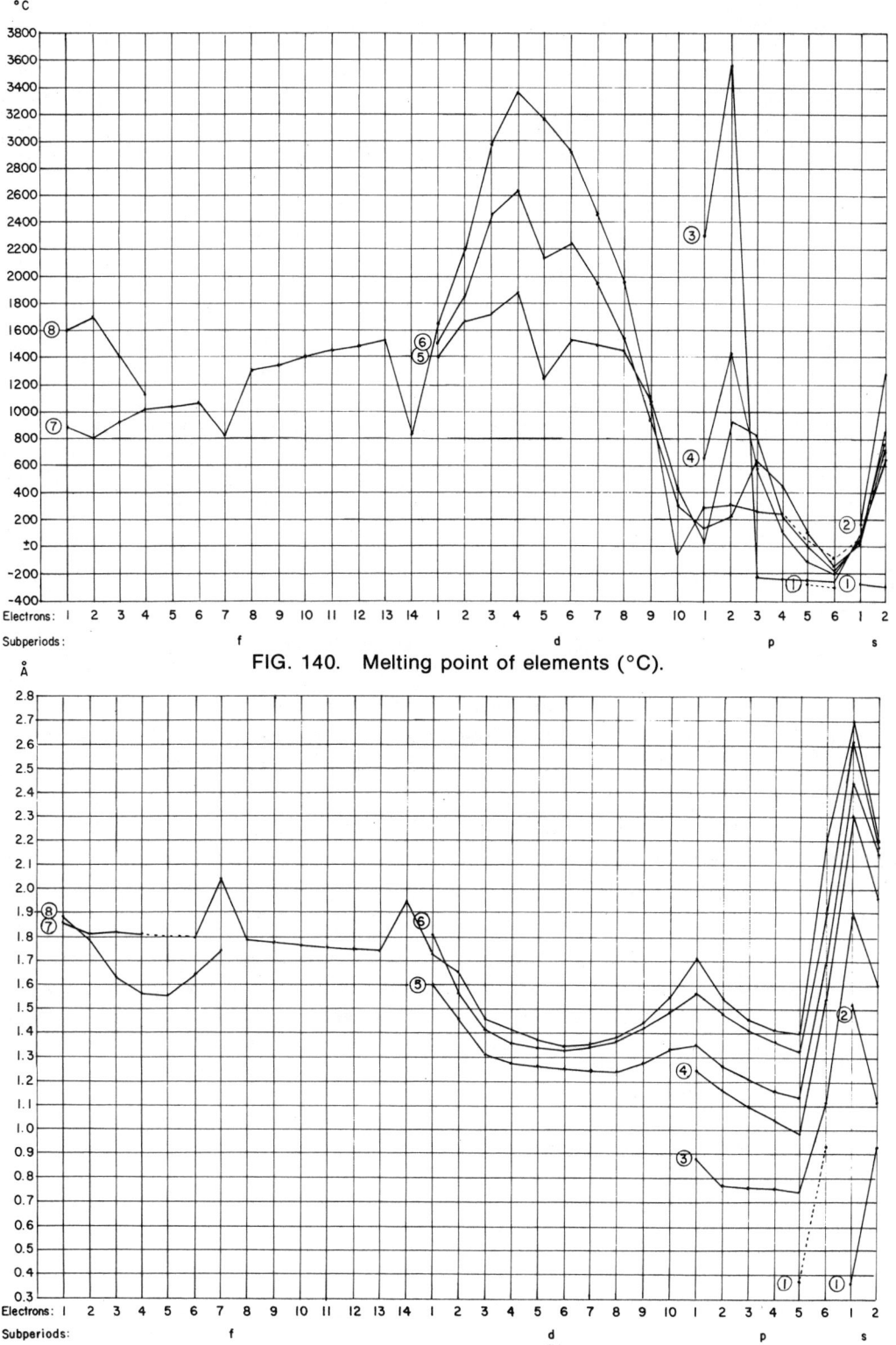

FIG. 140. Melting point of elements (°C).

FIG. 141. Atomic radius (Å).

140   GRAPHIC REPRESENTATIONS OF THE PERIODIC SYSTEM

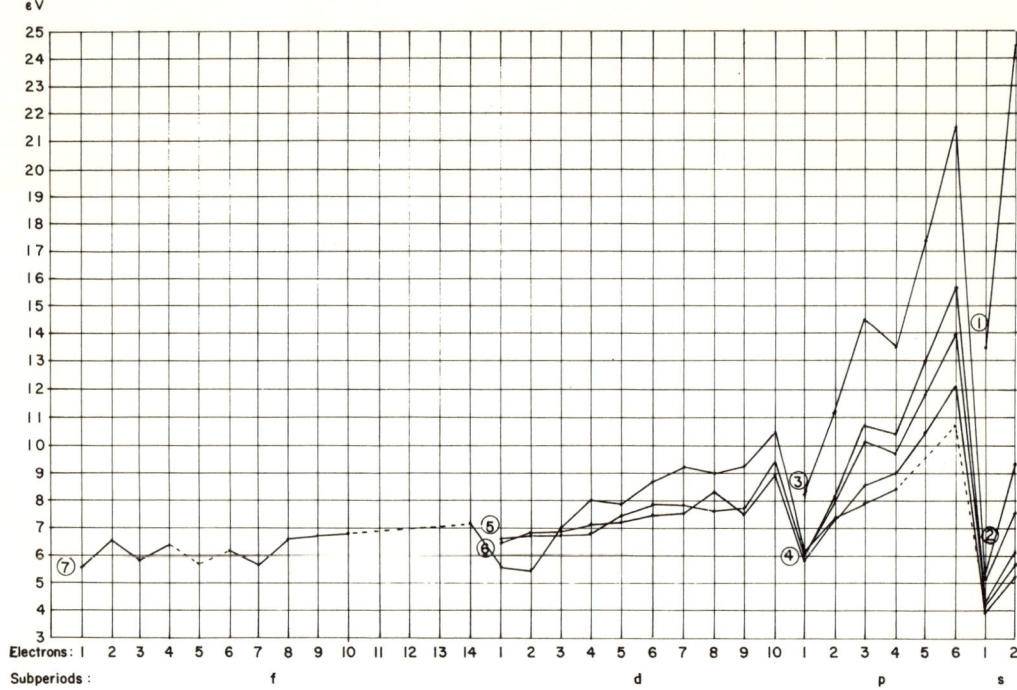

FIG. 142. First ionization potential (eV).

The chemical periodic tables have been constructed on the basis of the similarity in chemical and physical properties of the elements and they reflect this similarity. In order to retain this advantage, the proposed electronic configuration tables can be colored to show this relationship of elements. The nonmetals and the noble gases are colored with cool colors (green, blue, and violet) and the metals with warm colors (yellow, orange, red, brown, and also gray). By this coloration, the triangle of nonmetals in the *p* block is clearly shown. The distinction between nonmetals and metals is made according to Sanderson's rule (modified a little)(5):

elements are metals if the number of electrons in the outermost shell (always *s* and *p* subshells) is equal to or less than the shell number (principal quantum number) of the element; elements are nonmetals if the number of electrons in the outermost shell is greater than the shell number.

The *color legend* is as follows:
orange—inner transition metals (f^1–f^{14}) and rare metals (d^1);
light brown—acidic metals (d^2–d^5);
gray—Fe and Pt metals (d^6–d^8);
dark brown—coin and volatile metals (d^9 and d^{10});
yellow—metals of nonmetal groups (p^1–p^5);
light green—B and C nonmetals (p^1 and p^2);
dark green—N and O nonmetals (p^3 and p^4);

blue—halogens (p^5);
violet—noble gases (p^6) and helium (He);
carmine red—alkali metals (s^1);
scarlet red—alkaline earth metals (s^2);
white—hydrogen (H).

The colored Mazurs table appears at the back of this book as a separate foldout. On the bottom lines of this table, even the old group numbers and the maximum valences are shown.

In conclusion, mention of the four scientists who did most to establish good periodic tables is appropriate: Yeou Ta, Charles Janet, L. M. Simmons, and V. M. Klechkovskii. Yeou Ta clearly expressed the mathematical equation of the Periodic Law in 1946. Very early, in 1928, Janet drew a periodic table which already fit into the expression of this law, and first used the new division into periods. Simmons finally suggested the new division into periods in 1947. Both Simmons and Klechkovskii popularized the idea of basing the periodic table on electronic configuration and quantum numbers. Of these scientists, only the photographs of Janet, Simmons, and Klechkovskii were available. These are presented here with short biographical data. About Yeou Ta no information was obtainable.

Charles Janet was born in Paris in 1849 and died in Beauvais (France) in 1932. Although he graduated from an engineering school, he later received the degree of doctor of natural sciences. He spent all of his life working in geology and biology. Especially important are his investigation of ants and some fresh water algae. He devoted only the last five years of his life to the study of the periodic classification of chemical elements. His articles appeared in separate booklet form during the years 1927–1930.

L. M. Simmons was born in London (England) in 1905. He has been a professor of chemistry at Scots College in Sydney, Australia, since 1930. He published his two important articles in the *Journal of Chemical Education* in 1947 and 1948.

Vsevolod Mavrikievich Klechkovskii was born in Russia in 1900. In 1927 he graduated from Moscow Timiriazev Agriculture Academy. He is an agricultural chemist and doctor of science. In 1955 he became a professor at the same academy. In

1952 he received the Stalin prize for a work in agriculture. His articles appeared in various Russian journals during the years 1951–1971.

Finally, a comparison of the periodic tables of Janet and Simmons, in which the new periods are used, with the original chemical element tables of Dumas and Mendeleev is presented in Fig. 143. The comparison proves that the two geniuses, Dumas and Mendeleev, with their drawings were not far from the currently established good periodic tables.

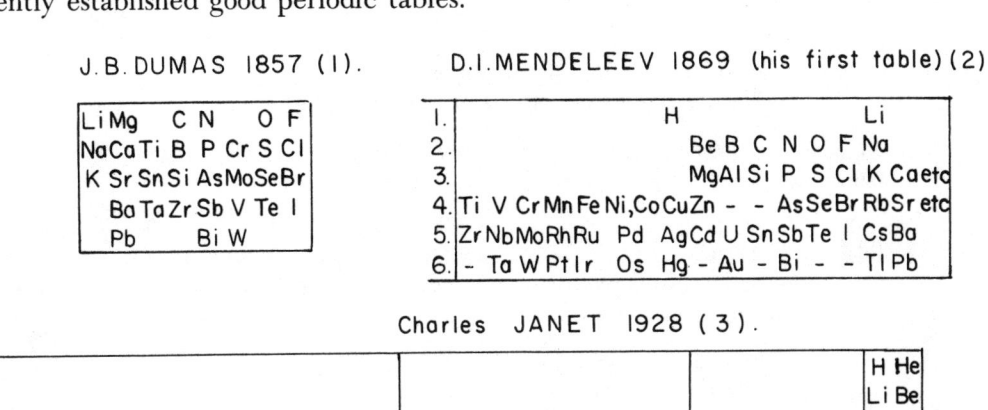

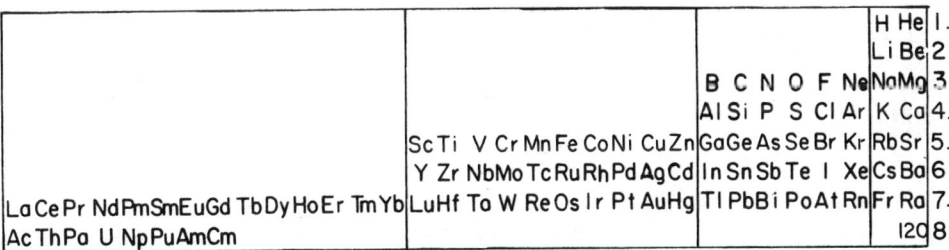

FIG. 143. Comparison of periodic tables.

5. Bibliography

(The references in each section are in chronological order.)

PREFACE
(REVIEWS OF PERIODIC TABLES)

(1) F. P. Venable. *The development of the Periodic Law.* Easton, Pa.: The Chemical Publishing Co., **1896**.

(2) George Rudorf. *The periodic classification and the problem of chemical evolution.* London: Whitaker & Co., **1900**.
——— *Das periodische System, seine Geschichte und Bedeutung für die chemische Systematik.* Hamburg u. Leipzig: L. Voss, **1904**.

(3) Henri Moissan. *Traité de chemie minérale.* P. I. Paris: **1904**.
——— *Classification des corps simples.* Paris: Masson et C$_{ie}$ **1904**.

(4) A. E. Garrett. *The Periodic Law.* London: Paul, Trench, Trübner & Co., **1909**.

(5) Curt Schmidt. *Das periodische System der chemischen Elemente.* Leipzig: J. A. Barth, **1917**.

(6) B. Smith Hopkins: *Chemistry of the rarer elements.* New York: Heath & Co., 14, **1923**.
——— *Chapters in the chemistry of the less familiar elements.* I. The periodic table. Champaign, Ill.: Stipes Publ. Co., **1939**.

(7) John D. Main Smith. *Chemistry and atomic structure.* London: Ernest Benn, **1924**.

(8) J. G. F. Druce. Some recent representations of the periodicity of the elements. *Chem. News* **130**, 322-326 (**1925**).

(9) W. Dahmen. Über das natürliche System der Elemente. *Z. physik.-chem. Unterricht* **40**, 106-111 (**1927**).

(10) M. A. Blokh. *Iubileinomu mendeleevskomu s'ezdu v oznamenovanie 100-letnei godovshchiny so dnia rozhdeniia D. I. Mendeleeva.* Leningrad: Akad. Nauk SSSR, **1934**.

(11) G. N. Quam and Mary Battell Quam. Types of graphic classifications of the elements. *J. Chem. Educ.* **11**, 27-32; 217-223; 228-297 (**1934**).

(12) A. von Antropoff. Les formes usuelles du système périodique des éléments. *Ann. Guébhard-Séverine* **13**, 161-174 (**1937**).

(13) W. F. Luder. Electron configuration as the basis of the periodic table. *J. Chem. Educ.* **20**, 21-26 (**1943**).

(14) Günter Langhammer. *Das Periodensystem.* Berlin u. Leipzig: Volk und Wissen, **1949**.

(15) Ryutaro Tsuchida. Thirty years of development of chemistry. Changes of periodic tables. *Kagaku (Chemistry)* **8**, 128-137 (**1953**).

(16) Robert Delhez. Aspects récents de la classification périodique. *Rev. Universelle Mines* [9] **10**(11), 679-683 (**1954**); **19**(7), 283-288 (**1963**).

(17) Edward G. Mazurs. Ķīmisko elementu periodiskās sistēmas tabulu veidi (in Latvian). *Technikas Apskats* (Lincoln, Nebr.) 2(7), 8-12 (**1955**).

———— Die wichtigsten Tabellen des Perioden-Systems. *Chem.-Ztg.* **80**, 195-203 (**1956**).

———— *Types of graphic representation of the Periodic System of chemical elements.* La Grange, Ill.: author's 1st edition, **1957**.

———— Up & downs of the periodic table. *Chemistry* (Washington) **39**(7), 6-12 (**1966**).

———— Ķīmisko elementu periodiskās tabulas izveidošanās (in Latvian). *Technikas Apskats* **16**(57), 4-8 (**1970**).

(18) Gil Chaverri-Rodríguez. *Historica grafica de la tabla periódica.* Univ. de Costa Rica, **1957**.

———— *Tabla periódica. Estructura Electronica y Enlace Químico.* Costa Rica: Ciudad Univers., Ser. 100, **1962**.

(19) Miroslav Zikmund. 90 rokov periodického zákona. *Chemické zvesti* (Bratislava) **13**(3), 137-162 (**1959**).

(20) V. I. Semishin. *Periodicheskii zakon i periodicheskaia sistema khimicheskikh elementov D. I. Mendeleeva v rabotakh russkikh uchenykh* (bibliography only). Moskva: Mosk. Inst. Khim. Mashinostroen., **1959**.

———— *Literatura po periodicheskomu zakonu D. I. Mendeleeva (1869-1969)* (bibliography only). Moskva: Vysshaia Shkola, **1969**.

———— O printsipakh postroeniia i formakh periodicheskoi sistemy, *Sto let periodich. zakona khimicheskikh elementov.* Moskva: Nauka, 71-98, **1969**.

(21) H. Iveković i Z. Balenović. Uvod u studij i bibliografiju periodnog sistema elemenata i periodiciteta fizikalnih i kemijskih svojstava elemenata i spojeva (bibliography only). *Spomenica u počast 40-godišnjice osnivanja SKJ*, Zagreb, 78-112, **1960**.

(22) J. W. van Spronsen. *The periodic system of chemical elements.* Chapter 6: Contributions to the development of the periodic system from 1871 to the present (p. 147-211). Amsterdam-London-New York: Elsevier, **1969**.

PREHISTORY OF THE DISCOVERY OF THE PERIODIC SYSTEM

(1) de Morveau. Sur les dénominations chymiques, la necéssité d'en perfectionner le système et le régles pour y parvenir. *J. phys., hist. nat. et arts* **19**, 370-382 (**1782**) (table p. 382).

(2) de Morveau, Lavoisier, Berthollet et de Fourcroy. *Méthode de nomenclature chimique.* Paris: Cuchet, Hassenfratz et Adet, **1787** (table p. 100/101).

(3) Lavoisier. *Traité élémentaire de chimie.* Paris: Cuchet, **1789** (table p. 192).

Law of Triads.

(0) Dumas. *Traité de chimie, appliqué aux arts.* Paris: Béchet Jeune, V. I., **1828** (table p. LXXVII).

(1a) Johann Wolfgang Döbereiner. Versuch zu einer Gruppirung der elementaren Stoffe nach ihrer Analogie. *Ann. Physik* (Pogg.) **15**, 301-307 (**1829**).

(1b) Auszug eines Briefes vom Hofrath Wurzer, Prof. der Chemie zu Marburg. *Ann. Physik (Gilbert)* **56**, 331-334 (**1817**).
Aus einem Schreiben des Herrn Prof. und Bergrat Döbereiner an den Prof. Gilbert. *Ibidem* **57**, 435-438 (**1817**).

(2) Leopold Gmelin. *Handbuch der Chemie.* Heidelberg: C. Winter, I., 4. Aufl., **1843** (table p. 52).

(3) Oliver Wolcott Gibbs. *An inaugural dissertation on a natural system of chemical classification.* Princeton, N.J.: John T. Robinson, **1845.**

(4) Max Pettenkofer. Über die regelmässigen Abstände der Äquivalentzahlen der sogenannten einfachen Radicale. *Gelehrten Anzeigen, Bayer. Akad. Wissensch, München,* **32** und **33,** 261-271 (**1850**).
――――― The same title. Eine Reclamation gegenüber Hrn. Dumas' Äquivalentgewichten der einfachen Körper. *Ann. Chem., Justus Liebigs,* **105,** 187-202 (**1858**).

(5) Jean Baptiste Dumas. Address before the British Association. *Proc. 21st meeting Brit. Assoc. at Ipswich,* **1851.**
――――― The same title. *The Athenaeum (J. Lit., Sci., fine Arts)* (London) 750, **1851.**
――――― Observations on atomic volumes and atomic weights, with considerations on the probability that certain bodies now considered as elementary may be composed. *Am. J. Sci.* **62** or [2] **12,** 275-277 (**1851**).
――――― Mémoire sur les équivalents des corps simples. *Compt. rend.* (Paris) **45,** 709-731 (**1857**).
――――― The same title. *Ann. chim. et phys.* [3] **55,** 129-210 (**1859**).

(6) J. H. Gladstone. *Phil. Mag.* [4] **5,** 573 (**1853**).

(7) Josiah P. Cooke. The numerical relations between the atomic weights, with some thoughts on the classification of the chemical elements. *Am. J. Sci.* **67** or [2] **17,** 387-407 (**1854**) (separate tables).
Cited by G. Elsen. Was het systeem van Mendelejew reeds voor 1869 bekend? *Chem. Weekblad* **27,** 378-379 (**1930**).

(8) William Odling. On the natural grouping of the elements. *Phil. Mag.* [4] **13,** 423-439; 480-497 (**1857**).
――――― The same subject. *Manual of Chemistry.* London: Green Longman and Robert Longman, **1861.**

(9) E. Lenssen. Über die Gruppierung der Elemente nach ihrem chemisch-physikalischen Charakter. *Ann. Chem., Justus Liebigs,* **103,** 121-131 (**1857**) (tables p. 122/123 and 124).

(10) J. A. R. Newlands. On relations among the equivalents. *Chem. News* **7,** 70-72 (**1863**).
――――― Relations between equivalents. *Ibidem* **10,** 59 (**1864**).

(11) Thénard. *Cited* by Oechsner de Coninck. *Notes et documents de chimie générale.* Paris: **1902.**

(12) E. Frémy. *Traité de chimie.* Paris: V. Masson et fils, **1865.**

(13) Anonymous author. The numerical relations of atoms. New elements predicted. *Chem. News (Am. Suppl.)* **4,** 217-218 (**1869**) (table p. 218).

(14) Wurtz. *Leçons élémentaires de chimie moderne.* **1870/71.**

Valence Tables.

(1) L. Meyer. *Die modernen Theorien der Chemie und ihre Bedeutung für die chemische Statistik.* Breslau: Maruschke und Berendt. I. Aufl., 135-139, **1864** (table p. 137, 138).
――――― The same subject. *Ostwald's Klass. d. exakt. Wissensch.* **68,** 3-5; 8 (**1895**).

(2) A. Naquet. *Principes de chimie fondés sur les théories modernes.* Paris: **1864.**

(3) A. W. Williamson. On the classification of the elements in relation to their atomicities. *J. Chem. Soc.* (London) [2] **17,** 211-222 (**1864**) (table p. 212/213).

(4) C. W. Blomstrand. *Die Chemie der Jetztzeit.* Heidelberg: C. Winter's Universitätsbuchhandl., 401, **1869**.

———— Bemerkungen über die Elemente. *Ber. deut. chem. Ges.* **3**, 533-539 (**1870**) (table p. 537).

(5) Anonymous author. The pairung of the elements. *Chem. News* (Am. Suppl.) **4**, 339-340 (**1869**) (table p. 340).

(6) Fr. Waechter. *Ber. deut. chem. Ges.* **11**, 11-16 (**1878**).

(7) William Sully Beebe. *Number and Matter.* New York: **1882** (table p. 5).

Law of Octaves.

(1a) A. E. Béguyer de Chancourtois. Mémoire sur un classement naturel des corps simples ou radicaux appelé vis tellurique. *Compt. rend.* (Paris) **54**, 757-761; 840-843; 967-971 (**1862**) (no table).

———— Tableau du classement naturel des corps simples, dit vis tellurique. *Ibidem* **55**, 600-601 (**1862**) (no table); **56**, 253-255 (**1863**).

(1b) ———— *Vis tellurique: classement naturel des corps simples ou radicaux, obtenu au moyen d'un système de classification hélicoïdal et numérique.* Paris: **1863** (table).

(1c) *Cited* by P. J. Hartog. A first foreshadowing of the Periodic Law. *Nature* **41**, 186-188 (**1889**) (shortened table p. 187).

Cited by Lecoq de Boisbaudran et A. de Lapparent. Sur une réclamation de priorité en faveur de M. de Chancourtois, relativement aux relations numériques des poids atomiques. *Compt. rend.* (Paris) **112**, 77-81 (**1891**) (shortened table p. 80).

———— A reclamation of priority on behalf of M. de Chancourtois referring to the numerical relations of the atomic weights. *Chem. News* **63**, 51-52 (**1891**) (no table).

Cited by Karl Seubert. Das natüraliche system der chemischen Elemente. *Ostwald's Klass. d. exakt. Wissensch.* **68**, 119-122 (**1895**) (shortened table p. 121).

(1d) *Cited* by J. W. van Spronsen. Prioriteit bij de ontdekking van het periodiek systeem. *Chem. Weekblad* **47**, 814-815 (**1951**) (complete table p. 815).

———— Le système périodique des éléments chimiques et l'apport de Béguyer de Chancourtois. *Nucleus* (Paris) 9(5), 295-306 (**1968**).

(2) J. A. R. Newlands. On the relations among equivalents. *Chem. News* **10**, 94 (**1864**) (table not complete).

———— On the Law of Octaves. *Ibidem* **12**, 83 (**1865**) (table); **13**, 113-114; 130 (**1866**).

———— On relation among the atomic weights of the elements when arranged in their natural order. *Ibidem* **32**, 21-22 (**1875**) (table p. 22).

———— On relations among the atomic weights of the elements, *Ibidem* **37**, 255-257 (**1878**) (tables p. 256/257).

Cited by W.A.T. (Tilden). John A. R. Newlands. *Nature* **58**, 395-396 (**1898**).

Cited by Wendell H. Taylor. J.A.R. Newlands: a pioneer in atomic numbers. *J. Chem. Educ.* **26**, 491-496 (**1949**) (table p. 493).

Cited by J. W. van Spronsen. One hundred years of the "Law of Octaves." *Chymia* **11**, 125-137 (**1966**).

Forerunners of the Periodic Law.

(1) William Odling. On the proportional numbers of the elements. *Quart. J. Sci.* **1**, 642-648 (**1864**) (table p. 643).

Cited by J. W. van Spronsen. Wegbereider on outdekken van het periodiek systeem der elementen 1864-1964. *Chem. Weekblad* **60**(50), 683-686 **(1964)**.

(2) Gustavus D. Hinrichs. *Programm der Atomechanik, oder die Chemie eine Mechanik der Panatome.* Iowa City, Iowa; **1867** (p. 7-8).

———— On the classification and the atomic weights of the so-called chemical elements, with reference to Stas' determinations. *Proc. Am. Assoc. Advanc. Sci.* **18**, 112-124 **(1869)** (table p. 115).

Cited by J. W. van Spronsen. Hinrichs as discoverer of the Periodic System (in Dutch). *Chem. Weekblad* **64**(10), 11-15 **(1968)**.

Cited by Carl A. Zapffe. Gustavus Hinrichs, precursor of Mendeleev. *Isis* **60**(4), 461-476 **(1969)**.

(3) Lothar Meyer. **1868**. Not published.

Cited by Karl Seubert. Zur Geschichte des periodischen Systems. *Z. anorg. Chem.* **9**, 334-338 **(1895)** (table p. 336/337).

———— Das natürliche System der chemischen Elemente. Abhandlungen von Lothar Meyer und D. Mendelejeff. *Ostwald's Klass. d. exakt. Wissensch.* **68**, 6-7 **(1895)**.

DISCOVERY OF THE PERIODIC LAW

D. I. Mendeleev and Lothar Meyer.

(1a) D. Mendeleev. Sootnoshenie svoistv s atomnym vesom elementov. *Zh. Russk. Khim. Obshch.* **1**, 60-77 **(1869)** (table p. 70).

Cited by Karl Seubert. Das natürliche System der chemischen Elemente. Abhandlungen von Lothar Meyer und D. Mendelejeff. *Ostwald's Klass. d. exakt. Wissensch.* **68**, 20-40 **(1895)**.

Cited by B. N. Menshutkin. Early history of Mendeleeff's Periodic Law. *Nature* **133**, 946 **(1934)**.

(1b) D. Mendelejeff. Über die Beziehungen der Eigenschaften zu den Atomgewichten der Elemente. *Z. f. Chem.* **12** or [2] **5**, 405-406 **(1869)** (table p. 405).

———— The same title. *Ostwald's Klass. d. exakt. Wissensch.* **68**, 18-19; 33 **(1895)**.

———— Versuche eines Systems der Elemente nach ihren Atomgewichten und chemischen Funktionen. *J. prakt. Chem.* [1] **106**, 251 **(1869)**.

(2) P. Walden. Dmitri Ivanowitsch Mendelejeff (1834-1907). *Ber. deut. chem. Ges.* **41**, 4741-4742 **(1908)**.

(3) B. M. Kedrov. Mendeleevskie formy Periodicheskoi Sistemy elementov 1869-1871. *Sto let per. zak. khim. elem.* Moskva: Nauka, 42-56, **1969**.

———— The same title. *Vopr. Istor. Estestvozn. Tekhn.* **4** (29), 59-70 **(1969)**.

(4) D. Mendeleev. *Osnovy Khimii.* St. Petersburg: 1st edit., V. II, **1871**.

(5) D. Mendeleev. Ob atomnom obiome prostykh veshchestv. *Trudy 2. s'ezda russk, estestvoispyt, i vrachei v Moskve* **1**, 62-71 **(1870)**.

Cited by B. N. Menshutkin. Early history of Mendeleeff's Periodic Law. *Nature* **133**, 946 **(1934)**.

(6a) D. Mendeleev. O kolichestve kisloroda v solianykh okislakh i ob atomnosti elementov. *Zh. Russk. Khim. Obschch.* **2**. 14-21 **(1870)**.

(6b) *Cited* by V. von Richter. Correspondenz aus St. Petersburg am 17. Oktober 1869. *Ber. deut. chem. Ges.* **2**, 553 **(1869)**.

(7) D. Mendeleev. *Osnovy Khimii*. St. Petersburg: 1st edit., V. I., **1869** (table p. IV).

(8) Lothar Meyer. Die Natur der chemischen Elemente als Function ihrer Atomgewichte. *Ann. Chem., Justus Liebigs,* **Suppl. 7**, 354-364 (**1870**) (table p. 356).

———— The same title. *Ostwald's Klass. d. exakt. Wissensch.* **68**, 9-17 (**1895**) (table p. 11).

(9) Heinr. Baumhauer. *Die Beziehungen zwischen dem Atomgewichte und der Natur der chemischen Elemente*. Braunschweig: F. Vieweg & Sohn, **1870** (separate table).

———— Über das natürliche System der chemischen Elemente. *Ber. deut. chem. Ges.* **6**, 652-655 (**1873**) (no table).

(10a) D. Mendelejew. Über die Stellung des Ceriums im System der Elemente (in German). *Bull. acad. imp. sci.* (St. Peterbourg) **16**, 45-51 (**1871**) (table p. 51).

(10b) ———— The same title. *Mélanges physiques et chimiques* **8**(4), 445-452 (**1872**) (table p. 447).

(10c) *Cited* by V. von Richter. Correspondenz aus Petersburg am 6/18. Dezember 1870. *Ber. deut. chem. Ges.* **3**, 990-992 (**1870**) (table p. 992).

(11) D. Mendeleev. Estestvennaia sistema elementov i primenenie ee k ukazaniiu svoistv neotkrytykh elementov. *Zh. Russk. Khim. Obshch.* **3**, 25-56 (**1871**) (table p. 31).

(12a) D. Mendeleeff. Die periodische Gesetzmässigkeit der chemischen Elemente. *Ann. Chem., Justus Liebigs,* **Suppl. 8**, 133-229 (**1871**) (tables pp. 149 and 151).

———— The same title. *Ostwald's Klass. d. exakt. Wissensch.* **68**, 41-118 (**1895**) (tables pp. 54 and 56).

(12b) D. Mendeleev (no author's name). The chemistry of the future. *Quart. J. Sci.* **14** [N.S.7] (55), 289-306 (**1887**) (table p. 304).

(12c) ———— The Periodic Law of the chemical elements. *Chem. News* **40**, 231-232; 243-244; 255-256; 267-268; 279-280; 291-292; 303-304 (**1879**) (table p. 268); **41**, 2-3; 27-28; 39-40; 49-50; 61-62; 71-72; 83-84; 93-94; 106-108; 113-114; 125-126 (**1880**).

(12d) ———— La loi périodique des éléments chimiques. *Monit. Sci. (Docteur Quesneville)* [3] **9**, 691 etc. (**1879**).

Question of Priority.

(1) D. Mendelejeff. Zur Frage über das System der Elemente. *Ber. deut. chem. Ges.* **4**, 348-352 (**1871**).

(2) John A. R. Newlands. On the Periodic Law. *Chem. News* **38**, 106-107 (**1878**).

———— Über das periodische System der Elemente. *Ber. deut. chem. Ges.* **11**, 516-517 (**1878**).

(3) Lothar Meyer. Zur Geschichte der periodischen Atomistik. *Ber. deut. chem. Ges.* **13**, 259-265 (**1880**).

(4) D. Mendelejeff. Zur Geschichte des periodischen Gesetzes. *Ber. deut. chem. Ges.* **13**, 1796-1804 (**1880**).

(5) J. A. Newlands. Zur Geschichte des periodischen Gesetzes. *Ber. deut. chem. Ges.* **17**, 1145-1150 (**1884**).

———— On the discovery of the Periodic Law, and on relations among the atomic weights. London: E. & F. N. Spon, **1884**.
Citation of this book. *Chem. News* **49**, 198-200 (**1884**).

———— On the Periodic Law. *Chem. News* **61**, 136 (**1890**).

(6) William Crookes. Telluric screw (priority of Chancourtois). *Chem. News* **63**, 51 **(1891)**.

(7) D. Mendeléeff. Comment j'ai trouvé le système périodique des éléments. *Rev. gén. chim. pur. appl.* **1**, 211-214; 510-512; **2**, 409-433 **(1899)**.

CLASSIFICATION OF PERIODIC TABLES ORIGINATED AFTER DISCOVERY OF THE LAW OF OCTAVES AND THE PERIODIC LAW

DIVISION I. SHORT TABLES.

Little Periodic Table: the Lanthanides in Two Series.

(1) Frank Austin Gooch and Claude Frederic Walker. *Outlines of organic chemistry.* New York, London: Macmillan Co., V. II, **1905** (table p. 9).

(2) Bohuslaw Brauner. Über die Stellung der Elemente der seltenen Erden im periodischen System. *Z. Elektrochem.* **14**, 525-527 **(1908)** (table p. 527).

(3) Emil Baur. *Z. physik. Chem.* **76**, 569-583 **(1911)**.

(4) R. J. Meyer. Die Stellung der Elemente der seltenen Erden im periodischen System. *Naturwissenschaften* **2**, 781-786 **(1914)** (table p. 781).

(5) Alfred W. Stewart. *Recent advances in physical and inorganic chemistry.* London: Longmans, Green & Co., 4th edit., **1920** (table p. 260).

(6) R. G. W. Norrish. *Chem. News* **124**, 16-22 **(1922)** (table p. 17).

(7) Carl Renz. Die seltenen Erden im periodischen System. *Z. anorg. allgem. Chem.* **122**, 135-145 **(1922)** (table p. 139).

(8) J. D. Main Smith. The rare earths. *Nature* **120**, 583-584 **(1927)**.

(9) Wilhelm Klemm. Eine Systematik der seltenen Erden, begründet auf periodischen Eigenschafsänderungen ihrer Ionen. *Z. anorg. allgem. Chem.* **184**, 345-351 **(1929)** (table p. 347).
———— Nachtrag zu der Mitteilung: "Eine Systematik der seltenen Erden". *Ibidem* **187**, 29-32 **(1930)**.
———— Bemerkung zur Systematik der seltenen Erden. *Ibidem* **209**, 321-324 **(1932)**.
G. Jantsch und W. Klemm. Über das Auftreten niederer Wertigkeit bei den Halogeniden der seltenen Erden. *Ibidem* **216**, 80-84 **(1934)** (scheme p. 81).
W. Klemm und H. Bommer. Zur Kenntniss der Metalle der seltenen Erden. *Ibidem* **231**, 138-171 **(1937)** (table p. 154).
W. Klemm. Einige Bemerkungen zur Atomvolumenkurve. *Chem.-Ztg.* **62**, 97-100 (1938) (table p. 99).
———— Zur Systematik der seltenen Erden. *Angew. Chem.* **51**, 575-581 **(1938)**.
———— The same title. *Chem.-Ztg.* **66**, 365 **(1942)**.
———— 100 let sistemy periodov khimicheskikh elementov. *Sto let per. zak. khim. elem.* Moskva: Nauka, 57-70, **1969**.

(10) Arnold Sommerfeld. Über paramagnetische Momente der seltenen Erden. *Sitzungsber. Akad. Wiss. Wien,* Abt. IIa, 11-17 **(1930)**.

(11) Herman Yagoda. Periodic classification of the rare earths. *J. Am. Chem. Soc.* **57**, 2329-2330 **(1935)** (table p. 2329).

(12) D. W. Pearce. Anomalous valence and periodicity within the rare earth group. *Chem. Rev.* **16**, 121-147 (**1935**) (table p. 135).

(13) Pierre van Rysselberghe. *J. Chem. Educ.* **12**, 474-475 (**1935**) (table p. 474).

(14) J. Gillis. *Natuurw. Tijdschr.* (Belg.) **17**, 218-220 (**1935**) (table p. 219).

(15) S. A. Shchukarev. Sovremennoe sostoianie periodicheskogo zakona D. I. Mendeleeva. *Trudy Iubilein. Mendeleev. s'ezda,* Moskva, Akad. *Nauk SSSR* **2**, 23-35 (in Russian); 37-49 (in French) (**1937**).

CLASS 1. TABLES WITH EIGHT GROUPS AND NO SUBGROUPS (LAW OF OCTAVES)

Type IA1-1.

(1) A. E. Béguyer de Chancourtois. *Vis tellurique: classement naturel des corps simples ou radicaux, obtenu au moyen d'un système de classification hélicoidal et numérique.* Paris: **1863** (table).
Cited by J. W. van Spronsen. Prioriteit bij de ontdekking van het periodiek systeem. *Chem. Weekblad* **47**, 814-815 (**1951**) (table p. 815).

(2) Lothar Meyer. *Die modernen Theorien der Chemie.* Breslau: 2. Aufl., 302 (**1872**).

(3) Lewis R. Gibbes. Synoptical table of the chemical elements. *Proc. Elliott Soc. Sci. Art, Charleston* **2**, 77-90 (**1875**) (table p. 85).
Cited by F. P. Venable. An early American arrangement of the elements. *J. Am. Chem. Soc.* **17**, 947-949 (**1895**) (table p. 948).
Cited by Wendell H. Taylor. Lewis Reeve Gibbes and classification of elements. *J. Chem. Educ.* **18**, 403-407 (**1941**).

(4) Elena Bogdan. O propunere de aranjere a elementelor în sistem periodic (in Rumanian). *Analele Stiint. Univ. "Al. I. Cuza", Iasi,* [N.S.] **10**(1), 1-16 (**1964**) (table p. 6).

Type IB1-1.

(0) V. Ia. Kurbatov. *Zakon D. I. Mendeleeva.* Leningrad: N. Kh. T. I., **1925**.

(1) B. H. Wallin. Untersuchung einiger periodischen Erscheinungen in der Reihe der chemischen Elemente. *Göteb. kungl. Vetenskaps-och Vitterhets-Sanhäll. Handl.* **30**(3), 5-28 (**1926**) (separate table Fig. I).

Type IC1-1.

(1) John A. R. Newlands. On relations among equivalents. *Chem. News* **10**, 94 (**1864**) (table not complete).
―――― On the Law of Octaves. *Ibidem* **12**, 83 (**1865**) (table).
―――― Proc. of Chem. Soc.: The Law of Octaves and the causes of numerical relations among the atomic weights. *Ibidem* **13**, 113-114 (**1866**) (table p. 113).
―――― On the Law of Octaves. *Ibidem* **13**, 130 (**1866**) (no table).
―――― On relations among the atomic weights of the elements. *Ibidem* **37**, 255-257 (**1878**) (table p. 256/257).

(2) D. Mendeleev. Sootnoshenie svoistv s atomnym vesom elementov. *Zh. Russk. Khim. Obshch.* **1**, 66-71 (**1869**) (fragment of the table p. 67).
Cited by V. von Richter. Correspondenz aus St. Petersburg am 17. Oktober 1869. *Ber. deut. chem. Ges.* **2**, 553 (**1869**).

(3) Thomas Carnelley. Mendelejeff's Periodic Law and the magnetic properties of the elements. *Chem. News* **40**, 183-184 (**1879**) (table p. 184).

———— Mendelejeff's periodisches Gesetz und die magnetischen Eigenschaften der Elemente. *Ber. deut. chem. Ges.* **12**, 1958-1961 (**1879**) (table p. 1960).

———— The Periodic Law, as illustrated by certain physical properties of inorganic compounds. *Phil. Mag.* [5] **18**, 1-22 (**1884**) (table p. 3).

———— Das periodische Gesetz, angewendet auf gewisse physikalische Eigenschaften anorganischer Körper. *Beibl. Ann. Phys. Chem.* **8**, 735-738 (**1884**) (no table).

(4) Lothar Meyer. *Die modernen Theorien der Chemie.* Breslau: Maruschke & Berendt, 4. Aufl., **1880** (table p. 138).

(5) E. S. Fedorov. O popytke podvesti atomnye vesa pod odin zakon. *Zh. Russk. Khim. Obshch.* **13**, 244-245 (**1881**).

Fedaroff. Relations between the atomic weights of elementary bodies. *Bull. Soc. Chim. France* [2] **36**, 559-560 (**1881**).

(6) A. Bazarov. Ob atomnykh vesakh elementov. *Zh. Russk. Fiz.-Khim. Obshch.* **19**, 61-73 (**1887**).

(7) B. N. Chicherin. *Sistema khimicheskikh elementov.* Petersburg: P. I., **1888** (table p. 11).

———— The same subject. *Zh. Russk. Fiz.-Khim. Obshch.*, Suppl. to **20**, 1-26 (**1888**).

(8) Geoffrey Martin. On a method of representing the properties of elements graphically by means of characteristic surfaces. *Chem. News* **90**, 175-177 (**1904**) (table p. 175).

———— *Researches on the affinities of the elements.* London: J. & A. Churchill, **1905** (table p. 22).

(9) Ferfando Sanford. The significance of the Periodic Law. *J. Am. Chem. Soc.* **33**, 1349-1353 (**1911**) (table p. 1350).

(10) Wilh. Palmaer. Om ett mått på grundämnenas kemiska frändskap. *Svensk.Kem. Tidskrift* **29**, 88-101 (**1917**) (table p. 96).

(11) Eduard Smiz. Betrachtungen über die Harmonie der Atomgewichte. *Ber. deut. pharm. Ges.* **29**, 504-518 (**1919**) (table p. 508).

(12) Heinrich Teudt. Ableitung des periodischen Systems der chemischen Elemente aus der Elektronentheorie. *Z. anorg. allgem. Chem.* **106**, 189-208 (**1919**) (table p. 192).

(13) A. F. Kapustinskii. Struktura periodicheskoi sistemy khimicheskikh elementov. *Dokl. Arad. Nauk SSSR* **81**(1), 47-50 (**1951**) (table p. 49).

(14) Jack Green. Geochemical table of the elements for 1953. *Bull. Geol. Soc. Amer.* **64**, 1001-1012 (**1953**) (separate table).

(15) Ana Maria Mojos, Hector Vicente Incolla. Ayer, hoy y mañana en la tabla periódica de los elementos químicos. *Saneamiento* (Buenos Aires) **33**(215), 265-268 (**1969**) (table p. 266).

Subtype IC1-1a.

(1) Lewis R. Gibbes. Synoptical table of the chemical elements. *Proc. Elliott Soc. Sci. Art, Charleston* **2**, 77-90 (**1875**) (table p. 80).

Cited by F. P. Venable. An early American arrangement of the elements. *J. Am. Chem. Soc.* **17**, 947-949 (**1895**).

Cited by Wendell H. Taylor. Lewis Reeve Gibbes and classification of elements. *J. Chem. Educ.* **18**, 403-407 (**1941**).

(2) H. Strache. The explanation of the Periodic System of the elements with the aid of the electron theory. *Ion* (London) **1**, 81-98 (**1908**) (table p. 89, 90).

───── Die Erklärung des periodischen Systems der Elemente mit Hilfe der Elektronentheorie. *Ber. deut. physik. Ges.* **6**, 798-803 (**1908**) (no table).

───── The same title. *Chem.-Ztg.* **32**, II, 931 (**1908**) (no table).

(3) Emil Kohlweiler. Konstitution und Konfiguration der Atome. *Z. phys. Chem.* **94**, 513-541 (**1920**) (table p. 520).

───── Neue Anschauungen über die chemischen Elemente und ihre Atome. *Physik. Z.* **21**, 203-208; 311-316 (**1920**) (table p. 313).

(4) P. Vinassi de Regny. La simmetria elettronica ed i minerali. *Periodico di mineralogia* **4**, 225-240 (**1933**) (table p. 236/237).

Subtype IA1-1b.

(1) A. G. Oppegaard. The Periodic System of the chemical elements in three dimensions (the "five-five" system), *J. Chem. Soc.* (London) **1948**, 318-321 (table p. 319).

Subtype IC1-1b.

(1) C. J. Reed. The graphical representation of the relation between valence and atomic weight. *Trans. St. Louis Acad. Sci.* **4**(4), 649-675 (**1885**).

───── A prediction of the discovery of Argon. *Chem. News* **71**, 213-215 (**1895**) (table p. 214).

(2) W. Preyer. Einige tatsächlige Grundlagen des genetischen Systems der Elemente. *Deut. Pharm. Ges.* **2**, 144-155 (**1892**) (table p. 146, 150).

───── *Das genetische System der chemischen Elemente.* Berlin: R. Friedländer & Sohn, **1893** (table p. 4).

(3) A. van den Broek. Das α-Teilchen und das periodische System der Elemente. *Ann. Physik* [4] **23**, 199-203 (**1907**) (table p. 201).

(4) Otto Reinmuth. The structure of matter. *J. Chem. Educ.* **5**, 1312-1320 (**1928**); **6**, 341-348 (**1929**) (separate table in vol. 6).

Subtype IC1-1c.

(1) N. P. Nechaev. Report at a meeting of the Society. *Zh. Russk. Fiz.-Khim. Obshch.* **26**(2)(1), 14-15 (**1894**).

(2) D. A. Goldhammer. Bemerkungen über die analytische Darstellung des periodischen Systems der Elemente. *Z. anorg. Chem.* **12**, 39-45 (**1896**).

───── Zametka ob analiticheskom vyrazhenii periodicheskoi sistemy elementov. *Zh. Russk. Fiz.-Khim. Obshch.* **29**, 53 (**1897**).

Subtype IC1-1d.

(1) Heinrich Biltz und Wilhelm Biltz. *Laboratory methods of inorganic chemistry.* New York: John Wiley & Sons, 2nd edit., **1928** (table p. 10).

Type IC1-2.

(1) C. T. Blanshard. Natural groups and cross-analogies. *Chem. News* **71**, 39-40 (**1895**) (table p. 39).

(2) J. D. Main Smith. The electronic structure of atoms. I. The periodic classification. *J. Chem. Soc.* (London) **1927** (Trans. II), 2029-2038) (tables pp. 2031, 2035, 2036).

(3) Marcel Oswald. Une présentation nouvelle de la classification des éléments, et son intérêt pour les recherches. *Compt. rend. 17. congr. chem. ind.* (Paris) **I**, 26-43 (**1937**) (table p. 30).
———— Mettalurgie et classification périodique des éléments. *Compt. rend. 18. congr. chim. ind.* (Nancy) **I**, 391-406 (**1938**) (table p. 392).

(4) Sidney J. French. The drama of chemistry. *Dow Diamond* 6(1), 21 (**1943**).

(5) S. T. Bowden. *Physicochemical periodic table.* Cardiff, England: University College, **1947**.

(6) Mircea V. Ionescu. Asupra clasificării periodice moderne a elementelor. *Acad. rep. popul. Romîne, Fil. cluj., Stud. cercet. chim.* (Bucharest), **7**(1/4), 7-21 (**1956**) (table p. 12).

(7) Emil Petrovici. Citeva observații privitoare la caracterul elementelor și poziția lor in sistemul periodic. *Bulet. stiint. și techn., Scoala politechn.* (Timisvara, Rumania) [N.S.] **1**(15)(1), 335-350 (**1956**) (table p. 338/339).

(8) R. T. Sanderson. A rational periodic table. *J. Chem. Educ.* **41**, 187-189 (**1964**) (table p. 188).

(9) V. K. Grigorovich. *Periodicheskii zakon i elektronnoe stroenie metallov.* Moskva: Nauka, **1966** (table p. 67).

(10) Eduard Pachmann. Periodická soustava prvků jako učební pomůcka na všcobecně vzdělávalí škole. *Chemické Listy* (Prague) **60**(10), 1385-1390 (**1966**) (table p. 1389).

(11) W. B. Guenther. An electronegativity spectrum for the periodic table. *J. Chem. Educ.* **47**, 317 (**1970**).

Subtype IC1-2a.

(1a) Orme Masson. *The classification of the chemical elements.* London, Melbourne: Melville, Mullen & Lade, **1896**.

(1b) William Ramsay. *The gases of atmosphere.* London: Macmillan & Co., **1896**, **1902** (table p. 218/219).
———— *Modern Chemistry.* London: I. M. Dent & Co., V. I., **1900** (table p. 50).

(2) K. B. McCutchon. A simplified periodic classification of the elements. *J. Chem. Educ.* **27**, 17-19 (**1950**).

(3) Prem Behari Mathur. Bi-planar periodic arrangement of elements. P. I. Classification on chemical grounds. *Proc. Natl. Acad. Sci* (India) **23A**(4/5), 125-132 (**1954**).

(4) Sumio Satou. A new periodic table of the chemical elements (in Japanese). *Mem. of the Kagawa Univ.* (Takamatsu, Japan) **II**(19), 1-5 (**1955**).
———— *Comparative Inorganic Chemistry* (in Japanese). Tokyo, Japan: Nikkan Kogyo Shinbunsha, **1959** (table p. 24/25).
———— Fundamental Inorganic Chemistry (in Japanese). Tokyo, Japan: Asakura Shoten, **1966**. (table p. 39/40).

Subtype IB1-2b.

(1) Darvin O. Lyon. *Das periodische System in neuer Anordnung.* Leipzig & Wien: Fr. Deuticke, **1928** (table p. 34 and 35).

Subtype IC1-2b.

(1) Eugen Rabinowitsch und Erich Thilo. *Periodisches System, Geschichte und Theorie.* Stuttgart: F. Enke, **1930** (table p. 40).

(2) Sidney French. Warping the periodic table. *J. Chem. Educ.* **14**, 571-573 (**1937**) (table p. 572).

(3) Reinh. Walter. Neuere Entwicklung des Periodischen Systems. *Chem.-Ztg.* **74**, 673-676 (**1950**) (table p. 673 Fig. 1).

(4) Tiberiu Golgoţiu, Julieta Linde şi Angela Luca. Cîteva consideraţiuni asupra poziţici elementelor tranziţionale în tabloul periodic (in Rumanian). *Bulet. Inst. Politechn., Jassy,* **6**, 109-114 (**1960**) (separate table p. 112/113).

Subtype IC1-2c.

(1) Lecoq de Boisbaudran. Classification des éléments chimiques. *Compt. rend.* (Paris) **120**, 1097-1103 (**1895**) (table p. 1100); **124**, 127-130 (**1897**).

———— Classification of the chemical elements. *Chem. News* **71**, 271-273 (**1895**) (table p. 272).

CLASS 2. TABLES WITH TWO SUBGROUPS "a" and "b"

Type IA2-1.

(1) Alfred W. Stewart. *Recent advances in physical and inorganic chemistry.* London, New York: Longmans, Green & Co., 3rd edit., **1919** (table pp. 270, 271).

Type IC2-1.

(0) D. Mendeleev. Sootnoshenie svoistv s atomnym vesom elementov. *Zh. Russk. Khim. Obshch.* **1**, 60-77 (**1869**) (footnote p. 70).

———— Ob atomnom obiome prostykh veshchestv. *Trudy 2. s'ezda russk. estestvoispyt. i vrachei v Moskve* **1**, 62-71 (**1870**) (table p. 65).

———— O kolochestve kisloroda v solianykh okislakh i ob atomnosti elementov. *Zh. Russk. Khim. Obshch.* **2**, 14-21 (**1870**) (table p. 19).

(1) D. I. Mendeleev (*cited* by V. von Richter). Correspondenz aus St. Petersburg am 6/18. Dezember 1870. *Ber. deut. chem. Ges.* **3**, 990-992 (**1870**) (table p. 992).

———— Über die Stellung des Ceriums im System der Elemente (in German). *Bull Acad. imp. sci.* (St. Petersburg) **16**, 45-51 (**1871**) (table p. 51).

———— The same title. *Mélanges physiques et chimiques* **8**(4), 445-452 (**1872**) (table at p. 447).

———— Estestvennaia sistema elementov i primenenie ee k ukazaniiu svoistv neotkrytykh elementov. *Zh. Russk. Khim. Obshch.* **3**, 25-56 (**1871**) (table p. 31).

———— Die periodische Gesetzmässigkeiten der chemischen Elemente. *Ann. Chem., Justus Liebigs,* **Suppl. 8**, 133-229 (**1871**) (table p. 151).

———— The same title. *Ostwald's Klass. d. exakt. Wissensch.* **68**, 41-118 (1895).

———— The same subject. *Osnovy Khimii.* St. Peterburg: 1st edit., V. II, **1871**.

———— Remarques à propos de la découverte du gallium. *Compt. rend.* (Paris) **81**, 969-972 (**1875**) (table p. 969).

———— The chemistry of the future. *Quart. J. Sci.* **14** [N.S.7] (55), 289-306 (**1877**) (table p. 304).

———— La loi périodiqúe des éléments chimiques. *Monit. sci.* (*Docteur Quesneville*) **9**, 691-737 (**1879**) (table p. 701).

———— The Periodic Law of the chemical elements. *Chem. News* **40**, 231-232; 243-244; 255-256; 267-268; 279-280; 291-292; 303-304 (**1879**) (table p. 268); **41**, 2-3; 27-28; 39-40; 49-50; 61-62; 71-72; 83-84; 93-94; 106-108; 113-114; 125-126 (**1880**).

(2) Lothar Meyer. *Die modernen Theorien der Chemie.* Breslau: Maruschke & Berendt, 4. Aufl., **1880** (table p. 185).

(3) Georg Rudorf. *The periodic classification and the problem of chemical evolution.* London: Whittaker & Co., **1900** (table p. 151).

(4) Bohuslav Brauner. Über die Stellung der Elemente der seltenen Erden im periodischen System von Mendelejeff. Z. anorg. Chem. **32**, 1-30 (**1902**) (table p. 18).

(5) L. A. Chugaev. *Periodicheskaia sistema khimicheskikh elementov.* St. Peterburg: Obrazovanie, **1913** (table p. 130/131 and 204).

Type IA2-2.

(1) O. I. Stewart. Another attempt to base a classification of the elements on atomic structure. *J. Chem. Educ.* **5**, 57-63 (**1928**) (table p. 62).

Type IB2-2.

(1) O. I. Stewart. Another attempt to base a classification of the elements on atomic structure. *J. Chem. Educ.* **5**, 57-63 (**1928**) (table p. 59).

Type IC2-2.

(1) V. von Richter. *Lehrbuch der anorganischen Chemie.* Bonn: M. Cohen & Sohn, 6th edit. **1889** (1st edit. 1857?) (table p. 272).

(2) Walther Nernst. *O. Dammer, Handbuch der anorganischen Chemie.* Stuttgart: F. Enke, B. I. **1892** (table p. 121).
———— *Theoretische Chemie.* Ibidem **1893**.
———— *Theoretical chemistry.* London: Macmillan & Co., **1895**.

(3) F. P. Venable. A modified arrangement of the elements under the natural law. *J. Am. Chem. Soc.* **17**, 75-84 (**1895**) (table p. 81).
———— *The development of Periodic Law.* Easton, Pa.: The Chemical Publishing Co., 269-271, **1896** (table p. 270).
———— and Jas. Lewis Howe. *Inorganic chemistry.* Ibidem **1898** (table p. 20).
———— *The study of the atom.* Ibidem **1904** (table p. 182).

(4) A. Reychler. *Les théories physico-chimiques.* Brussels: H. Lamertin, **1897**.
———— *Outlines of physical chemistry.* London, New York: **1899** (table p. 29).

(5) Jas. Lewis Howe. The eighth group of the Periodic System and some of its problems. *Chem. News* **82**, 15-17 (**1900**) (table p. 16).

(6) Georg Rudorf. *Das periodische System, seine Geschichte und Bedeutung für die chemische Systematik.* Hamburg, Leipzig: Voss, **1904** (table p. 247).

(7) Julius Kiss. *Das periodische System der Elemente und die Giftwirkung.* Wien, Leipzig: A. Hölder, **1909** (table p. 4).

(8) K. Fajans. Die Radioelemente und das periodische System. *Naturwissenshaften* **2**, 463-468 (**1914**) (table p. 465).

(9) R. J. Meyer. Die Stellung der Elemente der seltenen Erden im periodischen System. *Naturwissenschaften* **2**, 781-787 (**1914**) (table p. 781).

(10) G. Weissenberger. Über die Mannigfaltigkeit der Materie. *Vorträge des Vereins zur Verbreit. naturwiss. Kenntn.* **56**, 391-413 (**1916**) (table p. 398/399).

(11) Curt Schmidt. *Das periodische System der chemischen Elemente.* Leipzig: J. A. Barth, **1917** (table p. 130/131).

(12) Alfred W. Stewart. *Recent advances in physical and anorganic chemistry*. London, New York: Longmans, Green & Co., **1919** (table p. 269).

(13) J. Newton Friend. The periodic sphere and the position of rare earth metals. *Chem. News* **130**, 196-197 (**1925**) (table p. 196, 197).

(14) O. I. Stewart. Another attempt to base a classification of the elements on atomic structure. *J. Chem. Educ.* **5**, 9-24; 57-63 (**1928**) (table p. 58).

(15) Herbert Heribert. Über eine einfache Grundlage der Elemente. *Z. Elektrochem.* **36**, 687-688 (**1930**) (table p. 688).

(16) C. E. Fawsitt. The periodic arrangement of the chemical elements in 1930. *Chem. Engin. Mining Rev.* **23**, 152-153 (**1931**) (table p. 152/153).

(17) Robert M. Müller. *Über die Entfaltungsordnung und den Stammbau der chemischen Grundstoffe*. Halle: M. Niemeyer, **1944** (table p. 6).

(18) Niels Bjerrum. Remarques au sujet de la classification des éléments en groupes dans le système périodique. *Bull. Soc. Chim. Belges* **62**, 78-90 (**1953**) (table p. 80).

(19) Roderich Scheer. A variable periodic system. *J. Chem. Educ.* **32**, 590-591 (**1955**) (table p. 590).

(20) T. Pavolini. Nuova forma del sistema periodico degli elementi. *Chimica* (Milan) **33**, 7-9 (**1957**) (separate table p. 8/9).

(21) S. A. Shchukarev. *Lektsii po obshchemu kursu khimii*. Leningrad: Izdat. Leningr. Univ., V. I., **1962**.

(22) David Abbott. *An introduction to the periodic table*. London: J. M. Dent & Sons, **1966** (table p. 4).

Subtype IC2-2a.

(1) Heinrich Biltz. Zur Kenntniss des Perioden-Systems der Elemente. *Ber. deut. chem. Ges.* **35**, 562-568 (**1902**) (table p. 568).
_____ Berichtigung zu der Tabelle über das "Perioden-System der Elemente". *Ibidem* **35**, 4241 (**1902**) (correction of previous table).

(2) C. Zenghelis. Das periodische System und die methodische Einteilung der Elemente. *Chem.-Ztg.* **30**, I, 294-295; 316-317 (**1906**) (table p. 317).

(3) E. H. Büchner. De achtste groep van het periodiek systeem. *Chem. Weekblad* **12**, 336-339 (**1915**) (table p. 338).

(4) U. Sborgi. Per una teoria elettronica del comportamento anodico del metalli ed in particolare di quelli che presentano fenomeni di passività. *Atti Accad. Naz. Lincei* (Rome), Rendicoti [VI] **1**, 315-318; 388-392 (**1925**) (table p. 317).

Subtype IC2-2b.

(1) W. Preyer. Über das genetische System der chemischen Elemente. *Berlin Physik. Ges., Abh.* **10**, 85-88 (**1892**) (scheme p. 86/87).
_____ Einige tatsächliche Grundlagen des genetischen Systems der Elemente. *Deut. pharm. Ges.* **2**, 144-155 (**1892**) (scheme p. 147).
_____ *Das genetische System der Elemente*. Berlin: R. Friedländer & Sohn, **1893** (scheme p. 3).

(2) Saul Dushman. The Periodic Law. *Gen. Elec. Rev.* **18**, 614-621 (**1915**) (separate table p. 618/619).
——— The same title. A review of late developments and a revised form of Mendelejeff's table. *Sci. Amer. Suppl.* **81**, 44-46 (**1916**) (table p. 46).
——— Structure of the atom. P. I. *Gen. Elec. Rev.* **20**, 186-196 (**1917**) (table p. 188).

(3) Albert W. Hull. The crystal structures of the common elements. *J. Franklin Inst.* **193**, 189-216 (**1922**) (table p. 208).

(4) George W. Sears. A new form of periodic table as a practical means of correlating the facts of chemistry. *J. Chem. Educ.* **1**, 173-177 (**1924**) (table p. 175).
——— A theoretical point of view in the teaching of inorganic chemistry. *Ibidem* **10**, 430-432 (**1933**) (table p. 431).

(5) Robert M. Müller. *Über die Entfaltungsordnung und den Stammbau der chemischen Grundstoffe.* Halle: M. Niemeyer, **1944** (table p. 8).

Type IC2-3.

(1) Carl Benedicks. Über die Atomvolumina der seltenen Erden und deren Bedeutung für das periodische System. *Z. anorg. Chem.* **39**, 41-48 (**1904**) (table p. 42).

(2) Arthur A. Blanchard and Frank B. Wade. *Foundations of chemistry.* New York, Cincinnati, Chicago: American Book Co., **1914** (table p. 389).

(3) K. Fajans. Das periodische System der Elemente, die Radioaktiven Umwandlungen und die Struktur der Atome. *Physik. Z.* **16**, 456-486 (**1915**) (table p. 467).
——— *Radioaktivität und die neueste Entwickelung der Lehre von den chemischen Elementen.* Braunschweig: F. Vieweg & Sohn, **1919** (table p. 89, 90).

(4) William D. Harkins and R. E. Hall. The Periodic System and the properties of the elements. *J. Am. Chem. Soc.* **38**, 170-221 (**1916**) (table p. 202/203).
——— Das periodische System und die Eigenschaften der Elemente. *Z. anorg. Chem.* **97**, 175-240 (**1916**) (table p. 202/203).
William D. Harkins. The evolution of the elements and the stability of complex atoms. I. A new periodic system which shows a relation between the abundance of the elements and the structure of the nuclei of atom. *J. Am. Chem. Soc.* **39**, 856-879 (**1917**) (table p. 874/875).

(5) C. G. Bedreag. Remarques sur le système périodique des éléments. *Ann. sci. Univ. Jassy* **10**, 145-153 (**1916**) (separate table).

(6) G. Weissenberger. Über die Mannigfaltigkeit der Materie. *Vorträge des Vereins zur Verbreit. naturwiss. Kenntn.* **56**, 391-413 (**1916**) (table p. 411).

(7) Fritz Paneth. Das periodische System der chemischen Elemente. *Ergebn. exakt. Naturwissensch.* **1**, 362-403 (**1922**) (table p. 399).

(8) Karl Fehrle. Phasenverschiebung, Relativbewegung und radioaktiver Zerfall. *Z. Physik* **16**, 397-407 (**1923**) (table p. 405).

(9) W. Walter Meissner. *Chemischer Handatlas.* Braunschweig: G. Westermann, **1931** (table p. 1).

(10) Ernst Sommerfeldt. Beziehungen der Kristallgitter zur Atomchemie für das periodische System der Elemente. *Zentrbl. Mineral., Geol., Paläont.*, Abt. A. **1934**, 33-46 (table p. 34).

(11) Friedrich Kipp. Über das periodische System der Elemente. *Naturwissenschaften* **30**, 679-683 (**1942**) (table p. 680).

(12) F. H. Loring. Classification of the elements. I. Mainly stable elements. *Chem. Products* **6**. 51-58 (**1943**) (table p. 52).

(13) Anonymous author. Neue Tafel des periodischen Systems. *Physik. Blätter* **3**, 151-155 (**1947**) (table p. 152/153).

Type IB2-4.

(1) Heinr. Baumhauer. *Die Beziehungen zwischen dem Atomgewichte und der Natur der chemischen Elemente.* Braunschweig: F. Vieweg & Sohn, **1870** (separate table).
_____ Über das natürliche System der chemischen Elemente. *Ber. deut. chem. Ges.* **6**, 652-655 (**1873**) (no table).

(2) Thos. Carnelley. Suggestions as to the cause of the Periodic Law and the nature of the chemical elements. *Chem. News* **53**, 183-186 (**1886**) (table p. 184).

(3) L. Beaumont Tansley. Spiral classification of the elements. *Chem. News* **121**, 269 (**1920**); **122**, 121-122 (**1921**) (table p. 121).

(4) Pietro Saccardo. Rappresentazione periodica e tossicologica degli elementi. *Chimica* (Milan) **11**, 411-414 (**1955**) (table p. 411).

(5) N. P. Agafoshin. *Izbrannye glavy obshchei khimii.* Moskva: Gos. Uchebno-pedag. Inst., **1956** (separate table).

(6) Elena Bogdan. O propunere de aranjare a elementelor îu sistem periodic (in Rumanian). *Analele Stiint., Univ. "A. I. Cuza", Iasi,* [N.S.] **10**(1), 1-16 (**1964**) (table p. 6/7).

Type IC2-4.

(1) H. Gretschel und G. Bornemann. Das natürliche System der Elemente. *Jahrbuch der Erfindungen* **19**, 292-301 (**1883**) (table p. 293).

(2) Lothar Meyer. *Grundzüge der theoretischen Chemie.* Leipzig: Breitkopf & Härtel, **1892**.
_____ *Outlines of theoretical chemistry.* London: Longmans, Green & Co., **1892** (table p. 55).
_____ Über den Vortrag der anorganischen Chemie nach dem natürlichen System der Elemente. *Ber. deut. chem. Ges.* **20**, 1230-1250 (**1893**) (table p. 1232).

(3) Dan Radulescu. La théorie des radicaux comme interprétation du système de Mendeleeff. *Bulet. Soc. rom. stiinte Bukuresti* **21**, 39-71 (**1912**) (table p. 70/71).
_____ L'hypotèse d'une homologie nucléaire dans le système périodique. *Ibidem* **21**, 500-513 (**1912**) (table p. 512/513).

(4) J. R. Rydberg. Untersuchungen über das System der Grundstoffe. *Lunds Univ. Årsskr.* (Acta Universitatis Lundensis) [N.S.] **9**(18), 1-41 (**1913**) (table p. 20).
_____ Recherches sur le système des éléments. *J. chim. phys.* **12**, 585-639 (**1914**) (table p. 609).
_____ Le système des éléments chimiques. *Rev. gén. sci.* **25**, 734-741 (**1941**) (table p. 739).

(5) Kasimir Fajans. Die radioaktiven Umwandlungen und das periodische System der Elemente. *Ber. deut. chem. Ges.* **46**, 422-439 (**1913**) (table p. 437).

(6) R. Ladenburg. Atombau und periodisches System der Elemente. *Z. Elektrochem.* **26**, 262-274 (**1920**) (table p. 264).

(7) Franz Urbach. Periodisches System, Atombau und Radioaktivität. *Physik. Z.* **22**, 114-119 (**1921**) (table p. 116).

(8) Fritz Paneth. Über die heutige Schreibweise des periodischen Systems der Elemente. *Z. angew. Chem.* **36**, 407-410 (**1923**) (table p. 409); **37**, 421-422 (**1924**) (no table).

———— Die Entwicklung und der heutige Stand unserer Kenntnisse über das natürliche System der Elemente. *Naturwissenschaften* **18**, 964-976 (**1930**) (table p. 966).

(9) Iwan Bolin. Grundämnenas platser i det periodiska systemet enlight Bohrs senaste undersökningar. *Teknisk Tidskrift* **54**, 42-46 (**1924**) (table p. 43).

(10) Karl Mahler. *Atombau und periodisches System der Elemente.* Berlin: O. Salle, **1927** (table p. 117).

(11) John A. V. Butler. *The chemical elements and their compounds.* London: MacMillan Co., **1927** (table on the frontispiece).

(12) Stuart Graves. A periodic chart of the atoms. *J. Chem. Educ.* **6**, 553-555 (**1929**) (table p. 553).

(13) M. Haissinsky. Le système périodique des éléments. *Génie civil* **96**, 210-213 (**1930**) (table p. 210).

———— *Nuclear chemistry and its application* (in Russian). Moscow: **1961**.

———— The same title (in English). Reading, Mass.: Addison-Wesley Publ. Co., **1964** (table p. 236).

(14) C. H. Douglas Clark. A spectroscopic classification of the elements according to ground states. *Proc. Leeds Phil. Lit. Soc., Sci. Sect.*, **2**, 225-229 (**1931**) (table p. 226/227).

———— Some physical aspects of atomic linkages. *Chem. Rev.* **11**, 231-271 (**1932**) (table p. 234/235).

———— *The electronic structure and properties of matter; an introductory study of certain properties of matter in the light of atomic numbers.* London: Chapman & Hall, **1934** (table p. 54,55).

———— A simple electronic periodic table. *Proc. Leeds Phil. Lit. Soc., Sci. Sect.*, **3**, 281-292 (**1937**) (table p. 282/283).

———— *The fine structure of matter. The quantum theory and line spectra.* London: Chapman & Hall, V. III, **1938**.

(15) Harriett H. Fillinger. A "living" periodic chart of the elements. *J. Chem. Educ.* **9**, 1807-1810 (**1932**) (table p. 1807).

(16) C. J. Gorter en A. J. Rutgers. Het systeem der elementen. *Chem. Weekblad* **30**, 602-606; 632-635; 642-646; 654-657; 671-674; 682-684 (**1933**) (table p. 634).

(17) B. Smith Hopkins. The periodic table. *J. West. Soc. Eng.* **38**, 307-317 (**1933**) (table p. 309).

———— Recent developments in the chemistry of the rare-earth group. *J. Chem. Educ.* **13**, 363-368 (**1936**) (table p. 363).

(18) Henry D. Hubbard. *A key to the periodic chart of the atoms. Primer of the atoms.* Chicago: **1934**.

(19) Ia. K. Syrkin. Periodicheskaia sistema (K stoletiiu so dnia rozhdeniia D. I. Mendeleeva). *Uspekhi Khim.* **3**, 358-405 (**1934**) (table p. 374/375).

(20) B. W. Nekrassov. Les propréteés des ions. *Bull. Soc. Chim. France* [5] **3**, 151-159 (**1936**) (table p. 154).

(21) E. C. Payne. The Periodic System. *J. Chem. Educ.* **14**, 593-594 (**1937**) (table p. 594).

(22) Robert Höltje. Clemens Wikler und das periodische System der Elemente. *Deut. Mus., Abhandl. Ber.* **12**, 1-12 (**1940**) (table p. 11).

(23) Fahlenbach. Ein periodisches System der chemischen Elemente in neuer Anordnung. *Umschau Wissensch. Techn.* **44**, 403-404 (**1940**).

(24) Sidney J. French. The drama of chemistry. *Dow Diamond* **5**(4), 30 (**1942**).

(25) Antonio García Valcárcel. El sistema periódico de Mendelejeff en biologia. *Medicina Española* (Valencia) **14**, 558-562 (**1945**) (table p. 560).

(26) F. Oberhauser B. Sistema periódico o sistema natural de los elementos químicos. *An. facult. filos. y educ., Univ. Chile, Secc. de quim.* **3**, 131-154 (**1946**) (table p. 136).

(27) Włodzimierz Rodziewicz. Zmiany rozmieszczenia pierwiastków w ostatnim periodzie ich układu. *Przegląd Chemiczny* (Poland) **5**, 182-187 (**1947**) (table p. 185).

(28) Ugo Ventriglia. Contributo alla sistematica degli elementi cristallizzanti in reticoli ionici inquadrata nel sistema periodico, *Ricerca sci.* **18**, 1635-1640 (**1948**) (table p. 1637).

(29) R. Spence. Chemistry and atomic energy. *Research* (London) **2**, 115-119 (**1949**) (table p. 116).

(30) V. L. Albanskii. Novaia forma tablitsy periodicheskoi sistemy khimicheskikh elementov D. I. Mendeleeva, otrazhaiushchaia dostizheniia nauki po stroeniiu atoma. *Dokl. Akad. Nauk SSSR* [N.S.] **75**, 209-211 (**1950**) (separate table p. 210/211).
———— Tablitsa periodicheskoi sistemy khimicheskikh elementov D. I. Mendeleeva na osnove elektronnykh struktur atomov. *Zh. Obshch. Khim.* **21**, 1393-1395 (**1951**) (separate table p. 1394/1395).

(31) E. Hayek. Die neuen Elemente im periodischen System. *Pyramide* **2**, 9-12 (**1952**) (table p. 11).

(32) A. F. Kapustinskii. Periodichnost' v stroenii elektronnykh obolochek i iader atomov. I. Periodicheskaia sistema khimicheskikh elementov i ee sviaz' s teoriei chisel i s fiziko-khimicheskim analizom. *Izv. Akad. Nauk SSSR* **1953** (1), 3-4; 12-20 (table p. 4/5).

(33) Robert Delhez. Aspects récents de la classification périodique. *Rev. Universelle Mines* **10**, 679-683 (**1954**) (table p. 681).

(34) Simon Z. Roginskii. *Měndělejevova periodiská soustava prvků ve světle poslednich výzkumnů.* Praha: Státni naklad. techn. lit. **1954**.

(35) Lajos Kesztelyi. Az elemek periódusos rendszere. *Fizikai Szemle* (Budapest) **5**(2/3), 40-50 (**1953**).

(36) George K. Estok. An adjustable periodic chart for lecture purposes. *J. Chem. Educ.* **33**, 618-619 (**1956**).

(37) A. N. Nesmeyanov. La classification périodique de Mendéléev et la chimie organique. *Bull. Soc. Chim. France* **1960**, 987-1010 (table p. 990).

(38) G. E. Villar. Adjuste propuesto para la disposición actual de cuadro periódico. *Anales Real Soc. Españ. Fis. y Quim.*, Ser. **B 56** (5), 521-534 (**1960**) (table p. 534).

(39) S. A. Shchukarev. Cited by V. M. Vdovenko. *Khimiia urana i transuranovykh elementov.* Moskva-Leningrad: Izdat. Akad. Nauk SSSR **1960**.

(40) G. S. Vozdvizhenskii. Variant periodicheskoi sistemy, korrektiruiushchii polozhenie khimicheskikh elementov nulevoi, vosmoi i pervoi grupp. *Trudy Kazan. Khim.-Tekhnol. Inst.* **1960** (29), 33-35.

(41) Franz Matthes. *Atome, Elemente, Isotope.* Leipzig: Urania-Verlag, **1961** (separate table).

(42) John Albert Newton Friend. *Man and chemical elements.* London: Griffin, **1961** (table p. 5).

(43) V. I. Gol'danskii. *Novye elementy v periodicheskoi sisteme D. I. Mendeleeva.* Moskva: Atomizdat, **1964** (table p. 10/11).

(44) Pai Yen Loung. *Graphic handbook of chemistry and metallurgy.* New York: Chem. Publ. Co., **1965** (table I).

(45) N. S. Akhmatov i G. S. Vozdvizhenskii. Konstruktsiia VIII gruppy periodicheskoi sistemy khimicheskikh elementov. *Proc. 20th Congr. Intern. Un. Pure Appl. Chem.,* **1965**, (C-D), Report D-40, 87-88, Moscow.

(46) M. Kh. Karapet'iants, A. P. Kreshkov. Stoletie velikogo otkrytiia. *Plast. Massy* (Moscow), **1969** (3), 5-10 (table p. 7).

(47) Ana Maria Mojos, Hector Vicente Incolla. Ayer, hoy y mañana en la tabla periódica de los elementos químicos (in Spanish). *Saneamiento* (Buenos Aires) 33(215), 265-268 **(1969)** (table p. 267).

(48) Frank Dieter Leyh. Modell zur Erläuterung atomarer Elektronenkonfigurationen. *Chem. Schule* 17(6), 278-284 **(1970)** (tables pp. 280, 282, 283).

(49) D. N. Trifonov. Periodichnost' chetkaia i razmytaia. *Priroda* (Moscow) **1970** (5), 84-91.

(50) K. V. Astakhov. Periodicheskii zakon D. I. Mendeleeva v svete sovremennykh nauchnykh dannykh. *Zh. Vses. Khim. Obshch.* 16(3), 251-260 **(1971)** (table p. 256).

Subtype IC2-4a.

(1) A. Dauvillier. Analyse de la structure atomique. *Compt. rend.* (Paris) 173, 1077-1079 **(1921)** (table p. 1078).
───── Analyse de la structure électronique des éléments. *J. phys. radium* (Paris) 3, 154-177 **(1922)** (table p. 159).

Subtype IC2-4b.

(1a) Geoffrey Martin. *Researches on the affinities of the elements.* London: J. & A. Churchill, **1905** (table p. 35).

(1b) E. Molinari. *Trattato di chimica inorganica generale.* Milano: U. Hoepli, **1905** (table p. 666).

(2) Charles Baskerville. *General inorganic chemistry.* Boston, Mass.: Heath & Co., **1909** (table p. 110).

(3) A. van den Broek. β and γ rays and the structure of the atom (internal-charge numbers). *Nature* 93, 376-377 **(1914)** (table p. 377).

(4) D. Balarew. Über die Zusammenstellung der Molekularvolumen der Oxyde im periodischen System. *J. prakt. Chem.* [2] 102, 283-286 **(1921)** (table p. 284).

Subtype IC2-4c.

(1) W. Herz. *Leitfaden der theoretischen Chemie.* Stuttgart: F. Enke, **1912** (table p. 101).

(2) Frederick H. Getman. *Outlines of theoretical chemistry.* New York: John Wiley & Sons, **1913** (table p. 24).

(3) W. R. Cooper. The present position of the Periodic Law. *World Power* 1, 18-24 **(1924)** (table p. 18).

(4) B. W. Nekrassov. Les propriétés des ions. *Bull. Soc. Chim. France* [5] **3**, 151-159 (**1936**) (table p. 154).

(5) N. V. Sidgwick. *The chemical elements and their compounds.* Oxford: Clarendon Press, **1950** (table p. XXVIII).

Subtype IC2-4d.

(1) A. Morette. L'état actuelle de la classification des éléments. *J. pharm. chim.* [9] **1**, 437-454; 482-492 (**1941**) (table p. 452).

(2) Science Service. A new chart of the elements. Periodic table chart. *Chemistry* (Washington) **21**(14), 18 (**1947**) (separate table).

(3) Edu Guides. *The 102 elements. Building blocks of the Universe.* Chicago, Ill.: **1958** (table on the inside front cover).

Type IA2-5.

(1) William D. Harkins and R. E. Hall. The Periodic System and the properties of the elements. *J. Am. Chem. Soc.* **38**, 169-221 (**1916**) (tables pp. 170/171, 182, 183).
_____ Das periodische System und die Eigenschaften der Elemente. *Z. anorg. Chem.* **97**, 175-240 (**1916**) (tables pp. 179, 180, 181, 210).
William D. Harkins. The abundance of the elements in relation to the hydrogen-helium structure of the atoms. *Proc. Natl. Acad. Sci.* (Washington) **2**, 216-224 (**1916**) (table p. 218).

Type IB2-5.

(0) Gustavus Detlef Hinrichs. *Programm der Atomechanik, oder die Chemie eine Mechanik der Panatome.* Iowa City, Iowa: 7-8, **1867**.

(1) Ernst Huth. Das periodische Gesetz der Atomgewichte und das natürliche System der Elemente. *Samml. naturwiss. Vorträge* (Frankfurt) **I**, **1884** (separate table).

(2) F. V. Wells. Note on the Periodic System of the elements. *J. Wash. Acad. Sci.* **8**, 232-234 (**1918**) (table p. 233).

(3) I. R. Partington. Periodic classification of the elements. *Chem. News* **121**, 304 (**1920**).

(4) Arnaldo Piutti. Una rappresentazione didattica degli elementi. *Gazz. chim. ital.* **55**, 754-756 (**1925**) (separate table).
_____ The same title. *Rendic. Accad. Sci. Fis. Mat.* (Naples) [3] **31**, 142-143 (**1925**) (table p. 142/143).

(5) F. von Wolff. Die Kristallstrukturen der Elemente und ihre Beziehungen zum periodischen System. *Neues Jahrb. Mineral. Geol.*, Beil. Bd. A. **57**, 265-286 (**1928**) (table p. 267).

(6) Iu. K. Delimarskii. Periodicheskii zakon D. I. Mendeleeva (k stoletiiu so dnia otkrytiia). *Ukrain. Khim. Zh.* **35**(3), 227-232 (**1969**).
_____ and O. G. Zarubitskii. Spiralevidnaia forma Periodicheskoi Sistemy khimicheskikh elementov. *Zh. Obshch. Khim.* **39**(1), 11-15 (**1969**).

Type IC2-5.

(0) Ernst Huth. Das periodische Gesetz der Atomgewichte und das natürliche System der Elemente. *Samml, naturwiss. Vorträge* (Frankfurt) **I**, 12-16 (**1884**).

(1) Carl Arnold. *Repetitorium der Chemie.* Hamburg & Leipzig: L. Voss, **1885** (table p. 44/45).
———— *Abriss der allgemeinen oder physikalischen Chemie.* Ibidem **1903** (table p. 74/75).

(2) Nikolai Morozov. *Periodicheskie sistemy stroeniia veshchestva.* Moskva: **1907** (table p. 16/17 and separate table p. 328/329).
———— *D. I. Mendeleev i znachenie ego periodicheskoi sistemy.* Moskva: **1907.**

(3) Emil Baur. Über das periodische System der Elemente. *Z. physik. Chem.* **76**, 569-583 **(1911)** (table p. 583).

(4) K. Scheringa. Iets over het periodiek Systeem. *Chem. Weekblad* **8**, 868-869 **(1911)** (table p. 868).

(5) Edmond Bauer. *La théorie de Bohr: la constitution de l'atome et la classification périodique des éléments.* Paris: J. Hermann, **1922** (table p. 10).

(6) Wilh. Palmaer. Eine Aufstellung des periodischen Systems. *Z. physik. Chem.* **110**, 685-704 **(1924)** (table p. 696).
———— En ny upställning af det periodiska systemet. *Svensk. Kem. Tidskr.* **37**, 1-17 **(1925)** (table p. 10).

(7) J. D. Main Smith. *Chemistry and atomic structure.* London: Ernest Benn, **1924** (table p. 79).
———— The electronic structure of atoms. I. The periodic classification. *J. Chem. Soc.* (London) **1927** (Trans. II), 2029-2038 (table p. 2029).

(8) M. Centnerszwer. Über die Haupt- und Nebengruppen des Periodischen Systems. *Ber. deut. chem. Ges.* **59 B**, 786-788 **(1926)** (table p. 787).

(9) Arrigo Mazzuchelli. Sulla opportunità di introdurre i periodi quantici nella tavola di Mendeleieff. *Gazz. chim. ital.* **65**, 467-473 **(1935)** (table p. 469).

(10) B. N. Menshutkin. Glavnye momenty v razvitii periodicheskoi sistemy elementov (1869-1937). *Priroda* (Moscow) **26** (3), 116-126 **(1937)** (table p. 125).

(11) S. A. Shchukarev. Sovremennoe sostoianie periodicheskogo zakona D. I. Mendeleeva. *Trudy Iubilein. Mendeleev. s'ezda, Moskva, Akad. Nauk SSSR,* **2**, 23-35 (in Russian); 37-49 (in French) **(1937)** (table p. 26/27).

(12) C. J. Bakker. Het periodiek systeem der elementen. *Nederland. Tidschr. Natuurk.* **7**, 305-310 **(1940)** (separate table).

(13) Société Chimique de France, 1940. Cited by A. Morette. L'état actuel de la classification des éléments. *J. pharm. chim.* [9] **1**, 437-454 **(1941)** (table p. 450).

(14) G. B. Bokii. Raspredelenie elementov po podgruppam na osnove kristallokhimicheskikh predstavlenii. *Priroda* (Moscow) **31**, 34-47 **(1942)**.

(15) Walter Kwasnik. *Der Chemiker als Forscher.* München, Berlin: R. Oldenbourg, **1943** (table p. 29).

(16) Edgar I. Emerson. A chart based on atomic numbers showing the electronic structure of the elements. *J. Chem. Educ.* **21**, 254-255 **(1944)**.

(17) Pares Chandra Banerjee. A note on the classification of elements. *J. Indian Chem. Soc.* **22**, 130-131 **(1945)** (table p. 130).

(18) Ia. K. Syrkin i M. E. Diatkina. *Khimicheskaia sviaz' i stroenie molekul.* **1946.**
———— *Structure of molecules and the chemical bond.* New York, London: Interscience Publ., **1930** (table pp. 32, 33).

(19) T. S. Wheeler. The periodic table and electron configurations. *Chemistry & Industry* **1947**, 639-642 (table p. 640-641).

(20) I. P. Selinov. Periodicheskaia sistema elementov D. I. Mendeleeva i nekotorye voprosy atomnoi fiziki. *Uspekhi Fiz. Nauk* **44**, 511-526 (**1951**) (table p. 513 and separate table).

(21) V. I. Gol'danskii. Novye khimicheskie elementy v periodicheskoi sisteme D. I. Mendeleeva. *Priroda* (Moscow) **41** (7), 51-61 (**1952**) (table p. 52).

(22) B. M. Kedrov. *Nauchnyi arkhiv D. I. Mendeleeva* **1**, 838 (**1953**).

(23) R. T. Sanderson. One more periodic table. *J. Chem. Educ.* **31**, 481 (**1954**).
———— A new periodic chart, with electronegativities. *Ibidem* **33**, 443-445 (**1956**) (table p. 444).

(24) *Kratkaia Khimicheskaia Entsiklopediia*, V. 3, Moskva: Izdat. Sov. Entsikl., p. 967, **1964**. *Fizicheskii Entsiklopedicheskii Slovar'*, V. 3, Ibidem, p. 604/605, **1964**.

Subtype IC2-5a.

(1) Lothar Meyer. Die Natur der chemischen Elemente als Funktion ihrer Atomgewichte. *Ann. Chem., Justus Liebigs*, **Suppl. 7**, 354-364 (**1870**) (table p. 356).
———— *Ostwald's Klass. d. exakt. Wissensch.* **68**, 9-17 (**1895**) (table p. 11).
———— Zur Geschichte der periodischen Atomistik. *Ber. deut. chem. Ges.* **13**, 259-265 (**1880**).

CLASS 3. TABLES WITH THREE SUBGROUPS
"a," "b," and "c"

Type IA3-1.

(1) Henry Bassett. A tabular expression of the periodic relations of the elements. *Chem. News* **65**, 3-4; 19 (**1892**).

(2) Anonymous author. A novel arrangement of the table in three dimensions. *Chemistry* (Washington) **21**(4), 14-17 (**1947**) (tables pp. 15, 16, 17, and back cover).

(3) S. Horie. A new space model of the Periodic System of elements. *J. Chem. Educ.* **31**, 382 (**1954**).

Type IB3-1.

(0) Charles P. Steinmetz. The Periodic System of elements. *J. Am. Chem. Soc.* **40**, 733-739 (**1918**) (tables p. 738, 739).

(1) Frank O. Green and Bernard G. Jackson. A spherical arrangement of the chemical elements. *Trans. Illin. Acad. Sci.* **43**, 83-90 (**1950**) (table p. 84).

Type IC3-1.

(1a) Peter W. Schenk. Zur Einordnung der Lanthaniden und Transurane ins periodische System der Elemente. *Österr. Chem.-Ztg.* **50**, 52-54 (**1949**) (table p. 54).
———— The same title. *Angew. Chem.* **63**, 141-142 (**1951**) (table p. 142).

(1b) A. P. Faustov. Novyi sposob izobrazheniia sistemy elementov D. I. Mendeleeva. *Zh. Obshch. Khim.* **19**, 396-398 (**1949**).

(2) Frank O. Green and Bernard G. Jackson. A spherical arrangement of the chemical elements. *Trans. Illin. Acad. Sci.* **43**, 83-90 (**1950**) (table p. 88).

(3) Luis Hurtado Acera. Los cien elementos químicos. *Nucleo* (Madrid) **6**(8), 21-23 (**1951**) (table p. 23).
_____ La nueva tabla periódica y los elementos artificiales Atenio y Centurio. *Metalurg. y Electric.* (Spain) **15** (164), 36-40 (**1951**) (table p. 39).
_____ The same title. *Ion* (Madrid) **11**, 448-552 (**1951**).

(4) Ernst Grimsehl. *Lehrbuch der Physik.* Bd. 4. Struktur der Materie. Leipzig: Teubner, **1959** (table p. XII).

(5) Stefan Amsterdamski. *Rozvój pojęcia pierwiastka chemicznego.* Warsaw: Panstwowe Wyd. Naukowe, **1961**, 152/153.

(6) V. K. Grigorovich. K voprosu o vliianii stroeniia vnutrennikh elektronnykh obolochek atomov na razmeshchenie elementov v periodicheskoi sisteme D. I. Mendeleeva. *Trudy Inst. Metall. Baikova, Gos. Kom. po chernoi i tsvetnoi metallurg., Akad. Nauk SSSR,* **14**, 155-187 (**1963**) (table p. 176/177).
_____ O meste lantanidov i aktinidov v periodicheskoi sisteme elementov D. I. Mendeleeva i o sdvigakh elementov-analogov v sootvetstvii s khimicheskimi svoistvami. *Ibidem* **14**, 188-211 (**1963**) (table p. 196).
_____ Sviaz' telplot obrazovaniia okislov, sul'fidov i galogenidov metallov s periodicheskim zakonom D. I. Mendeleeva. *Izv. Akad. Nauk SSSR, Metall. i Gornoe Delo,* **1964** (4), 91-105 (table p. 95).
_____ O sviazi kristallicheskikh struktur soedinenii perekhodnykh metallov tipa NaCl i NiAs s ikh elektronnym stroeniem. *Vysokotemp. Neorgan. Soedinen., Akad, Nauk Ukrain. SSR, Inst. Probl. Materialoved.,* **1965**, 5-24 (table p. 22).
_____ *Periodicheskii zakon i elektronnoe stroenie metallov.* Moskva: Nauka, **1966** (tables p. 21 and 23).

(7) D. I. Petriichuk. Novyi variant razmeshcheniia redkozemel'nykh i drugikh elementov v tablitse periodicheskoi sistemy D. I. Mendeleeva. *Izsled. po khim. redkikh i soputstv. elem., Akad. Nauk Kirgiz. SSR, Inst. neorg. i fizich. khim.* **1966**, 194-200 (table p. 197).

(8) S. M. Ali. A new periodic table based on electronic configuration (in English). *Pakistan J. Sci. Res.* **18**(4), 210-213 (**1966**).

Subtype IC3-1a.

(1) V. M. Chistiakov. K razvitiiu korotkoi formy tablitsy Mendeleeva. *Uch. Zap. Kuibysh. Gos. Pedagog. Inst.* **42**, 101-123 (**1964**).

Subtype IB3-1b.

(1) N. P. Agafoshin. Modernizatsiia periodicheskoi sistemy elementov Mendeleeva. *Zh. Obshch. Khim.* **22**, 177-184 (1952) (separate table p. 180/181).

Subtype IC3-1b.

(1) F. M. Shemiakin. K voprosu o vkl'uchenii redkikh zemel' v periodicheskuiu sistemu. *Zh. Obshch. Khim.* **2**, 62-64 (**1932**) (table p. 63).

(2) A. I. Mashentsev. O Mendeleevskoi sistematike elementov. *Zh. Obshch. Khim.* **24**(86), 1094-1095 (**1954**) (table p. 1094/1095).

(3) Pai Yen Loung. *The electronic evolution in the atoms of the elements and the construction of a new periodic table.* Palo Alto, Calif.: Periodex Sci. Co., **1965** (table I).

Type IA3-2.

(1) Rudolf Vogel. Über die Beziehungen der seltenen Erden zum periodischen System. *Z. anorg. allgem. Chem.* **102**, 117-200 (**1918**) (table p. 197).

(2) E. C. Payne. A Periodic System of the elements. *J. Chem. Educ.* **15**, 180-183 (**1938**) (table p. 182).

(3) William E. Rice. A helical periodic table. *J. Chem. Educ.* **33**, 492 (**1956**).

Type IC3-2.

(1) Edward G. Mazurs. Ķīmisko elementu periodiskās sistemas tabulu veidi (in Latvian). *Technikas Apskats* (Lincoln, Nebr.) **2**(7), 8-12 (**1955**) (table p. 8).
———— Die wichtigsten Tabellen des Perioden-Systems. *Chem.-Ztg.* **80**, 195-203 (**1956**) (table p. 196, Fig. 6).

(2) A. A. Clifford. Periodic classification of elements. *Nature* **184** (**suppl. 26**), 2012 (**1959**).

Subtype IC3-2a.

(1) G. M. Murashov. Kvantovo-mekhanicheskaia forma periodicheskoi sistemy elementov. *Zh. Obshch. Khim.* **19**, 399-403 (**1949**) (table p. 402).

Subtype IC3-2b.

(1) Fritz Scheele. Die Einordnung der Lanthaniden und Actiniden in das periodische System. *Z. Naturforsch.* **4a**, 137-139 (**1949**) (table p. 139).

(2) I. A. Lebedev. K voprosu o polozhenii transaktinievykh elementov v Periodicheskoi Sisteme D. I. Mendeleeva. *Radiokhimiia* **14** (4), 618-622 (**1972**) (table p. 619, erroneously marked as Table 1, should be Table 2).

DIVISION II. MEDIUM TABLES

CLASS 1. TABLES WITH SIXTEEN GROUPS

Type IIA1-1.

(1) Karl Hack. *Angriffe auf verschiedene Grundanschauungen in der Physik und der Chemie. III. Die Genesis der Elemente und das periodische System.* Mittenberg a./M.: G. Volkhardt'sche Druckerei, **1910** (table No. 14).

Type IIB1-1.

(1) Flavian Flavitskii. O funktsii vyrazhaiushchei periodichnost' svoistv khimicheskikh elementov. *Trudy Kazan. Fiz.-Matem. Nauk* **5**, 214 (**1887**).
———— Über eine Funktion, welche der Periodizität der Eigenschaften der chemischen Elemente entspricht. *Z. Physik. Chem.* **2**, 102 (**1888**) (no table).
———— The same title. *Z. anorg. Chem.* **11**, 264-267 (**1896**) (no table).

(2) G. Johnstone Stoney. On a logarithmic law of the atomic weights. *Chem. News* **57**, 163 (**1888**) (no table).

---------- The same title. *Proc. Roy. Soc.* (London) **44**, 115-117 (**1888**) (no table).

---------- On the law of atomic weights. *Phil. Mag.* [6] **4**, 411-416; 504-505 (**1902**) (separate table).

Cited by Lord Rayleigh. On Dr. Johnstone Stoney's logarithmic law of atomic weights. *Proc. Roy Soc.* (London) **85 A**, 471-473 (**1911**) (table p. 472 taken from original manuscript 1888).

(3) H. Erdmann. *Lehrbuch der anorganischen Chemie*. Braunschweig: F. Vieweg, 727, **1902** (separate table).

(4) Sir William Ramsay. Einige Betrachtungen über das periodische Gesetz der Elemente. *Verh. Ges. deut. Naturforsch. und Ärzte* **75**, 62-74 (**1903**) (table p. 63).

---------- Quelques considérations sur la loi périodique des éléments. *Rev. gén. chim. pure appl.* **6**, 449-455 (**1903**) (no table).

(5) W. Simon and Daniel Base. *Manual of chemistry*. Philadelphia, New York: Lea & Febiger, **1909** (table p. 131).

(6) Karl Hack. *Angriffe auf verschiedene Grundanschauungen in der Physik und der Chemie. III. Die Genesis der Elemente und das periodische System*. Mittenberg a./M.: G. Volkhardt'sche Druckerei, **1910** (table No. 15).

(7) Ingo W. D. Hackh. *Das synthetische System der Atome*. Hamburg: Hephaestos, **1914** (table p. 4).

IIC1-1.

(1) D. Mendeleev. Sootnoshenie svoistv s atomnym vesom elementov. *Zh. Russk. Khim. Obshch.* **1**, 60-77 (**1869**) (fragment of the table foot-note p. 71).

---------- Die Beziehungen zwischen den Eigenschaften der Elemente und ihren Atomgewichten. *Ostwald's Klass. d. exakt. Wissensch.* **68**, 20-40 (**1895**) (foot-note p. 32).

---------- The same subject. *Osnovy Khimii*. St. Peterburg: 1st edit. V. II. **1871**; 4th edit., **1881**; 5th edit., **1889**.

---------- Zur Geschichte des periodischen Gesetzes. *Ber. deut. chem. Ges.* **13**, 1796-1804 (**1880**) (table p. 1804).

---------- Periodicheskaia sistema elementov. *Zh. Russk. Khim. Obshch.* **13**, 517-520 (**1881**) (table p. 519).

Cited by Jawein. Protok. d. J. d. russ. phys,-chem. Gesellsch. 1881, 517. *Ber. deut. chem. Ges.* **14**, 2821-2823 (**1881**) (table p. 2882).

---------- Comment j'ai trouvé le système périodique des éléments. *Rev. gén. chim. pur. appl.* **1**, 211-214; 510-512 (**1899**).

(2) Lothar Meyer. *Die modernen Theorien der Chemie*. Breslau: Maruschke & Berendt, 4. Aufl., **1880** (table p. 184).

(3) B. N. Chicherin. *Sistema khimicheskikh elementov*. Peterburg: V. I., **1888** (table p. 9).

---------- The same subject. *Zh. Russk. Fiz.-Khim. Obshch.* **Suppl. to 20**, 1-26 (**1888**).

(4) Sir William Crookes. On the position of helium, argon, and krypton in the scheme of elements. *Chem. News* **78**, 25-26 (**1898**) (table p. 25).

---------- The same title. *Proc. Roy. Soc.* (London) **63**, 408-411 (**1898**) (table p. 409).

---------- Die Stellung von Helium, Argon und Krypton im System der Elemente. *Z. anorg. Chem.* **18**, 72-76 (**1898**) (table p. 74).

(5) A. J. Batschinski. Ein Versuch, die periodische Gesetzmässigkeit der chemischen Elemente physikalisch zu erklären. *Z. physik. Chem.* **43**, 372-375 (**1903**) (table p. 374).

(6) G. Bourgerel. Série de Mendéléeff ordonnée et mise a jour en 1917 d'aprés les derniers poids atomiques publiés. *Monit. sci. (Docteur Quesneville)* [5] **10**, 241-242 (**1920**).

(7) F. W. Rixon. A new diagram of the periodic table. *Chemistry & Industry* **52**, 260-261 (**1933**) (table p. 261).

(8) J. Guzmán. Tabla periódica de los elementos químicos. *Anales Real Soc. Espan̄. Fis.y Quím.* (mem. y not.) **35**, 104-106 (**1937**) (table p. 162).

(9) J. G. Vogel. Gemoderniseerd periodiek systeem. *Chem. Weekblad* **38**, 529-531 (**1941**).

Subtype IIC1-1a.

(0) F. Kirchhof. Das periodische System der Elemente im Lichte der Theorie des radioaktiven Zerfalls. *Z. physik. Chem.* **94**, 257-262 (**1920**).

(1) I. A. Vaisman. Opyt periodizatsii elementov na osnove stroeniia atomnogo iadra. *Dokl. Akad. Nauk SSSR* [N.S.] **62**, 211-214 (**1948**) (table p. 213).

Subtype IIC1-1b.

(1) E. Loew. Versuch einer graphischen Darstellung für das periodische System der Elemente. *Z. physik. Chem.* **23**, 1-12 (**1897**) (table p. 5).

(2) A. E. Fersman. *Geokhimiia*. T. I., **1933**.

(3) E. N. Dobrocvetov. Fizičke konstante hemiskih elemenata kao funkcije rednor broja i položaja u periodnom sistemu. *Glasn. hem. društ. Beograd (Bull. soc. chim. Belgrade)* **13**, 145-160 (**1948**) (table p. 146).

Type IIA1-2.

(1) Sir William Crookes. On the position of helium, argon, and krypton in the scheme of elements. *Chem. News* **78**, 25-26 (**1898**) (table p. 25).
⎯⎯⎯ The same title. *Proc. Roy. Soc.* (London) **63**, 408-411 (**1898**) (table p. 409).
⎯⎯⎯ Die Stellung von Helium, Argon und Krypton im System der Elemente. *Z. anorg. Chem.* **18**, 72-76 (**1898**) (table p. 73).

(2) Ernest Beutel. Über ein Unterrichtsmodell des periodischen Systems der Elemente. *Z. physik. chem. Unterricht* **26**, 13-19 (**1913**).

Type IIC1-2.

(1) J. Emerson Reynolds. Note on a method of illustrating the Periodic Law. *Chem. News* **54**, 1-4 (**1886**) (table p. 2).
Citation. *Ber. deut. chem. Ges.* **19**, 647-648 (**1886**) (no table).
⎯⎯⎯ Argon and the Periodic System. *Nature* **51**, 486-487 (**1895**).

(2) William Crookes. Address to the Chemical Section of the British Association. *Chem. News* **54**, 115-126 (**1886**) (table p. 120).
⎯⎯⎯ Genesis of elements. *Ibidem* **55**, 83-88; 95-99 (**1887**) (table p. 96).

(3) Samuel Haughton. Geometrical illustrations of Newlands' and Mendeleeff's Periodic Law of the atomic weights of the chemical elements. *Chem. News* **58**, 93-95; 102-103 (**1888**) (table p. 94).

(4) W. Preyer. *Das genetische System der Elemente*. Berlin: R. Friedländer & Sohn, **1893** (separate table).

(5) W. F. Kemble and C. R. Underhill. *The Periodic Law and the hydrogen spectrum*. New York: D. Van Nostrand Co., **1909** (separate table).

CLASS 2. TABLES WITH DISPOSITION OF ELEMENTS: 2, 8, AND 18.

Type IIA2-1.

(00) D. Mendeleev. Sootnoshenie svoistv s atomnym vesom elementov. *Zh. Russk. Khim. Obshch.* **1**, 60-77 (**1869**) (fragment of table foot-note p. 70).
———— Die Beziehungen der Elemente und ihren Atomgewichten. *Ostwald's Klass. d. exakt. Wissensch.* **68**, 20-40 (**1895**) (foot-note p. 32).

(0) William Ramsay. *Modern Chemistry*. London: I. M. Deut & Co., V. I., 49-51 (**1900**).

(1) B. K. Emerson. Helix chemica. A study of the periodic relations of the elements and their graphic representation. *Am. Chem. J.* **45**, 160-210 (**1911**) (tables pp. 160/161, 162/163).
———— Concerning a new arrangement of the elements on a helix, and the relationships which may be usefully expressed thereon. *Science* **34**, 640-652 (**1911**) (table p. 644).
———— The helix chemica. *Chem. Rev.* **5**, 215-229 (**1928**) (table p. 216).

(2) M. Courtines. A model of periodic table. *J. Chem. Educ.* **2**, 107-109 (**1925**) (table p. 107).

(3) The Staff of the Museum of Science and Industry. *The periodic table of the elements*. Chicago: **1946**.

(4) Selig Hecht. *Explaining the atom*. New York: The Viking Press, **1947** (table p. 26, 27).
Cited by L. S. Foster. The periodic table and arrangement of extranuclear electrons. *J. Chem. Educ.* **26**, 283-285 (**1949**) (table p. 284).

Type IIB2-1.

(1) B. K. Emerson. Helix chemica. A study of the periodic relations of the elements and their graphic representation. *Am. Chem. J.* **45**, 160-210 (**1911**) (tables pp. 164/165, 174/175, 190/191 etc.).
———— Concerning a new arrangement of the elements on a helix, and the relationships which may be usefully expressed thereon. *Science* **34**, 640-652 (**1911**) (tables pp. 646, 647, 649, 650, 651).
———— The helix chemica. *Chem. Rev.* **5**, 215-229 (**1928**) (table p. 218).

(2) Jacob Kunz. On the present theory of magnetism and the Periodic System of chemical elements. *8th Intern. Congr. Appl. Chem.* **22**, 187-203 (**1912**) (table p. 192).

(3) John D. Clark. A new periodic chart. *J. Chem. Educ.* **10**, 675-677 (**1933**) (table p. 675).

(4) K. Gordon Irvin. A periodic arrangement of the elements to meet modern chemistry needs. *School Sci. Math.* **38**(6), 654-655 (**1938**) (table p. 655).
———— Periodicity paterns of the elements. *J. Chem. Educ.* **16**, 335-340 (**1939**) (table p. 335).

(5) Exhibition of Science at South Kensington (England).
Cited by A. J. Garratt. Physics in the festival of Britain. *Research* (London) **4**, 320-323 (**1951**) (table p. 320).

(6) S. I. Tomkeieff. The periodic table and its application. *Nature* **173**, 393-395 (**1954**) (tables pp. 394, 395).
———— *A new periodic table of the elements, based on the structure of the atom*. London: Chapman & Hall, **1954** (table p. 28).
———— A new approach to the periodic table. *Discovery* **15** (9), 375-378 (**1954**) (table p. 377).

_____ Atomic sizes and bond types in their relation to the Periodic System and to the structure of the atom. *Sci. Progress* **43**, 28-44 **(1955)** (table p. 31), **44**, 38 **(1956)**.

_____ The periodic table of the elements as a basis of geochemistry. *Ibidem* **46**, 46-62 **(1958)**.

Type IIC2-1.

(1) D. Mendeleev. Die periodische Gesetzmässigkeit der chemischen Elemente. *Ann. Chem., Justus Liebigs*, **Suppl. 8**, 133-229 **(1871)** (table p. 149).
Cited by Karl Seubert. *Ostwald's Klass. d. exakt. Wissensch.* **68**, 41-118 **(1895)** (table p. 54).
_____ *Osnovy Khimii*. St. Peterburg: 1st edit. V. II., **1871**; 2nd edit. V. I. **1872**.
_____ (no author's name) The chemistry of the future. *Quart. J. Sci.* **14**, [N.S.7] (55), 289-306 **(1877)** (table p. 304).
_____ La loi périodique des éléments chimiques. *Monit. Sci. (Docteur Quesneville)* **9**, 691-737 **(1879)** (tables pp. 692/693, 700).
_____ The Periodic Law of the chemical elements. *Chem. News* **40**, 231-232; 243-244; 255-256; 267-268; 279-280; 291-292; 303-304 **(1879)** (table p. 231, 268); **41**, 2-3; 27-28; 39-40; 49-50; 61-62; 71-72; 83-84; 93-94; 106-108; 113-114; 125-126 **(1880)**.

(2) James Walker. On the periodic tabulation of the elements. *Chem. News* **63**, 251-253 **(1891)**.

(3) Oscar Scarpa. Alcune rappresentazioni grafiche del sistema periodico degli elementi. *9. Congr. Intern. Quim. Pura Aplic.* (Madrid) **2**, 158-164 **(1934)** (table p. 162).

Subtype IIC2-1a.

(1) F. H. Loring. *Studies in valency*. London: Simpkin, Marshall, Hamilton, Kent & Co., **1913** (table p. 23).
_____ Is H composed of a whole number part (A) plus an auxiliary part (B) and a rotating electron (C)? *Chem. News* **121**, 315-318 **(1920)** (table p. 316).
_____ *Atomic theories*. London: Methuen & Co., **1923** (table p. 102).
_____ *The chemical elements*. Ibidem **1923** (table p. 14).

Type IIA2-2.

(1) Frederick Soddy. La table périodique des éléments. *Radium* **11**, 6-8 **(1914)** (table p. 8).
_____ *The chemistry of the radioelements*. II. The radioelements and the Periodic Law. London: Longmans, Green & Co., **1914** (table p. 11).

(2) William D. Harkins and R. E. Hall. The Periodic System and the properties of the elements. *J. Am. Chem. Soc.* **38**, 169-221 **(1916)** (table p. 192).
_____ Das periodische System und die Eigenschaften der Elemente. *Z. anorg. Chem.* **97**, 175-240 **(1916)** (table p. 206).

(3) Marc Chauvierre. Sur une nouvelle classification périodique des éléments chimiques. *Bull. Soc. Chim. France* [4] **25**, 297-305 **(1919)** (table p. 299).

Type IIB2-2.

(1) Friedrich Kipp. Über das periodische System der Elemente. *Naturwissenschaften* **30**, 679-683 **(1942)** (table p. 681).

Type IIC2-2.

(1) Geo Woodiwiss. The chemical elements. A new classification. *Chem. News* **93**, 214-215 **(1906)** (table p. 214).

(2) J. R. Rydberg. Untersuchungen über das System der Grundstoffe. *Lunds Univ. Årskr.* (*Acta Univ. Lundensis*) [N.S.] **9**(18), 1-41 (**1913**) (table p. 21).
_____ Recherches sur le système des éléments. *J. chim. phys.* **12**, 585-639 (**1914**) (table p. 610).

(3) W. M. Hicks. High-frequency spectra and the periodic table. *Phil. Mag.* [6] **28**, 139-142 (**1914**) (table p. 141).

(4) H. A. Geauque. A classification of the elements with respect to their properties. *J. Chem. Educ.* **2**, 464-466 (**1925**) (table p. 465).

(5) B. K. Emerson. The helix chemica. *Chem. Rev.* **5**, 215-229 (**1928**) (table p. 227).

(6) A. K. Dmitriev. Variant periodicheskoi sistemy elementov D. I. Mendeleeva. *Khimiia v shkole* **1**(3), 24-27 (**1937**) (tables pp. 25, 27).

(7) E. S. Sarkisov. Raschet mezhatomnykh rasstoianii inertnykh gazov v kristallakh. *Dokl. Akad. Nauk SSSR* [N.S.] **62**, 231-234 (**1948**) (table 233).
_____ O periodicheskoi sisteme elementov D. I. Mendeleeva. *Zh. Fiz. Khim.* **24**, 487-502 (**1950**) (table p. 500/501).

Subtype IIC2-2a.

(0) D. Mendeleev. Sootnoshenie svoistv s atomnym vesom elementov. *Zh. Russk. Khim. Obshch.* **1**, 60-77 (**1869**) (fragment of a table foot-note p. 70).

(1) R. M. Deeley. A new diagram and periodic table of the elements. *J. Chem. Soc.* (London) (Trans.) **63**, 852-867 (**1893**) (table p. 865).

(2) Arnaldo Piutti. Sopra una rappresentazione degli elementi chimici mediante punti nello spazio ordinario. *Atti R. Accad. Lincei* (Rome) [5] **22**, I, 569-575 (**1913**) (tables pp. 571, 574).
_____ A representation of the chemical elements by means of points in ordinary space. *Chem. News* **108**, 76-78 (**1913**) (table p. 77).
_____ Sur une mode de représentation des éléments chimiques au moyen de points dans l'espace ordinaire. *J. chim. phys.* **12**, 58-65 (**1914**) (tables pp. 60, 63).

Subtype IIC2-2b.

(0) Gustavus Hinrichs. On the classification and the atomic weights of the so-called chemical elements, with reference to Stas' determinations. *Proc. Am. Assoc. Advanc. Sci.* **18**, 112-124 (**1869**) (table p. 115).

(1) W. Spring. *Tableau représentant la loi périodique des éléments chimiques.* Liege: **1881** (table).
_____ Über eine Methode, das periodische Gesetz zu erläutern. *Ber. deut. chem. Ges.* **19**, 3092-3093 (**1886**) (no table).

(2) Paul Sabatier. Sur la classification des corps simples par la loi périodique. *Ann. fac. sci. Toulouse* **4**, B1-B14 (**1880**) (table p. B7).

(3) S. A. Shchukarev. Zakony D. I. Mendeleeva i periodicheskaia sistema. *20th Intern. Congr. Pure Appl. Chem.*, Report **D-33**, (Moscow) 80-81, **1965**.

Subtype IIB2-2c.

(1) E. Bindel und A. Blickle. *Zahlengesetze in der Stoffewelt und in der Erdenentwicklung.* Stuttgart: Freies Geistesleben, **1952**.

Type IIA2-3.

(1) S. I. Tomkeieff. *A new periodic table of the elements based on the structure of the atom.* London: Chapman & Hall, **1954** (tables p. 26 No. 3 and separate table).
———— A new approach to the periodic table. *Discovery* **15**(9), 375-378 (**1954**) (table p. 377).

Type IIB2-3.

(1) Giuseppe Oddo. La mia classificazzione periodica degli elementi e la constituzione elettrica degli atomi e delle valenza. *Gazz. chim. ital.* **55**, 149-174 (**1925**) (scheme p. 165).

Type IIC2-3.

(1) Thos. Carnelley. Suggestions as to the cause of the Periodic Law, and the nature of the chemical elements. *Chem. News* **53**, 157-159; 169-172; 183-186; 197-200 (**1886**) (table p. 198/199).

(2) V. von Richter. *Lehrbuch der anorganischen Chemie.* Bonn: M. Cohen & Sohn, 6. edit., **1889** (table p. 270).

(3) Walther Nernst. *Theoretische Chemie.* Stuttgart: F. Enke, **1893**.

(4) Theodore Williams Richards. A table of atomic weights. *Chem. News* **78**, 182-183; 193-195 (**1898**) (table p. 194).
———— The same title. *Am. Chem. J.* **20**, 543-554 (**1898**) (table p. 554).

(5) Stefan Meyer. Über die magnetischen Eigenschaften der Elemente. *Monatsh. Chem.* **20**, 369-382 (**1899**) (table p. 382/383).
———— Magnetisierungszahlen anorganischer Verbindungen. *Sitzungsber. Akad. Wiss. Wien, Math.-Naturw. Kl.* **108**, 861-898 (**1899**) (table p. 886/887).
———— Periodische Systeme der Elemente. *Physik. Z.* **19**, 178-179 (**1918**) (table p. 179).

(6) H. E. Roscoe and C. Schorlemmer. *A treatise on chemistry.* London, New York: Macmillan & Co., V. II, **1900** (table p. 48).

(7) George Rudorf. Vergleichende Studien im periodischen System. Die verschiedenen Verbindungsstufen der Elemente. *Z. anorg. Chem.* **37**, 177-198 (**1903**) (tables pp. 178, 180, 190).

(8) D. Mendeleev. *Osnovy Khimii.* St. Peterburg: 8th edit., **1906** (table p. 255).

(9) S. M. Losanitsch. Die Grenzen des periodischen Systems der chemischen Elemente. *Serb. Akad. Wiss.* (Belgrad) **1906** (table p. 13).
———— The limits of the Periodic System of the chemical elements. *Ion* (London) **1**, 259-274 (**1909**) (table p. 264).

(10) Sir William A. Tilden. *The elements.* London, New York: Harper & Brothers, **1900** (separate table).

(11) Arthur John Hopkins. The specific gravities of the elements considered in their relation to the Periodic System. *J. Am. Chem. Soc.* **33**, 1005-1027 (**1911**) (table p. 1025).

(12) B. K. Emerson. Concerning a new arrangement of the elements on a helix, and the relationships which may be usefully expressed thereon. *Science* **34**, 640-652 (**1911**) (table p. 643).

(13) C. I. Istrati et G. G. Longinescu. *Cours élémentaire de chimie.* Paris: Gauthier-Villars, **1913** (table p. 258).

(14) Ivan D. Margary. The periodic table. A modification more in accord with atomic structure. *Phil. Mag.* [6] **42**, 287-288 (**1921**) (table p. 288).

(15) John A. V. Butler. *The chemical elements and their compounds.* London: Macmillan Co., **1927** (table append. II).

(16) B. W. Nekrassow. Les propriétés des ions. *Bull. Soc. Chim. France* [5] **3**, 151-159 (**1936**) (table p. 153).

(17) Michele Ragno. Osservazioni al sistema periodico degli elementi. *Chimica* (Milan) **14**, 147-148 (**1938**) (table p. 147).

(18) S. I. Tomkeieff. *A new periodic table of the elements based on the structure of the atoms.* London: Chapman & Hall, **1954** (separate table).

Subtype IIC2-3a.

(0) Stefan Meyer. Observations on the Periodic System of the elements, and an attempt at classifying the radio-elements in the same. *Ion* (London) **1**, 249-259 (**1909**) (table p. 256).

(1) Andreas von Antropoff. *Cited* by Mark von Stackelberg. Der Einfluss der Gebiete des "inneren Aufbaus" der Atome im periodischen System auf die Ionenradien. *Z. physik. Chem.* **118**, 342-346 (**1925**) (table p. 343).
——— Eine neue Form des periodischen Systems der Elemente. *Z. angew. Chem.* **39**, 722-725; 725-728 (**1926**) (table p. 724, 725).
——— Zusammenhänge zwischen den physikalischen und chemischen Eigenschaften des Siliciums und seiner Stellung im periodischen System. *Z. Elektrochem.* **32**, 423-428 (**1926**) (table p. 425, 426).
——— A new form of the Periodic System of the elements. *J. Am. Chem. Soc.* **49**, 888-889 (**1927**) (table p. 888).

(2) Mark von Stackelberg. *Atlas der physikalischen und anorganischen Chemie.* Dissertation, Berlin: Verlag Chemie, **1929** (separate table).

(3) P. E. Cleator. *The periodic problem.* Wallasey, Cheshire (England): **1950**.

(4) Roderich Scheer. A variable periodic system. *J. Chem. Educ.* **32**, 590-591 (**1955**) (table p. 590).

(5) D. G. Cooper. *The periodic table.* London: Butterworths Sci. Publ., **1958** (table on the frontispiece).

Subtype IIC2-3b.

(1) V. K. Grigorovich. K voprosu o vliianii stroeniia vnutrennikh elektronnykh obolochek atomov na razmeshchenie elementov v periodicheskoi sisteme D. I. Mendeleeva. *Trudy Inst. Metall. Baikova, Gos. Kom. po chernoi i tsvetnoi metall., Akad. Nauk SSSR* **14**, 155-187 (**1963**) (table p. 184).
——— *Periodicheskii zakon i elektronnoe stroenie metallov.* Moskva: Nauka, **1966** (table p. 61).

Subtype IIC2-3c.

(1) F. W. Schenk. Zur Einordnung der Lanthaniden und Transurane in das periodische System der Elemente. *Angew. Chem.* **63**, 141-142 (**1951**) (table p. 142).

Type IIA2-4.

(1) Eduard von Stackelberg. Versuch einer neuen tabellarischen Gruppierung der Elemente auf Grund des periodischen Systems. *Z. physik. Chem.* **77**, 75-81 (**1911**).

Type IIB2-4.

(1) J. F. Tocher. Note on periodicity of properties of the elements: new arrangement. *Pharm. J.* **85**, or [4] **31**, 159-160 (**1910**) (table p. 160).

(2) William D. Harkins and R. E. Hall. The Periodic System and the properties of the elements. *J. Am. Chem. Soc.* **38**, 169-221 (**1916**) (table p. 189).
———— Das periodische System und die Eigenschaften der Elemente. *Z. anorg. Chem.* **97**, 175-240; 336 (**1916**) (table p. 200).

(3) Karl Hack. *Das natürliche System der Elemente in Form der eutropischen Spirale.* Würzburg: **1926**.
———— Eutropisches System der Elemente. *Standesz. deut. Apoth. (Deut. Apotheke)* **3**, 6-9 (**1934**) (table p. 7).

(4) Darvin O. Lyon. *Das periodische System in neuer Anordnung.* Leinzig, Wien: Deuticke, **1928** (table p. 36).

(5) A. E. Caswell. A new graphical arrangement of the periodic table. *Phys. Rev.* [2] **34**, 543 (**1929**).

(6) Iu. Ia. Bilibin. Ob odnom sposobe izobrazheniia Mendeleevskoi tablitsy. *Izv. Akad. Nauk SSSR, otd. geol.* **1939** (5), 172-175 (table p. 173).

(7) Edgar I. Emerson. A new spiral form of the periodic table. *J. Chem. Educ.* **21**, 111-115 (**1944**) (table p. 112).

(8) Gerhard Hübner. Das Periodensystem im Chemie-Unterricht. *Chem.-Ztg.* **75**, 63-65 (**1951**) (table p. 63).

(9) V. P. Sokoloff. Diagrams relating the periodic table to geochemistry. *J. Chem. Educ.* **31**, 15-17 (**1954**) (table p. 16).

Type IIC2-4.

(1) D. Mendeleev. Sootnoshenie svoistv s atomnym vesom elementov. *Zh. Russk. Khim. Obshch.* **1**, 60-77 (**1869**) (fragment of the table foot-note p. 69).
———— Die Beziehungen zwischen den Eigenschaften der Elemente und ihren Atomgewichten. *Ostwald's Klass. d. exakt. Wissensch.* **68**, 20-40 (**1895**) (foot-note p. 31).

(2) J. A. Groshans. Sur des points de contact entre la loi des périods de M. Mendelejeff et la loi des nombres de densité. *Recueil Trav. Chim. Pay-Bas* **3**, 310-330 (**1884**) (table p. 311).

(3) A. Horstmann. 2. Theoretische Chemie. I. A. Horstmann, H. Landolt. A. Winkelmann. Lehrbuch der physikalischen und theoretischen Chemie. *Graham-Otto. Lehrbuch der Chemie.* Braunschweig: F. Vieweg & Sohn, 164-165, **1885** (table p. 162/163).

(4) B. N. Chicherin. *Sistema khimicheskikh elementov.* Peterburg: P. I. **1888** (table p. 17).
———— The same title. *Zh. Russk. Fiz.-Khim. Obshch.*, **Suppl. to 20**, 1-26 (**1888**).

(5) Charles Skeele Palmer. The nature of the chemical elements. *Proc. Colo. Sci. Soc.* **3**, 287-307 (**1888-1890**) (table p. 296); **4**, 56-74 (**1891-1893**) (table p. 69).
———— The same title. Argon and helium in the periodic sequence. *Ibidem*, **1897**.

(6) P. J. F. Rang. The periodic arrangement of the elements. *Chem. News* **67**, 178 (**1893**).
—— The period-table. *Ibidem* **72**, 200-201 (**1895**) (table p. 201).

(7) Walther Nernst. *Theoretische Chemie.* Stuttgart: F. Enke, **1893**.

(8) James Walker. *Introduction to physical chemistry.* London, New York: Macmillan & Co., **1899** (table p. 44).

(9) Iu. M. Radik. O preryvnoi periodicheskoi sisteme khimicheskikh elementov. *Zh. Russk. Fiz.-Khim. Obshch.* **33**(3), 195-196 (**1901**).
Cited by V. I. Semishin. Rabota Iu. M. Radika po Periodicheskoi Sisteme elementov (tezisy). *Nauka v Pribaltike v XVIII—nachale XX veka,* Akad. Nauk Latv. SSR, Riga, **1962**, 91-93.

(10) H. Staigmüller. Das periodische System der Elemente. *Z. physik. Chem.* **39**, 245-248 (**1901**) (table p. 247).

(11) James Monckman. On a natural system of arranging the chemical elements, in which they fall into the periodic groups, based solely upon the atomic volumes and the combining weights. *Chem. News* **95**, 5-9 (**1907**) (table p. 6, 8).

(12) Eduard von Stackelberg. Versuch einer neuen tabellarischen Gruppierung der Elemente auf Grund des periodischen Systems. *Z. physik. Chem.* **77**, 75-81 (**1911**) (table p. 78, 79).

(13) B. K. Emerson. Helix chemica. A study of the periodic relations of the elements and their graphic representation. *Am. Chem. J.* **45**, 160-210 (**1911**) (table pp. 173, 180, 183).

(14) Toribio Cáceres. La clasificación de los elementos. *An. Soc. Españ. Fis. Quim.* **9**, 121-124 (**1911**) (table p. 124).

(15) Ceka Emu. *Die chemischen Elemente.* Linburg a./L. **1912**.

(16) P. P. fon Veimarn. Graficheskie izobrazheniia zavisimosti mezhdu atomnym vesom i skorost'iu dvizheniia atomov elementov pri ikh temperaturakh plavl'eniia i neposredstvennyi vyvod estestvennoi sistemy elementov iz etikh grafikov. *Zh. Russk. Fiz.-Khim. Obshch.* **47**, 481-489 (**1915**) (table p. 488/489).

(17) Marc Chauvierre. Sur une nouvelle classification périodique des éléments chimiques. *Bull. Soc. Chim. France* [4] **25**, 297-305 (**1919**) (table p. 300).

(18) Alfred W. Stewart. *Recent advances in physical and inorganic chemistry.* London: Longmans, Green & Co., 3. edit., **1919** (table p. 269).
—— The classification of the atoms. *Scientia* **37**, 373-382 (**1925**) (table p. 379).

(19) Paul Pfeiffer. Die Befruchtung der Chemie durch die Röntgenstrahlenphysik. *Naturwissenschaften* **8**, 984-991 (**1920**) (table p. 991).
—— Über die Stellung von Beryllium und Magnesium im periodischen System der Elemente. *Z. angew. Chem.* **37**, 41 (**1924**).

(20) R. G. W. Norrish. Transitional elements and the octet theory. *Chem. News* **124**, 16-22 (**1922**) (table p. 17).

(21) Fritz Paneth. Über die heutige Schreibweise des periodischen Systems der Elemente. *Z. angew. Chem.* **36**, 407-410 (**1923**) (table p. 409).
—— *Radio-elements as indicators.* New York, London: McGraw-Hill Book Co., **1928** (table p. 129).
—— Die Entwicklung und der heutige Stand unserer Kenntnisse über das natürliche System der Elemente. *Naturwissenschaften* **18**, 964-976 (**1930**) (table p. 965).
—— Radioactivity and the completion of the Periodic System. *Nature* **149**, 565-568 (**1942**) (table p. 565).

(22) Constantin G. Bedreag. Système physique des éléments. *Compt. rend.* (Paris) **179.** II, 766-768 (**1924**) (table p. 767, 768); **180,** 653-655 (**1925**) (no table).
———— The same title (in French). *Bull. Acad. Roumaine, Sect. Sci.* **9,** 8-14; 158-164 (**1925**).
———— Le système physique des éléments. *Ann. sci. Univ. Jassy* **13,** 62-76; 315-345 (**1926**) (tables p. 65, 318).
———— Le groupement physique des éléments (in French). *Bull. fac. stiint. Cernauti* **1,** 14-32 (**1927**) (table p. 16, 17).
———— Sur la configuration électronic des éléments. *Ibidem* **2,** 44-64 (**1928**) (table 50).
———— Systématique naturelle des éléments. *Ibidem* **6,** 197-203 (**1932**) (tables pp. 199, 200, 202).
———— The same title. *Bull. Acad. Roumaine, Sect. Sci.* **16,** 27-32 (**1932**).
———— Système physique des éléments. *Compt. rend.* (Paris) **197.** II, 838-840 (**1933**) (tables pp. 839, 840).
———— La place des protons et des neutrons dans la systématique naturelle des éléments (in French). *Bull. fac. stiint. Cernauti* **8,** 160-166 (**1934**) (table p. 161).
———— Place des uranides 93, 94 dans la systématique naturelle des éléments. *Compt. rend.* (Paris) **215,** 537-539 (**1942**) (table p. 539).
———— Structure et systématique naturelle des éléments. *Ann. sci. Univ. Jassy* **28,** 143-148 (**1942**) (table p. 146).
———— Die Systematik der Elektronenhülle der Elemente. *Naturwissenschaften* **31,** 490 (**1943**).
———— Die natürliche elektronische Systematik der Elemente (in German). *Bull. Acad. Roumaine, Sect. Sci.* **25,** 407-409 (**1943**) (table p. 408).
———— Système harmonique des éléments (in French). *Bul. Inst. Politechn. "Gh. Asachi", Iasi (Bull. de l'ecole polytechn. de Jassy)* **3,** 317-343 (**1948**) (table p. 320).
———— Periodic table of elements from 1 to 100. *Acad. Rep. Populare Romîne, Jaşi, Studii Cercet. Stiint.* 3(1-4), 71-81 (**1952**).
———— Représentation physique de la systématique des éléments. *Compt. rend.* (Paris) **252,** 1604-1606 (**1961**) (table p. 1606).
———— Physical presentation of the system of elements. II. *Comm. Acad. Rep. Populare Romîne* 12(2), 149-153 (**1962**).
———— Physical systematics of the elements (in English). *Anales stiint. Univ. "A. I. Cuza", Iaşi,* Sect. I, 8(2), 451-463 (**1962**).

(23) Edgar T. Wherry. Further notes on atomic volume isomorphism. *Am. Mineral.* **9,** 165-169 (**1924**) (table p. 168).

(24) Wilhelm Prandtl und Albert Grimm. Über die Aufsuchung des Elements No. 61. *Z. anorg. allgem. Chem.* **136,** 283-288 (**1924**) (table p. 238).

(25) Worth H. Rodebush. A compact arrangement of the Periodic System. *J. Chem. Educ.* **2,** 381-383 (**1925**) (table p. 382).

(26) Richard Swinne. Das periodische System der chemischen Elemente im Lichte des Atombaus. *Z. techn. Physik* **7,** 166-180; 205-216 (**1926**) (table p. 209).
———— Periodicheskaia sistema khimicheskikh elementov v svete teorii stroeniia atoma. *Uspekhi Fiz. Nauk* **6**(4/5), 330-374 (**1926**).

(27) Karl Mahler. *Atombau und periodisches System der Elemente.* Berlin: O. Salle, **1927** (table p. 118).

(28) Darvin O. Lyon. *Das periodische System in neuer Anordnung.* Leipzig & Wien: F. Deuticke, **1928** (table p. 19).

(29) Eugen Rabinowitsch und Erich Thilo. *Periodisches System, Geschichte und Theorie.* Stuttgart: F. Enke, **1930** (table p. 246).

(30) W. Walter Meissner. *Chemischer Handatlas.* Braunschweig: G. Westermann, **1931** (table p. 2).

(31) C. H. Douglas Clark. *The electronic structure and properties of matter; an introductory study of certain properties of matter in the light of atomic numbers.* London: Chapman & Hall, **1934** (table p. 6).

(32) Harvey Elliott White. *Introduction to atomic spectra.* New York, London: McGraw-Hill Book Co., **1934** (table p. 85).

(33) Ida Noddack. Das periodische System der Elemente und seine Lücken. *Angew. Chem.* **47**, 301-305 (**1934**) (table p. 301).
———— The same title. *Uspekhi Khim.* **4**, 11-21 (**1935**).

(34) A. E. Fersman. Periodicheskii zakon Mendeleeva v geokhimii. *Trudy Iubilein. Mendeleev. s'ezda,* Moskva. *Akad. Nauk SSSR,* V. I. **1936**, 383-417 (in Russian); 419-454 (in French) (table p. 388/389).

(35) Walter Noddack. Über den Ausbau des periodischen Systems. *Trudy Iubil. Mendeleev. s'ezda,* Moskva, *Akad. Nauk SSSR,* V. II, **1937**, 53-59 (in German); 61-67 (in Russian) (table p. 55).

(36) Laurence S. Foster. The advantages of the "long form" of the periodic table. *Rep. New Engl. Assoc. Teachers* **39**, 23-30 (**1937**) (table p. 25).
———— Why not modernize the textbooks also? I. The periodic table. *J. Chem. Educ.* **16**, 409-412 (**1939**) (table p. 410).
———— Periodic classification of the elements. *Ibidem* **23**, 602-603 (**1946**).

(37) Michele Ragno. Osservazioni al sistema periodico degli elementi. *Chimica* (Milan) **14**, 147-148 (**1938**) (table p. 147).

(38) H. G. S. Snijder. Een nieuve uitgave van het periodiek systeem. *Chem. Weekblad* **36**, 676-677 (**1939**) (table p. 676).
———— *Periodiek Systeem der Elementen met Toelichtingen.* Gröningen: R. Nordhoff, **1939**.

(39) Georges Chaudron. La classification périodique des éléments. *Métaux et Corrosion* **15**, 86-93 (**1940**) (table p. 92).

(40) G. B. Bokii. Raspredelenie elementov po podgruppam na osnove kristallokhimicheskikh predstavlenii. *Priroda* (Moscow) **31**, 38-47 (**1942**).

(41) Thomas H. Hazlehurst and Frank J. Fornoff. Representation of periodic properties of the elements. *J. Chem. Educ.* **20**, 77-79 (**1943**).

(42) Raymond Daudel. Mouvement scientifique. Les agrégats atomiques. *Rev. sci.* (Paris) **81**, 397-408 (**1943**) (table p. 406/407).

(43) Paul Renaud. Répartition des points de fusion dans le tableau de Mendeléeff. *Bull. Soc. Chim. France* [5] **12**, 1060-1062 (**1945**) (table p. 1061).

(44) Glenn T. Seaborg. The chemical and radioactive properties of the heavy elements. *Chem. Eng. News* **23**, 2190-2193 (**1945**) (table p. 2191).
———— The transuranium elements. *Science* **104**, 379-386 (**1946**) (table p. 385).
———— Place in periodic system and electronic structure of the heaviest elements. *Nucleonics* **5**(5), 16-36 (**1949**) (table p. 30/31).
———— The transuranium elements. *Endeavour* **18**, 5-13 (**1959**) (table p. 6).
———— Recent work with transuranium elements. *Proc. Natl. Acad. Sci.* **45**(4), 471-482 (**1959**).
G. Siborg. Evoliutsiia Periodicheskoi Sistemy elementov so vremen D. I. Mendeleeva do

nashikh dnei. *Sto let per. zak. khim. elem.* Moskva: Nauka, **1969**, 136-157 (tables pp. 139, 140, 141, 145, 149).

———— Rasshirenie predelov Periodicheskoi Sistemy. *Sto let per. zak. khim. elem. Dokl. plenarn. zased. Iubil. Mendeleev. s'ezda, 10, 1969.* Moskva: Nauka 21-39, **1941** (table p. 22).

(45) Ia. K. Syrkin i M. E. Diatkina. *Khimicheskaia sviaz' i stroenie molekul.* **1946.**

———— *Structure of molecules and the chemical bond.* New York, London: Interscience Publ. Co., **1950** (table pp. 34, 35).

(46) F. Oberhauser B. Sistema periódico o sistema natural de los elementos químicos. *Anales Fac. Filos. y Educ., Univ. Chile, Secc. de quim.* 3, 131-154 (**1946**) (table pp. 149, 150).

(47) J. G. Ryss. Obshchie svoistva perekhodnykh elementov. *Uspekhi Khim.* **17**, 372-388 (**1948**) (table p. 387).

(48) C. C. Addison. The grouping of the elements. *Chemistry & Industry* **1948**, 227-229 (table p. 227).

(49) A. N. Wrigley, W. C. Mast, and T. P. McCutcheon. A laminar form of the periodic table. *J. Chem. Educ.* **26**, 216-218; 248-250 (**1949**).

(50) William Q. Hull. The transuranium elements. *Chem. Eng. News* **30**, 232-237 (**1952**) (table p. 232).

(51) Carlos Lopez-Bustos. The chemical elements and their biological functions (in Spanish). *Farmacia Nueva* (Madrid) **17**, 541-546; 593-602 (**1952**).

———— Die chemischen Elemente und deren biologische Functionen. *Mitt. chem. Forschungsinst. Wirt Österr.* **8**, 7-11 (**1954**) (table p. 7).

(52) V. P. Sokoloff. Diagrams relating the periodic table to geochemistry. *J. Chem. Educ.* **31**, 15-17 (**1954**) (table p. 16).

(53) Milton T. Heald. A periodic table of elements for geologists. *J. Geol. Educ.* **2**, 19-23 (**1954**).

(54) B. A. Fickers. The chemist's "Magna Charta". *Sci. Couns.* **18**(1), 11; 28 (**1955**) (table p. 11).

(55) George K. Estok. An adjustable periodic chart for lecture purposes. *J. Chem. Educ.* **33**, 618-619 (**1956**).

(56) Gilbert Gordon. *The chemical elements and their isotopes.* **1960** (table p. 10/11).

(57) J. P. Redfern and J. E. Salmon. Periodic classification of the elements. *J. Chem. Educ.* **39**, 41 (**1962**).

(58) George T. Austin and Helen F. Austin. The Periodic Law correlates properties of the elements. *Chem. Eng.* (London) **70**(26), 87-92 (**1963**) (table p. 88).

(59) G. E. Villar. La estructura electronica y la posición de los elementos en la clasificación periódica. *Boletin Fac. Ing. y Agrimens., Univ. Montevideo* **8**(12), 493-510 (**1964**) (table p. 499).

———— A suggested modification to the periodic chart. *J. Inorg. Nucl. Chem.* **28**(1), 25-29 (**1966**) (table p. 27).

———— Las configuraciónes electronicas y la hibridización de los orbitales en los elementos de transición. *Boletin Fac. Ingen. y Agrimens., Montevideo,* **10**(3), 49-67 (**1967**) (tables pp. 50, 57).

———— El centenario de la tabla periódica. Estado actual, posibilidades de desarrollo y algunos ajustes propuestos. *Revista Real Acad. Cienc. Exact., Fiz. Natural.* (Madrid) **63**(3), 517-532 (**1969**) (table p. 529).

(60) Pai Yen Loung. *Graphic handbook of chemistry and metallurgy*. New York: Chem. Publ. Co., **1965** (table II).

(61) David Abbott. *An introduction to the periodic table*. London: J. M. Dent & Sons, **1966** (table p. 8).

(62) Eduard Pachmann. Periodická soustava prvků jako učební pomůcke na všcobecně vzdělávaeí škole. *Chemické Listy* (Prague) **60**(10), 1385-1390 (**1966**) (table p. 1387).

(63) Gerhardt Heist. Bau einer Klaptafel "Periodensystem der Elemente". *Chem. Schule* **16**(12), 435-437 (**1969**) (table p. 436).

(64) Ana Maria Mojos, Hector Vicente Incolla. Ayer, hoy y mañana en la tabla periódica de los elementos químicos. *Saneamiento* (Buenos Aires) **33**(215), 265-268 (**1969**) (table p. 268).

(65) William M. Allen. The diagonal periodic relationship. *Chemistry* (Washington) **43**(4), 22-24 (**1970**).

(66) Dieter Nebel. Zur Position der Aktiniden- und Transaktinidenelemente im Periodensystem der Elemente. *Z. f. Chem.* **10**(7), 251-260 (**1970**) (table p. 257).

(67) K. V. Astakhov. Periodicheskii zakon D. I. Mendeleeva v svete sovremennykh nauchnykh dannykh. *Zh. Vses. Khim. Obshch.* **16**(3), 251-260 (**1971**) (table p. 255).

Subtype IIC2-4a.

(0) R. G. W. Norrish. Transitional elements and the octet theory. *Chem. News* **124**, 16-22 (**1922**) (table p. 17).

(1) V. K. Grigorovich. K voprosu o vliianii stroeniia vnutrennikh elektronnykh obolochek atomov na razmeshchenie elementov v Periodicheskoi Sisteme D. I. Mendeleeva. *Trudy Inst. Metall. Baikova, Gos. Kom. po chernoi i tsvetnoi metall.*, Akad. Nauk SSSR **14**, 155-187 (**1963**) (table p. 177/178).
———— *Periodicheskii zakon i elektronnoe stroenie metallov*. Moskva: Nauka, **1966** (separate table p. 40/41 and p. 65).

(2) Pai Yen Loung. *The electronic evolution in the atoms of the elements and the construction of a new periodic table*. Palo Alto, Calif.: Periodex Sci. Co., **1965**.

Type IIC2-5.

(1) Curt Schmidt. Periodisches System und Genesis der Elemente. *Z. anorg. allgem. Chem.* **103**, 79-118 (**1918**) (table p. 93).

(2) H. G. Grimm. Periodisches System der Atomionen. *Z. physik. Chem.* **101**, 410-413 (**1922**) (table p. 411).

(3) F. P. Worley. A new view of atomic structure and relationship of the chemical elements. *Rep. Austral. Ass. Adv. Sci.* **16**, 212-219 (**1923**) (table p. 219).

(4) Heinrich Remy, *Lehrbuch der anorganischen Chemie*. Leipzig: Akademische Verlagsgesellschaft, Bd. I, **1931** (table p. 105).

(5) Karl Erik Zimens. Periodiska systemet av i dag. *Festskr. tillägnad J. Arvid Hedvall* (Göteborg) 635-650, **1948** (table p. 649).
———— The same title. *Chalmers Tekn. Högsk. Handl.* (Göteborg) **78**, 3-18 (**1948**) (table p. 17).

(6) N. V. Agaev. Periodicheskii zakon Mendeleeva i metallicheskie splavy. *Izv. sekt. Fiz.-Khim. Anal., Inst. Obshch. Neorg. Khim., Akad. Nauk SSSR*, **19**, 97-102 (**1949**).

(7) Reinh. Walter. Neuere Entwicklung des periodischen Systems. *Chem.-Ztg.* **74**, 673-676 (**1950**) (table p. 674 Fig. 3).

(8) Samuel Ruben. *The electronics of materials.* Indianopolis: Bobbs-Merrill Co., **1964** (table p. 25).
_____ *The elements.* Indianopolis: H. W. Sams, **1965** (table on the last page).

Subtype IIC2-5a.

(1) A. van den Broek. Das Mendelejeffsche "kubische" periodische System und die Einordnung der Radioelemente in dieses System. *Physik. Z.* **12**, 490-497 (**1911**) (table p. 495).

(2) Giuseppe Oddo. Nuova classificazione periodica degli elementi. *Gazz. chim. ital.* **50.** II, 213-245 (**1920**) (separate table).
_____ La mia classificazione periodica degli elementi e la constituzione elettrica degli atomi e della valenza. *Ibidem* **55**, 149-174 (**1925**) (table p. 152/153, 164/165).
_____ Sulla mia classificazione periodica degli elementi. *Ibidem* **61**, 694-698 (**1931**) (table p. 696/697).

(3) P. Vinassa de Regny. Il "numero elettronico" ed i constituenti terrestri. *Atti accad. naz. Lincei* (Rome) [6] **5**, 940-945 (**1927**) (table p. 943).

Subtype IIC2-5b.

(1) G. F. Horsley. 1893. Cited by George Rudorf. *The periodic classification and the problem of chemical evolution.* London: Whittaker & Co., 152-153, **1900** (table p. 153).
_____ *Das periodische System, seine Geschichte und Bedeutung für die chemische Systematik.* Hamburg, Leipzig: Voss, **1904** (table p. 239).

(2) C. Cuthbertson and E. Parr Metcalfe (on suggestion of A. W. Porter). On the refractive indices of gaseous K, Zn, Cd, Hg, As, Se, and Te. *Phil. Trans. Roy. Soc.* (London) **A 207**, 135-148 (**1908**) (table p. 147).

(3) Eduard von Stackelberg. Versuch einer neuen tabellarischen Gruppierung der Elemente auf Grund des periodischen Systems. *Z. physik. Chem.* **77**, 75-81 (**1911**) (table p. 81).

(4) Stefan Meyer. Periodische Systeme der Elemente. *Physik. Z.* **19**, 178-179 (**1918**) (table p. 179).

(5) M. Courtines. A model of periodic table. *J. Chem. Educ.* **2**, 107-109 (**1925**) (table p. 108).

(6) Charles A. Kraus. *Contemporary developments in chemistry. Radicals as chemical individuals.* New York: Columbia Univ. Press, **1927** (table p. 3).

(7) Ernst Sommerfeldt. Beziehungen der Kristallgitter zur Atomchemie für das periodische System der Elemente. *Zentrbl. Mineral., Geol., Paläont.,* **Abt. A. 1934**, 33-46 (table p. 43).

(8) A. E. Fersman. Periodicheskii zakon Mendeleeva v geokhimii. *Trudy Iubil. Mendel. s'ezda,* Moskva, *Akad. Nauk SSSR,* **1936**, V. I., 383-417 (in Russian); 419-454 (in French) (table p. 387).

(9) Zoltan G. Szabó and Béla Lakatos. A periodic system based on the fine distribution of the electrons. *Research* (London) **5**, 590-592 (**1952**) (table p. 591).
_____ A new form of the Periodic System and new periodic functions. *Acta Chim. Acad.*

Sci Hung. 4(2-4), 129-149 (**1954**) (tables pp. 133, 134, and separate table 134/135).

——— A periodic table and new periodic functions. *J. Chem. Educ.* **34**, 423-432 (**1957**) (table p. 430).

Z. G. Sabo. Periodicheskaia Sistema i periodicheskie funktsii. *Sto let per. zak. khim. elem.* Moskva: Nauka, **1969**, 244-255 (scheme of table p. 248/249).

Type IIC2-6.

(00) William Odling. On the proportional numbers of the elements. *Quart. J. Sci.* **I**, 642-648 (**1864**) (table p. 643).

(0) Lothar Meyer. **1868**, not published. *Cited* by Karl Seubert. Zur Geschichte des periodischen Systems. *Z. anorg. Chem.* **9**, 334-338 (**1895**) (table p. 336/337).

——— Das natürliche System der chemischen Elemente. *Ostwald's Klass. d. exakt. Wissensch.* **68**, 6-7 (**1895**).

(1) D. Mendeleev. Sootnoshenie svoistv s atomnym vesom elementov. *Zh. Russk. Khim. Obshch.* **1**, 60-77 (**1869**) (table p. 70).

——— Über die Beziehungen der Eigenschaften zu den Atomgewichten der Elemente. *Z. f. Chem.* **12**, or [2] **5**, 405-406 (**1869**) (table p. 405).

——— Versuche eines Systems der Elemente nach ihren Atomgewichten und chemischen Funktionen. *J. prakt. Chem.* [1] **106**, 251 (**1869**).

——— The same subject. *Osnovy Khimii*. St. Peterburg: V. I., 1st edit., **1869** (table p. IV). *Cited* by Karl Seubert. *Ostwald's Klass. d. exakt. Wissensch.* **68**, 18-19; 20-40 (**1895**) (table p. 33).

Cited by B. N. Menshutkin. Early history of Mendeleeff's Periodic Law. *Nature* **133**, 946 (**1934**).

(2) T. Silbermann. Das Gesetz der Periodizität der Elemente und das natürliche periodische System. *Ber. deut. chem. Ges.* **49**, 2219-2222 (**1916**).

(3) Darvin O. Lyon. *Das periodische System in neuer Anordnung.* Leipzig, Wien: F. Deuticke, **1928** (table pp. 17, 18).

(4) Robert F. Bacher and Samuel A. Goudsmit. *Atomic energy states.* New York: McGraw-Hill, **1932** (table p. XIII).

(5) Émile Carrière et Henri Guiter. Remarques sur la classification périodique des éléments suggérées par des preparations de sels par voie sèche on humide. *Bull. Soc. Chim. France* [5] **10**, 259-261 (**1943**).

(6) Paul Pascal. *Chimie générale.* Paris: Masson, **1949** (table p. 16).

Type IIC2-7.

(0) Geo Woodiwiss. The chemical elements. A new classification. *Chem. News* **93**, 214-215 (**1906**) (table p. 215).

(1) Egon Wiberg. Zur Systematik der chemischen Elemente im Unterricht. *Angew. Chem.* **49**, 480-481 (**1936**) (table p. 481).

Subtype IIC2-7a.

(1) Ingo W. D. Hackh. *Das synthetische System der Atome.* Hamburg: Hephaestos, **1914** (table pp. 9, 10).

——— Synthetic system of the elements. *Weltwissen* 3(9), 63 (**1915**).

——— Arc and spark spectra and Periodic System. *Astrophys. J.* **48**, 241-255 (**1918**) (table p. 245).
——— A modification of the periodic table. *Am. J. Sci.* **46**, 481-501 (**1918**) (table p. 489).
——— A new table of the Periodic System. *J. Am. Chem. Soc.* **40**, 1023-1026 (**1918**) (table p. 1025).
——— *The romance of the chemical elements*. Lancaster, Pa.: The New Era Printing Co., **1918** (table p. 2).
——— The same title. *Am. J. Pharm.* **90**, 478-492; 565-579 (**1918**).
——— The classification of the elements. *Sci. Amer. Suppl.* **87**, 146-149 (**1919**) (table p. 148).
——— The new periodic table and atomic structure. *Chem. News* **139**, 275-278 (**1929**) (table pp. 275, 276).
——— *A chemical dictionary*. Philadelphia: Blakiston Co., **1929** (table p. 545).

(2) Gerhard Hübner. Das Periodensystem im Chemieunterricht. *Chem.-Ztg.* **75**, 63-65 (1951) (table p. 63); **77**, 574-575 (**1953**) (table p. 575).

DIVISION III. LONG TABLES

SUBDIVISION IIIA: CHEMICAL TABLES

CLASS 1. TABLES OF ONE REVOLUTION AND OF ONE ROW

Type IIIB1-1.

(0) E. Loew. Versuch einer graphischen Darstellung für das periodische System der Elemente. *Z. physik. Chem.* **23**, 1-12 (**1897**) (table p. 3).

(1) Nicolas Opolonick. Chemical elements and their atomic numbers as points on a spiral. *J. Chem. Educ.* **12**, 265-267 (**1935**) (table pp. 266, 267).

Type IIIC1-1.

(1) M. Carey Lea. On the color relations of atoms, ions and molecules. *Am. J. Sci.* [5] **49**, 357-374 (**1895**) (separate table p. 360/361).

(2) A. von Antropoff. Les formes usuelles du système périodique des éléments. *Ann. Guébhard-Séverine* **13**, 161-174 (**1937**) (table p. 163).

Subtype IIIC1-1a.

(1) D. Mendeleev. The Periodic Law of the chemical elements (Faraday lecture on June 4, 1889). *J. Chem. Soc.* (London) [Trans.] **55**, 634-656 (**1889**) (separate table p. 656/657).
——— Periodicheskaia zakonnost' khimicheskikh elementov. *Zh. Russk. Fiz.-Khim. Obshch.* **21**, 233-257 (**1889**).
——— La loi périodique des éléments chimiques. *Monit. Sci.* [4] 3(572), 899-904 (**1889**).
——— The same subject. *Osnovy Khimii*. St. Peterburg: 5th. edit. **1889** (table p. 464/465).
——— *Grundlagen der Chemie*. St. Petersburg: C. Ricker, 666-703, **1892** (table p. 684/685).
——— *The principles of chemistry*. London, New York: Longmans, Green & Co., **1897** (separate table, V. II).

(2) F. H. Loring. *Studies in valency*. London: Simpkin, Marshall, Hamilton, Kent & Co., **1913** (table p. 17).
——— Valency and atomic number numeric. *Chem. Products* **5**, 21-23; 39-41 (**1942**) (table p. 22).

(3) Alois Bilecki. *Gedanken über das periodische System der chemischen Elemente.* Troppau: Buchholz & Diebel, **1915**.

(4) Ingo W. D. Hackh. A new table of Periodic System. *J. Am. Chem. Soc.* **40**, 1023-1026 (**1918**) (table p. 1024).

———— The classification of the chemical elements. *Sci. Amer. Suppl.* **87**, 146-149 (**1919**) (table p. 147).

(5) Vladimir Karapetoff. A chart of consecutive sets of electronic orbits within of chemical elements. *J. Franklin Inst.* **210**, 609-624 (**1930**) (separate table.)

(6) Chin Fang Hsueh and Ming Chien Chiang. Periodic properties of elements. *J. Chinese Chem. Soc.* **5**, 253-275 (**1937**) (table p. 254).

(7) Émile Rinck et Pierre Feschotte. Classification périodique et loi de Proust. *Bull. Soc. Chim. France* **1962**(4), 856-862 (table p. 859).

Subtype IIIB1-1b.

(1) K. Stoye. Eine neue Darstellung des Periodensystems. *Chem. Schule* **1**(1), 32-34 (**1954**).

Subtype IIIC1-1b.

(1) H. P. Vincent. A graphic form of the periodic chart. *Can. Chem. Educ.* **5**(1), 9-11 (**1969**) (table p. 10).

———— Further investigations on the graphic arrangement of the periodic chart. *Ibidem* **7**(1), 14-18 (**1971**).

CLASS 2. TABLES WITH DISPOSITION OF ELEMENTS: 4, 16, 36, AND 64 (CYCLES).

Type IIIA2-1.

(1) A. F. Kapustinskii. Periodichnost' v stroenii elektronnykh obolochek i iader atomov. I. Periodicheskaia sistema khimicheskikh elementov i ee sviaz' s teoriei chisel i s fiziko-khimicheskim analizom. *Izv. Akad. Nauk SSSR* **1953**, 3-11; 12-20 (table pp. 9, 10).

(2) Ernst Bindel. *Die geistigen Grundlagen der Zahlen.* Stuttgart: Freies Geistesleben, **1958**.

Type IIIB2-1.

(1) J. R. Rydberg. Untersuchungen über das System der Grundstoffe. *Lunds Univ. Årsskrift (Acta Univers. Lundensis)* [N.S.] **9**(18), 1-41 (**1913**) (table p. 23).

———— Recherches sur le système des éléments. *J. chim. phys.* **12**, 585-639 (**1914**) (table p. 611).

———— Le système des éléments chimiques. *Rev. gén. sci.* **25**, 734-741 (**1914**) (table p. 739). Cited by W. Pauli. Rydberg and the Periodic System of the elements. *Lunds Univ. Årsskrift, Kgl. Fysiograf. Sällskap. Handl.* **65** (21), 22-26 (**1954/55**) (table p. 25).

———— *Proc. of the Rydberg centen. conf. on atom. spectrosc., 1954.* Lund: C. W. K. Gleerup, 22-26, **1955** (table p. 25).

Type IIIC2-1.

(1) Edward G. Mazurs, **1965**. Not published.

Type IIIA2-2.

(1) Alois Bilecki. *Gedanken über das periodische System der chemischen Elemente.* Troppau: Buchholz & Diebel, **1915** (separate table).

CLASS 3. TABLES WITH DISPOSITION OF ELEMENTS: 2, 8, 18, AND 32 (PERIODS).

Type IIIA3-1.

(1) Hugo Stintzing. Eine neue Anordnung des periodischen Systems der Elemente. *Z. physik. Chem.* **91**, 500-507 (**1916**) (table p. 504).

Type IIIB3-1.

(1) Ingo W. D. Hackh. *Das synthetische System der Atome.* Hamburg: Hephaestos, **1914** (table p. 5).
_____ Arc and spark spectra and Periodic System. *Astrophys. J.* **48**, 241-255 (**1918**) (table p. 244).
_____ The classification of the chemical elements. *Sci. Amer. Suppl.* **87**, 146-149 (**1919**) (table p. 147).
_____ The new periodic table and atomic structure. *Chem. News* **139**, 275-278 (**1929**) (table p. 276).
_____ *A chemical dictionary.* Philadelphia: Blakiston Co., 544-545, **1929** (table p. 544).

(2) Hugo Stintzing. Eine neue Anordnung des periodischen Systems der Elemente. *Z. physik. Chem.* **91**, 500-507 (**1916**) (table p. 503).

(3) William Havens (of the Columbia Univ. staff). The atom. *Life* **26**(20), 68-88 (**1949**).

(4) John D. Clark. A modern periodic chart of chemical elements. *Science* **111**, 661-663 (**1950**) (table p. 662).

Type IIIC3-1.

(1) Henry Bassett. A tabular expression of the periodic relations of the elements. *Chem. News* **65**, 3-4; 19 (**1892**) (table p. 4).

(2) B. D. Steele. The place of the rare earth metals among the elements. *Chem. News* **84**, 245-247 (**1901**).

(3) Thomas Midgley jr. (arrangement suggested by R. E. Wilson). From the periodic table to production. *Chemistry & Industry* **56**, 133-136 (**1937**) (table p. 134).

(4) George Dubpernell. A periodic chart for electroplaters. *Proc. Amer. electroplaters' Soc.* **33** sess., 244-257 (**1946**) (table p. 248).

(5) Dorothy Ransom. Psychobiologic periodic table of chemical elements. *Sci. Monthly* **74**, 358-365 (**1952**) (table p. 359).

(6) A. Klemenc. *J. Thomsen und N. Bohr Periodensystem der chemischen Elemente.* Reprint, **1953**.

(7) S. V. Tikhomirov. O forme vyrazheniia zakona periodichnosti. *Izv. Vyssh. Uchebn. Zaveden., Geolog. Razved.* **1959**(6), 47-55 (separate table p. 48/49).

(8) Torolf Ternström. Ny uppställning av periodiska systemet. *Svensk Kem. Tidskr.* **75**, (11), 579-583 (**1963**) (table p. 580).
_____ A periodic table. *J. Chem. Educ.* **41**, 190-191 (**1964**) (table p. 190).

(9) A. A. Chaikhorskii. O nekotorykh zakonomernostiakh Periodicheskoi Sistemy D. I. Mendeleeva i predskazanii khimicheskikh svoistv sverkhtiazholykh elementov. *Radiokhimiia* **12**(6), 801-808 (**1970**) (table p. 802).

———— K voprosu teorii aktinidov. *Ibidem* **13**(1), 3-9 (**1971**) (table p. 5).

Subtype IIIA3-1a.

(1) D. F. Stedman. A periodic arrangement of the elements. *Can. J. Research* **25 B**, 199-210 (**1947**) (tables pp. 200, 201).

Subtype IIIB3-1a.

(1) Charles Janet. *La classification hélicoidale des éléments chimiques.* Beauvais: Novembre, **1928** (table VII-17).

———— The helicoidal classification of the elements. *Chem. News* **138**, 372-374; 388-393 (**1929**) (table p. 391).

Cited by A. Quintana y Mari. La classificación helicoidal de los elementos químicos. *Quim. e Industr.* (Barcelona) **8**(94), 287-292 (**1931**) (table p. 289).

(2) D. F. Stedman. A periodic arrangement of the elements. *Can. J. Research* **25 B**, 199-210 (**1947**) (table p. 202).

(3) S. A. Napol'skii. Novaia forma raspolozheniia elementov v tablitse D. I. Mendeleeva. *Uch. zap. Kirov. Pedagog. Inst.* (USSR) **1948**(4), 3-19.

(4) Gert G. Schlesinger. A summer short course in coordination chemistry. *Chemistry* (Washington) **39**(6), 8-13 (**1966**) (table p. 9).

(5) Z. Sh. Garaiev. Contemporary structure of Periodic System of chemical elements (in Azerbaidzhan). *Azerb. Khim. Zh.* (Baku) **1969**(5), 9-17 (table 17).

Subtype IIIA3-1b.

(1) Georg Gamov. *The birth and death of the sun.* New York: The Viking Press, **1940** (tables pp. 46, 47).

(2) Staff of the Museum of Science and Industry. *The periodic table of the elements.* Chicago, **1949**.

(3) Jennie E. Clauson. A space model of the Periodic System of elements. *J. Chem. Educ.* **29**, 250-251 (**1952**) (table p. 251).

———— A cut-out chart of the Periodic System. *Ibidem* **31**, 550-552 (**1954**) (table p. 552).

(4) Gilbert Gordon. *The chemical elements and their isotopes.* **1960** (table p. 10/11).

(5) A. B. Sidle. Presentation and construction of a helical periodic table. *Educ. in Chem.* (London) **3**(2), 107-108 (**1966**).

Subtype IIIA3-1c.

(1) R. Tremlelt. A model of the periodic classification of the elements. *School Sci. Rev.* **44**, 414-416 (**1963**) (table 414).

Subtype IIIC3-1d.

(1) J. E. Stareck. A natural periodic system including rare earths. *J. Chem. Educ.* **9**, 1625-1635 (**1932**) (table p. 1625).

Subtype IIIC3-1e.

(1) W. Denham Verschoyle. The Periodic Law. *Chem. News* **97**, 226-228 (**1908**).

(2) Irving Langmuir. The arrangement of electrons in atoms and molecules. *J. Am. Chem. Soc.* **41**, 868-934 (**1919**) (table p. 874).

(3) F. H. Loring. Missing elements in Periodic System. *Chem. News* **125**, 309-311; 386-388 (**1922**) (table p. 387); **126**, 1-4 (**1923**).
_____ *Atomic theories.* London: Methuen & Co., **1923** (separate table).
_____ *The chemical elements.* Ibidem, **1923** (table p. 34).

(4) Thomas Midgley jr. (arrangement suggested by R. E. Wilson). From the periodic table to production. *Chemistry & Industry* **56**, 133-136 (**1937**) (table p. 134).

(5) Osvaldo Baca-Mendoza. Leyes genéticas de los elementos químicos. Nuevo sistema periódico. *Univ. Nacl. del Cuzco* (Peru) 1-23 (in Spanish); 27-42 (in English) **1953** (tables II, III).
_____ The same title. *Revista Universit., Univ. Nacl. Cuzco* (Peru) **43**(106), 162-167 (**1954**) (no table).
_____ The same title. *Bol. Soc. Quim. Peru* (Lima) **21**, 5-23 (**1955**) (separate table II p. 24/25).
_____ Potenciales de ionización y radios iónicos en el "Nuevo Sistema Periódico". *Revista Univer., Univ. Nacl. Cuzco* (Peru) **46**(112), 277-291 (**1957**) (separate table).

(6) Sumio Satou. A new periodic table of the chemical elements (in Japanese). *Mem. of Kagawa Univ.* (Takamatsu, Japan) **II**(19), 1-5 (**1955**).
_____ *Comparative inorganic chemistry* (in Japanese). Tokyo, Japan: Nikkan Kogyo Shinbunsha, **1959** (table p. 20).
_____ *Fundamental inorganic chemistry* (in Japanese). Tokyo, Japan: Asakura Shoten, **1966** (table p. 34).

(7) K. V. Astakhov. Periodicheskii zakon D. I. Mendeleeva v svete sovremennykh nauchnykh dannykh. *Zh. Vses. Khim. Obshch.* **16**(3), 251-260 (**1971**) (table p. 253).

Type IIIA3-2.

(1) Karl Schirmeisen. Zur Ausgestaltung des periodischen Systems der chemischen Elemente. *Z. physik. Chem.* **33**, 223-236 (**1900**) (table I).

(2) Mulk Raj Verma and Anoop Krishen. A modified arrangement for the periodic classification of elements. *J. Sci. Industr. Res.* (India) **11 A**, 138-145 (**1952**) (table pp. 141, 144).

(3) Octavia S. Sell. A three-dimensional periodic chart. *J. Chem. Educ.* **32**, 524 (**1955**).

Type IIIB3-2.

(1) Charles Janet. *Essais de classification hélicoidale des éléments chimiques.* Beauvais: Avril, **1928** (table IV-6).

(2) V. Romanoff. Le système périodique de Mendeléeff par représentation graphique. *Rev. sci.* (Paris) **72**, 661-665 (**1934**) (table p. 664).

Type IIIC3-2.

(1) Eugenio Saz. La valencia química y el sistema periódico de los elementos. *Iberica* **17**(427), 296-301 (**1922**) (table pp. 296, 297).
_____ La théorie des valences positives et négatives. *Rev. gén. sci. appl.* **36**, 455-460; 503-509 (**1925**) (table p. 459).

———— El sistema periódico de los elementos. *Iberica* **35**(870), 186-190 (**1931**) (table pp. 186, 187).

———— Fórmulas de estructura de los compuestos de nitrógeno según la teoría de las valencias positivas y negativas. *9. Congr. Intern. quim. pura y aplic.* (Madrid) **3**, 268-279 (**1934**).

Subtype IIIA3-2a.

(1) Frank Austin Gooch and Claude Frederic Walker. *Outlines of inorganic chemistry.* London, New York: Macmillan Co., **II**, 3-11, **1905** (table p. 8/9).

(2) Carl Renz. Die seltenen Erden im periodischen System. *Z. anorg. allgem. Chem.* **122**, 135-145 (**1922**) (table p. 139).

Subtype IIIB3-2a.

(1) Émile Rinck et Pierre Feschotte. Classification périodique et loi de Proust. *Bull. Soc. Chim. France* **1962**(4), 856-862 (table p. 858).

Type IIIA3-3.

(1) I. Aucken. The Pagoda of the elements. *Chemistry & Industry* **1951**, 912 (table).
———— Valency and the Pagoda. *Ibidem* **1952**, 40 (no table).

Type IIIC3-3.

(1) Thomas Bayley. On the connexion between the atomic weight and the chemical and physical properties of elements. *Phil. Mag.* [5] **13**, 26-37 (**1882**) (scheme p. 31 and separate table).
———— Abstract of the article. *J. Chem. Soc.* (London) **42**, 359-360 (**1882**).
———— The cyclical law of the elements. *J. Am. Chem. Soc.* **20**, 927-934; 935-948 (**1898**) (no table).
———— The same title. *Chem. News* **77**, 157-160 (**1898**).

(2) Julius Thomsen. Systematische Gruppierung der chemischen Elemente. *Z. anorg. Chem.* **9**, 190-193 (**1895**) (table p. 192).
———— Über die mutmassliche Gruppe inaktiver Elemente. *Ibidem* **9**, 283-288 (**1895**) (scheme p. 284).
———— O sistematicheskoi gruppirovke khimicheskikh elementov. *Zh. Russk. Fiz.-Khim. Obshch.* **29**(2)(1), 49-50 (**1897**).
———— O predlagaemoi gruppe inaktivnykh elementov. *Ibidem* **29**, 50-51 (**1897**).

(3) Elliott Quincy Adams. A modification of the periodic table. *J. Am. Chem. Soc.* **33**, 684-688 (**1911**) (table p. 686).

(4) Georg Schaltenbrand. Die Gliederung des periodischen Systems der Elemente. *Z. anorg. allgem. Chem.* **115**, 127-130 (**1921**) (scheme p. 129).

(5) Niels Bohr. *Drei Aufsätze über Spektren und Atombau.* Braunschweig: F. Vieweg & Sohn, **1922** (table p. 132).
———— Über den Bau der Atome. *Z. Physik* **9**, 1-67 (**1922**) (no table).
———— The same title. *Naturwissenschaften* **11**, 606-624 (**1923**) (table p. 607).
———— *The theory of spectra and atomic constitution.* Cambridge: **1922**.
———— Om atomernes bygning. *Fys. Tidsskr.* **21**(1-2), 6-44 (**1923**) (table p. 9).
———— The structure of the atom. *Nature* **112**, 29-44 (**1923**) (table p. 30).
———— and D. Coster. Röntgenspektren und periodisches System der Elemente. *Z. Physik.* **12**, 342-374 (**1923**) (table p. 345).

(6) Darvin O. Lyon. *Das periodische System in neuer Anordnung.* Leipzig, Wien: Fr. Deuticke, **1928** (table p. 13).

(7) J. E. Stareck. A natural periodic system including rare earths. *J. Chem. Educ.* **9**, 1625-1635 (**1932**) (table p. 1626).

(8) Laurence L. Quill. The transuranium elements. *Chem. Rev.* **23**, 87-155 (**1938**) (table p. 105).

(9) Ia. I. Mikhailenko. *Periodicheskaia sistema khimicheskikh elementov Mendeleeva kak klassifikatsiia atomov po stroeniiu ikh elektronnoi obolochki.* Moskva: Izd. M. Kh. T. I., **1940**.

(10) E. I. Akhumov. O periodicheskoi sisteme khimicheskikh elementov. *Zh. Obshch. Khim.* **17**, 1241-1246 (**1947**) (table p. 1245).

———— Periodic systems of elements and antielements. *Ibidem* **31**, 1661-1664 (in English); 1777-1780 (in Russian) (**1961**).

(11) Włodzimierz Rodziewicz. Zmiany rozmieszczenia pierwiastków w ostatnim periodzie ich układu. *Przegląd Chem.* (Poland) **5**, 182-187 (**1947**) (table p. 186).

(12) Frank H. Spedding. The rare earths. *Sci. Amer.* **185**(5), 26-30 (**1951**) (table p. 27).

(13) George Glocker and Alexander I. Popov. Valency and the periodic table. *J. Chem. Educ.* **28**, 212-213 (**1951**) (table p. 213).

———— The long form of the periodic table. *Ibidem* **29**, 358 (**1952**) (no table).

(14) Charles D. Coryell. The place of the synthetic elements in the Periodic System. *Record. Chem. Progr.* **12**, 55-64 (**1951**).

———— The periodic table: the $6d$-$5f$ mixed transition group. *J. Chem. Educ.* **29**, 62-64 (**1952**) (table p. 62).

(15) A. F. Kapustinskii. Struktura periodicheskoi sistemy khimicheskikh elementov. *Dokl. Akad. Nauk SSSR* **81**(17), 47-50 (**1951**) (table p. 48).

(16) Umberto Sborgi. Quelques aspects de l'enseignement de la chimie inorganique basés sur le développement de la chimie depuis Mendeleieff jusqu'à la découverte de la structure de l'atome. *Ann. Guébhard-Séverine* **27/28**, 93-97 (**1951/1952**) (table p. 95).

(17) L. M. Marson e U. Zucchi. Rappresentazione matematica del sistema periodico degli elementi chimici. *Chimica* (Milan) **11**, 369-374 (**1955**) (table p. 370).

(18) Alberto Ghiorso and Glenn T. Seaborg. The newest synthetic elements. *Sci. Amer.* **195** (6), 66-80 (**1956**) (table p. 69).

Glenn T. Seaborg and Arnold R. Fritsch. The synthetic elements. III. *Ibidem* **208**(4), 68-70 (**1963**) (table p. 71).

G. Siborg. Evoliutsiia Periodicheskoi Sistemy elementov so vremen D. I. Mendeleeva do nashikh dnei. *Sto let per. zak. khim. elem.* Moskva: Nauka, **1969**, 136-157 (table p. 147).

———— Rasshirenie predelov Periodicheskoi Sistemy. *Sto let per. zak. khim. elem. Dokl. plenarn. zased. Iubil. Mendel. s'ezda, 10, 1969.* Moskva: Nauka, **1971**, 21-39 (table p. 28).

(19) J. E. Spice. Some comments on the classification of the elements. *School Sci. Rev.* (England) **41**, 244-260 (**1960**) (table p. 258).

(20) G. E. Villar. Nota sobre la clasificación periódica de los elementos. *An. Soc. Españ. Fis. Quim., Ser. B.* **60**(2/3), 103-106 (**1964**) (table p. 105).

(21) Thomas Chr. Frassares. *Chem. Chronika* B **31**(11-12), 169-171 (**1966**) (table p. 170) (in Greek).

(22) Mieczyslaw Taube. The Periodic System and superheavy elements (in English). *Nukleonika* (Warsaw) **12**(4), 309-312 (**1967**).

(23) V. Klemm. 100 let sistemy periodov khimicheskikh elementov. *Sto let per. zak. khim. elem.* Moskva: Nauka, **1969**, 57-70.

(24) Dieter Nebel. Zur Position der Actiniden- und Transactinidenelemente im Periodensystem der Elemente. *Z. f. Chem.* **10**(7), 251-260 (**1970**) (table p. 259).

(25) M. Gaisinskii (Haissinsky). Vliianie 4f- i 5f-elektronov na khimicheskie svoistva elementov. *Sto let per. zak. khim. elem., Dokl. plenarn. zased. Iubil. Mendel. s'ezda, 10, 1969.* Moskva: Nauka, **1971**, 68-74.

Subtype IIIA3-3a.

(1) K. K. Sugathan and T.C.K. Menon. Periodic classification and electronic configuration of elements. *Current Sci.* (India) **25**, 85 (**1956**).

Subtype IIIC3-3b.

(1) Ingo W. D. Hackh. *Das synthetische System der Atome.* Hamburg: Hephaestos, **1924** (table p. 7).

(2) Andreas von Antropoff. Eine neue Form des periodischen Systems der Elemente, *Z. Angew. Chem.* **39**, 722-725; 725-728 (**1926**) (table p. 724).
⎯⎯⎯⎯ Zusammenhänge zwischen den physikalischen und chemischen Eigenschaften des Siliciums und seiner Stellung im periodischen System. *Z. Elektrochem.* **32**, 423-428 (**1926**) (table p. 424).

(3) Mark von Stackelberg. *Atlas der physikalischen und anorganischen Chemie.* Dissertation, Berlin: Verlag Chemie, **1929** (table p. 3 Fig. 5).

(4) Fritz Scheele. Die Einordnung der Lanthaniden und Actiniden in das Periodische System. *Z. Naturforsch.* **4a**, 137-139 (**1949**).

Subtype IIIC3-3c.

(0) Emil V. Zmaczynski. Periodic System of the elements in a new form. *J. Chem. Educ.* **14**, 232-235 (**1937**) (table p. 233).

(1) Henry A. Wagner and Harold Simmons Booth. A new periodic table. *J. Chem. Educ.* **22**, 128-129 (**1945**) (table p. 128).

Type IIIA3-4.

(1) Emil V. Zmaczynski. Periodic System of the elements in a new form. *J. Chem Educ.* **14**, 232-235 (**1937**) (table p. 233).

(2) George A. Scherer. Models of a spiral periodic chart. *J. Chem. Educ.* **26**, 133-134 (**1949**) (table p. 133).
⎯⎯⎯⎯ New aids for teaching the Periodic Law. *School Sci. Math.* **49**, 91-97 (**1949**).

(3) D. Shireby. Presentation of the periodic table. *Educ. in Chem.* 3(4), 217-218 (**1966**) (table p. 217).

Type IIIB3-4.

(1) Charles Janet. *La classification hélicoidale des éléments chimiques.* Beauvais: Novembre, **1928** (table VIII-18, IX-19).

_____ The helicoidal classification of the elements. *Chem. News* **138**, 372-374; 388-393 (**1929**) (table p. 373, 389).

_____ *Concordance de l'arrangement quantique, de base, des électrons planétaires des atomes, avec la classification scalariforme, hélicoidale, des éléments chimiques.* Beauvais: Novembre, **1930** (table VI-15).

Cited by A. Quintana y Mari. La clasificación helicoidal de los elementos químicos. *Quím. e Industr.* (Barcelona) 8(94), 287-292 (**1931**) (table p. 291).

(2) Harold P. Brown. Periodic tables. *Trans. Kansas Acad. Sci.* **43**, 211-212 (**1940**) (table—chart 1).

(3) H. Strack. Das Perioden-System der Elemente mit Elektronenkonfigurationen und Quantenzahlen. *Chem.-Ztg.* **76**, 42-44 (**1952**).

(4) Frederick C. Strong III. The atomic form periodic table. *J. Chem. Educ.* **36**, 344-345 (**1959**).

(5) A. E. Dwight. Position of thorium and uranium in the periodic table. *Nature* **187**, 505-506 (**1960**) (table p. 505).

(6) Helen K. Griff. A clockwise spiral system of the chemical elements. *J. Chem. Educ.* **41**, 191 (**1964**).

(7) Thomas Chr. Frassares. *Chem. Chronika* **B 31**(11-12), 169-171 (**1966**) (table p. 171) (in Greek).

Type IIIC3-4.

(1) A. Werner. Beitrag zum Aufbau des periodischen Systems. *Ber. deut. chem. Ges.* **38**, 914-921; 2022-2027 (**1905**) (table p. 916).

_____ *Neuere Anschauungen auf dem Gebiete der anorganischen Chemie.* Braunschweig: **1905**, 7.

(2) Richard Swinne. Das periodische System der chemischen Elemente im Lichte der jüngsten Kanalstrahlenforschung. *Naturwissenschaften* **8**, 727-730 (**1920**).

(3) M. Kahanovicz. Le constanti elastiche in rapporto al sistema periodico degli elementi. *Atti R. Accad. Lincei* (Rome), *Rend.* [6] **8**, 584-590 (**1928**) (table p. 585).

(4) Jack Claude Rhodes. *The Rhodes Periodic System.* Willbur, Wash.: F. A. Bantz, **1933**.

(5) V. Romanoff. Le système périodique de Mendeléeff par représentation graphique. *Rev. sci.* (Paris) **72**, 661-665 (**1934**) (table p. 662).

_____ Représentation synoptique du sistème périodique de Mendeléeff et de la théorie de Bohr. *Compt. rend. 17. Congr. Chim. Ind.* (Paris) **II**, 2059-2062 (**1937**) (table p. 2060).

(6) Emil V. Zmaczynski. Periodic System of the elements in a new form. *J. Chem. Educ.* **14**, 232-235 (**1937**) (table p. 233).

(7) Edgar I. Emerson. A chart based on atomic numbers showing the electronic structure of the elements. *J. Chem. Educ.* **21**, 254-255 (**1944**) (table p. 254).

(8) A. N. Wrigley, W. C. Mast, and T. P. McCutcheon. A laminar form of the periodic table. *J. Chem. Educ.* **26**, 216-218; 248-250 (**1949**).

(9) K. E. Zimens. Eine Darstellung des periodischen Systems und der Isotopenzusammensetzung der Elemente für Unterrichtszwecke. *Svensk Kem. Tidskr.* **62**, 187-190 (**1950**).

(10) J. W. van Spronsen. De twaalf nieuwe elementen. *Chem. Weekblad* **47**, 55-60 (**1951**) (table p. 57).

(11) Richard Lepsius und S. K. Asunmaa. Prinzipielle Betrachtungen im periodischen System der Elemente. *Naturwissenschaften* **39**, 477-478; 490-491; 491 (**1952**) (scheme p. 491).

(12) Rudolf Reuber. *Das Periodensystem der Elemente mit Electronenanordnung.* Braunschweig: G. Westermann Verlag, **1954**.

(13) R. F. Wheeler. A new arrangement of the periodic table. *Chemistry & Industry* **45**, 1441-1443 (**1955**) (table p. 1441).

(14) George K. Estok. An adjustable periodic chart for lecture purposes. *J. Chem. Educ.* **33**, 618-619 (**1956**).

(15) Mircea V. Ionescu. Asupra clasificării periodice moderne a elementelor. *Acad. Rep. Popul. Romîne, Fil. Cluj., Stud. Cercet. Chim.* (Bucharest) **7**(2/4) 7-21 (**1956**) (table p. 13).

(16) A. A. Clifford. Periodic classification of elements. *Nature* **186**, 153-155 (**1960**) (table p. 155).

(17) G. E. Villar. Las tierras raras no son elementos de transición d. *Boletin, Fac. Ing. y Agrimens., Univ. Montevideo* **8**(4), 119-135 (**1962**) (table p. 130).
———— El centenario de la tabla periódica. Estado actual, posibilidades de desarrollo y algunas ajustes propuestos. *Revista Acad. Cienc. Exact., Fisic. Natur.* (Madrid) **63**(3), 517-532 (**1969**) (table p. 591).

(18) Ivan Baialovich. Periodni sistem sa potpunim periodama (in Serbokroatian). *Glasn. Hemisk. Druš.* (Beograd) **27**, 173-175 (**1962**) (separate table).

(19) V. K. Grigorovich. K voprosu o vliianii stroeniia vnutrennikh elektronnykh obolochek atomov na razmeshchenie elementov v periodicheskoi sisteme D. I. Mendeleeva. *Trudy Inst. Metall. Baikova, Gos. Kom. po chernoi i tsvetnoi metall., Akad. Nauk SSSR* **14**, 155-187 (**1963**) (table p. 172/173).

(20) V. M. Chistiakov. K razvitiiu korotkoi formy tablitsy Mendeleeva. *Uch. Zap. Kuibysh. Gos. Pedag. Inst.* **42**, 101-123 (**1964**) (table p. 105).

(21) A. V. Pitter. The periodic classification of the elements: the "PSD" periodic table. *School Sci. Rev.* (England) **46**(159), 412-416 (**1965**).

(22) S. S. Chissick and J. Baldwin. Nomenclature of the periodic table. *Educ. in Chem.* (London) **2**(4), 181-184 (**1965**) (table p. 182).

(23) Pai Yen Loung. *The electronic evolution in the atoms of the elements and the construction of the new periodic table.* Palo Alto, Calif.: Periodex Sci. Co., **1965** (table 3).

(24) Horst-Dietrich Hardt. Zur hundertjährigen Geschichte des periodischen Systems der chemischen Elemente. *Naturwiss. Rundschau* (Stuttgart) **19**(8), 313-316 (**1966**).
———— Eingliederung der Lanthaniden und Actiniden in das Periodensystem. *Ibidem* **22**(7), 302-305 (**1969**) (table II p. 303).
———— Das vollständige Periodensystem. *Allg. prakt. Chemie* **21**(3), 68-72 (**1970**) (table p. 68).

(25) Thomas Chr. Frassares. *Chem. Chronika* B **31**(11-12), 169-171 (**1966**) (table p. 170) (in Greek).

(26) B. J. Stokes. *Nuffield Foundation Science Teaching Project.* **1966**.

(27) V. I. Gol'danskii. O stroenii vos'mogo perioda sistemy D. I. Mendeleeva. *Dokl. Akad. Nauk SSSR* **180**(6), 1360-1362 (**1968**).
———— Samye liogkie i samye tiazholye atomy. *Priroda* (Moscow) **1969**(2), 19-27.
———— Periodicheskaia Sistema i problemy iadernoi khimii. *Sto let per. zak. khim. elem.* Moskva: Nauka, **1969**, 158-177 (table p. 167).
———— and S. M. Polikanov. *Tiazhelee urana.* Moskva: Nauka, **1969** (table p. 8/9).

V. I. Gol'danskii. The Periodic System of D. I. Mendeleev and problems of nuclear chemistry. *J. Chem. Educ.* **47**, 406-416 (**1970**) (scheme of table p. 411).

(28) M. Kh. Karapet'iants, A. P. Kreshkov. Stoletie velikogo otkrytiia. *Plast. Massy* (Moscow) **1969**(3), 5-10 (table p. 9).

(29) K. V. Astakhov. Periodicheskii zakon D. I. Mendeleeva v svete sovremennykh nauchnykh dannykh. *Zh. Vses. Khim. Obshch.* **16**(3), 251-260 (**1971**) (table p. 255).

Type IIIC3-5.

(1) D. Mendeleev. Sootnoshenie svoistv s atomnym vesom elementov. *Zh. Russk. Khim. Obshch.* **1**, 60-77 (**1869**) (table p. 70).
———— Über die Beziehungen der Eigenschaften zu den Atomgewichten der Elemente. *Z. f. Chem.* **12** or [2] **5**, 405-406 (**1869**) (table p. 405).
———— Versuche eines Systems der Elemente nach ihren Atomgewichten und chemischen Funktionen. *J. prakt. Chem.* [1] **106**, 251 (**1869**).
———— The same subject. *Osnovy Khimii.* St. Peterburg: V. I. 1st edit. **1869** (table p. IV). *Cited* by Karl Seubert. *Ostwald's Klass. d. exakt. Wissensch.* **68**, 18-19; 20-40 (**1895**) (table p. 33).

(2) C. J. Monroe and W. D. Turner. A new periodic table of the elements. *J. Chem. Educ.* **3**, 1058-1065 (**1926**) (table p. 1063).

(3) Royce H. LeRoy. A modified periodic classification of elements adapted to the teaching of elementary chemistry. *J. Chem. Educ.* **8**, 2052-2056 (**1931**).

(3) Jean Perrin. Grain de matière et de lumière II. Structure des atomes. *Actual. sci. et ind.* **191**, 31 (**1935**).

(4) Raymond Lautié. De quelques considérations sur les volumes atomiques. *Bull. Soc. Chim. France* [5] **5**, 695-711 (**1938**) (table p. 699).

(6) G. E. Villar. Existe en el sistema periódico una pléyada de 15 elementos radioactivos semejante a la de las tierras raras? *Boletin Fac. Ingen. Agrimens., Univ. Montevideo* **3**(5), 231-234 (**1938**) (table p. 231).

(7) Harold P. Brown. Periodic tables. *Trans. Kansas Acad. Sci.* **43**, 211-212 (**1940**) (table—chart II).

Subtype IIIB3-5a.

(1) C. R. Nodder. A convenient form of the periodic classification of the elements. *Chem. News* **121**, 269 (**1920**).

Subtype IIIC3-5a.

(1) Curt Schmidt. Studien über das periodische System. *Z. physik. Chem.* **75**, 651-664 (**1911**) (table p. 652).
———— *Das periodische System der chemischen Elemente.* Leipzig: **1917** (table p. 103).

Subtype IIIC3-5b.

(1) Edward G. Mazurs, **1957**. Not published.

Subtype IIIC3-5c.

(1) W. F. Sheehan. *Physical Chemistry.* Boston: Allyn & Bacon, **1961**.

Type IIIC3-6.

(1) Charles Janet. *La structure du noyau de l'atome, considérée dans la classification périodique des éléments chimiques.* Beauvais: Novembre, **1927** (table I and scheme p. 15).

———— *Essais de classification hélicoidale des éléments chimiques.* Ibidem: Avril, **1928** (table VI-11).

———— *La classification hélicoidale des éléments chimiques.* Ibidem: Novembre, **1928** (table I-1, X-22).

———— *Considérations sur la structure du noyau de l'atome.* Ibidem: Decembre, **1929** (table II-5, III-3).

———— The helicoidal classification of the elements. *Chem. News,* **138,** 372-374; 388-393 (**1929**) (table pp. 372, 374, 390, 393).

———— *Concordance de l'arrangement quantique, de base, des électrons planétaires des atomes, avec la classification scalariforme, hélicoidale, des éléments chimiques.* Beauvais: Novembre, **1930** (table 2, 14, 16).

Cited by A. Quintana y Mari. La clasificación helicoidal de los elementos químicos. *Quím. e Industr.* (Barcelona) **8**(94), 287-292 (**1931**) (table Fig. II).

(2) Royce H. LeRoy. Teaching the periodic classification of elements. *School Sci. Math.* **27,** 793-799 (**1927**) (table p. 796).

(3) S. A. Korff. A periodic classification of the hardness and meltingpoints of the elements. *Science* **67,** 370-371 (**1928**) (table p. 370).

(4) V. Romanoff. Le système périodique de Mendeléeff par représentation graphique. *Rev. sci.* (Paris) **72,** 661-665 (**1934**) (table p. 662).

———— Représentation synoptique du système périodique de Mendeléeff et de la théorie de Bohr. *Compt. rend. 17. Congr. Chim. Ind.* (Paris) **II,** 2059-2062 (**1937**) (table p. 2060).

(5) Germán E. Villar. Existe en el sistema periódico una pléyada de 15 elementos radioactivos semejante a la de las tierras raras? *Boletin Fac. Ing. Agrimens., Univ. Montevideo* **3**(5), 231-234 (**1938**) (table p. 232).

———— La incorporación de electrones f en las envolventes extranucleares de los Lantánidos y de los Actínidos. *Ibidem* **6**(5), 149-162 (**1957**) (table p. 151).

(6) Raymond Lautié. Sur la classification et la correspondance des éléments chimiques et de leurs composés. *Bull. Soc. Chim. France* (Mém.) [5] **6,** 677-683 (**1939**) (table p. 680).

(7) V. S. Sadikov. Novaia klassifikatsiia khimicheskikh elementov. *Priroda* (Moscow) **29**(7), 66-68 (**1940**).

(8) L. M. Simmons. A modification of the periodic table. *J. Chem. Educ.* **24,** 588-591 (**1947**) (table p. 589).

———— The display of electronic configuration by a periodic table. *Ibidem* **25,** 658-661 (**1948**) (table p. 659).

(9) O. von Auwers. Periodisches System nach Quantenzahlen. *Physik. Blätter* **4,** 423-425 (**1948**) (table p. 424).

(10) Reino W. Hakala. Letter to the editor. *J. Chem. Educ.* **25,** 229 (**1948**) (scheme).

———— The Periodic Law in mathematical form. *J. Phys. Chem.* **56,** 178-181 (**1952**) (table p. 179).

(11) V. M. Klechkovskii. $(n+l)$-gruppy v posledovatel'nom zapolnenii elektronnykh konfiguratsii atomov. *Dokl. Akad. Nauk SSSR* **80**(4), 603-606 (**1951**) (table p. 605).

———— O posledovatel'nosti v zapolnenii elektronnykh urovnei atomov. *Zh. Eksper. i Teoret. Fiz.* **23**(1), 115-122 (**1952**) (table p. 121).

_____ Konfiguratsionnyi indeks elektronnoi struktury normal'nykh atomov. *Izv. Timir. Sel'skokh. Akad.* **40**(3), 161-176 (**1961**) (table p. 173).

_____ Raspredelenie atomnykh elektronov i pravilo posledovatel'nogo zapolneniia ($n+l$)-grupp. Moskva: Nauka, **1968** (table p. 15).

_____ K teorii Periodicheskoi Sistemy elementov D. I. Mendeleeva. *Vopr. Istor. Estestvozn. Tekhn.* **4**(29), 3-16 (**1969**) (scheme p. 8).

(12) Richard Lepsius und S. K. Asunmaa. Prinzipielle Betrachtungen im Periodischen System der Elemente. *Naturwissenschaften* **39**, 490-491 (**1952**) (scheme p. 491); **41**, 221-227 (**1954**) (table p. 224); **43**, 324 (**1956**) (scheme).

(13) Neville Smith. Letter to the editor: The periodic table. *Chemistry & Industry* 1673-1674, **1955** (scheme p. 1674).

(14) Knud Aage Nissen. *Comments on the representation of the periodic table.* Copenhagen: Niels Bohr legatet, **1956**.

(15) Ortega G. Barraza, Calero A. De Lope, F. Farré-Torá. *Clasificación electronica de los elementos,* **1964** (only the table).
Francisco Farré-Torá. Classificación cuantica de los elementos químicos. *Univ. Nacl. Auton. de Mexico, Escuela Nacl. Preparatoria, Dept. de Química,* **1966**/**67** (table p. 8 and separate table).
_____ Estructuración de los elementos y clasificación periódica. *Ibidem* **1967**/**68**, 2.

(16) B. V. Tokarev. Regulo de $n+l$ grupoj kaj ĉefaj interrilatoj en Perioda Sistemo de D. I. Mendelejev (in Esperanto). *Scienca Revuo* **16**(3/4), 121-130 (**1966**) (table p. 126/127).

(17) D. Neubert. Double shell structure of the Periodic System of the elements (in English). *Z. Naturforsch.* **25 A**, 210-217 (**1970**) (table p. 211).

(18) Robert Mills. Systematic table of the elements. *Am. J. Phys.* **40**(8), 1169-1170 (**1972**) (table p. 1169).

Subtype IIIC3-6a.

(00) D. Balarew. Gedanken eines Ausbaues des Systems der Elemente auf breiterer Grundlage. *Z. anorg. allgem. Chem.* **121**, 22-24 (**1922**).

(0) A. Janek. Zur Statistik der Wertigkeiten. *Z. anorg. Chem.* **252**, 354-356 (**1944**) (table p. 355).

(1) Edward G. Mazurs. Ķīmisko elementu periodiskās sistemas tabulu veidi (in Latvian). *Technikas Apskats* (Lincoln, Nebr.) **2**(7), 8-12 (**1955**) (table p. 10).
_____ Die wichtigsten Tabellen des Perioden-Systems. *Chem.-Ztg.* **80**, 195-203 (**1956**) (table p. 200).

Subtype IIIC3-6b.

(0) V. Romanoff. Le système périodique de Mendeléeff par représentation graphique. *Rev. Sci.* (Paris) **72**, 661-665 (**1934**) (table p. 662).
_____ Représentation synoptique du système périodique de Mendeléeff et de la théorie de Bohr. *Compt. rend. 17. Congr. Chim. Ind.* (Paris) **II**, 2059-2062 (**1937**) (table p. 2060).

(1) Edward G. Mazurs. Ķīmisko elementu periodiskās sistemas tabulu veidi (in Latvian). *Technikas Apskats* (Lincoln, Nebr.) **2**(7), 8-12 (**1955**) (table p. 9).
_____ Die wichtigsten Tabellen des Perioden-Systems. *Chem.-Ztg.* **80**, 195-203 (**1956**) (table p. 200).

INTRODUCTION TO ELECTRONIC CONFIGURATION TABLES

(1) Mendeleeff. The Periodic Law of the chemical elements. *J. Chem. Soc. (London)* [Trans.] **55**, 649 (**1889**).

(2) Alois Bilecki. Periodisches System und Atomgewichte. *Z. physik. Chem.* **82**, 249-252 (**1913**).

(3) A. Sommerfeld. Electronic structure of the atom and the quantum theory. *Mem. a. Proc., Manchester Lit. a. Phil. Soc.* **70**, 141-151 (**1925/26**) (scheme p. 144).

(4) Charles Janet. *La structure du noyau de l'atome, considérée dans la classification périodique des éléments chimiques.* Beauvais: Novembre **1927** (scheme p. 15 a. separate table No. I).
 _____ The helicoidal classification of the elements. *Chem. News* **138**, 372; 388-393 (**1929**) (see the denoting of the tables p. 372).
 _____ *Concordance de l'arrangement quantique, de base, des électrons planétaires des atomes, avec la classification scalariforme, hélicoidale, des éléments chimiques.* Beauvais: Novembre **1930** (scheme table 12).

(5) Eugen Rabinowitsch und Erich Thilo. *Periodisches System, Geschichte und Theorie.* Stuttgart: F. Enke, **1930** (scheme p. 163).

(6) Heinrich Remy. *Lehrbuch der anorganischen Chemie.* Bd. I, **1931**.

(7) K. V. Nikol'skii. Periodicheskaia sistema elementov D. I. Mendeleeva v svete sovremennoi fiziki. *Priroda* (Moscow) **23** (3), 48-65 (**1934**).

(8) E. Madelung. *Die mathematischen Hilfsmittel des Physikers.* Berlin: J. Springer, **1936** (p. 359 a. 360).

(9) Harry J. Eméleus and J. S. Anderson. *Modern Aspects of Inorganic Chemistry.* London: G. Routledge & Sons, 1. edit., **1938**.

(10) Linus Pauling. *The nature of the chemical bond and the structure of molecules and crystals.* Ithaca, N.Y.: Cornell University Press, **1939**, **1940** (scheme p. 26).

(11) Thomas H. Hazlehurst. Quantum numbers and the periodic table. *J. Chem. Educ.* **18**, 580-581 (**1941**) (scheme p. 581).

(12) Raymond Daudel. Mouvement scientific. Les agrégats atomiques. *Rev. sci.* (Paris) **81**, 379-408 (**1943**) (scheme p. 405).

(13) Don DeVault. A method of teaching the electronic structure of atoms. *J. Chem. Educ.* **21**, 526-534; 575-581 (**1944**) (diagrams pp. 531 and 576/577).

(14) William J. Wiswesser. The periodic system and atomic structure. I. An elementary physical approach. *J. Chem. Educ.* **22**, 314-322 (**1945**) (scheme p. XXI a. 318).

(15) Yeou Ta. Une nouvelle représentation du tableau périodique des éléments. *Ann. de phys.* (Paris) [12] **1**, 88-99 (**1946**).

(16) Pao-Fang Yi. To the editor. *J. Chem. Educ.* **24**, 567 (**1947**) (scheme p. 567).

(17) L. M. Simmons. A modification of the periodic table. *J. Chem. Educ.* **24**, 588-591 (**1947**).
 _____ Display of electronic configuration by a periodic table. *Ibidem* **25**, 658-661 (**1948**).

(18) Reino W. Hakala. Letter to the editor. *J. Chem. Educ.* **25**, 229 (**1948**).
 _____ The periodic law in mathematical form. *J. Phys. Chem.* **56**, 178-181 (**1952**).

(19) Benjamin Carroll and Alexander Lehrman. The electron configuration and the ground state of the elements. *J. Chem. Educ.* **25**, 662-666 (**1948**).

(20) O. von Auwers. Periodisches System und Quantenzahlen. *Physik. Blätter* **4**, 423-425 (**1948**).

(21) J. A. Campbell. Interpreting electronic structure. *J. Chem. Educ.* **26**, 477-480 (**1949**) (scheme p. 479 a. 480).

(22) Sigge Hähnel. Grundämnenas periodiska system. *Teknisk Tidskrift* (Stockholm) **11**, 181-190 (**1949**).

(23) V. M. Klechkovskii. $(n+l)$-gruppy v posledovatel'nom zapolnenii elektronnykh konfiguratsii atomov. *Dokl. Akad. Nauk SSSR* **80** (4), 603-606 (**1951**) (scheme p. 604).

———— K voprosu o zakonomernosti v poriadke zapolneniia elektronnykh urovnei atoma s uvelicheniem atomnogo nomera elementa. *Ref. Dokl. Moskov. Sel'skokh Akad. Timir.* **13**, 127-133 (**1951**).

———— Ob elektronnykh gruppakh v atomakh. *Dokl. Akad. Nauk SSSR* **83**, 411-414 (**1952**).

———— O posledovatel'nosti v zapolnenii elektronnykh urovnei atomov. *Zh. Eksper. i. Teoret. Fiz.* **23**, (1), 115-122 (**1952**) (scheme table I).

———— K formulirovke pravil zapolneniia elektronnykh urovnei. *Dokl. Akad. Nauk SSSR* **92** (5), 923-926 (**1953**).

———— K voprosu o posledovatel'nosti termov v spektrakh mnogoelektronnykh atomov. *Zh. Eksper. i Teoret. Fiz.* **25**, 179-187 (**1953**) (scheme Fig. 1).

———— K teorii voprosa o chisle elementov v periode sistemy D. I. Mendeleeva. *Zh. Fizich. Khim.* **27**, 1251-1255 (**1953**).

———— O zakonomernosti, opredeliaiushchei dlinu periodov sistemy D. I. Mendeleeva. *Izv. Moskov. Timir. Sel'skokh. Akad.* **1**, 159-170 (**1953**).

———— O zapolnenii n, l-podgrupp v predelakh $(n+l)$-gruppy kvantovykh urovnei. *Ref. Dokl. Moskov. Sel'skokh. Akad. Timir.* **18**, 297-301 (**1954**).

———— O pravilakh, formuliruemykh pri pomoshchi $(n+l)$-gruppirovki kvantovykh urovnei. *Izv. Moskov. Timir. Sel'skokh. Akad.* **2** (6), 205-218 (**1954**) (scheme p. 211 Fig. 3).

———— O zavisimosti mezhdu Z i chislom s-, p-, d-, i f-elektronov v atome. *Dokl. Akad. Nauk SSSR* **95** (6), 1173-1176 (**1954**).

———— O pervykh elektronakh s dannym l v neitral'nom atome. *Zh. Eksper. i Teoret. Fiz.* **26** (6), 760-761 (**1954**).

———— O pervom poiavlenii atomnykh elektronov s dannymi l, n, n_r, i $n+l$. *Ibidem.* **30**, 199-201 (**1956**).

———— O nachale i okonchanii zapolneniia elektronami nekotorykh sovokupnostei kvantovykh urovnei. *Dokl. Moskov. Sel'skokh. Akad. Timir.* **23**, 200-207 (**1956**).

———— O zavisimosti mezhdu Z i raspredeleniem atomnykh elektronov. *Optika i Spektrosk.* **2** (1), 3-9 (**1957**).

———— K kharakteristike tipa elektronnoi konfiguratsii atomov. *Dokl. Akad. Nauk SSSR* **135**, 655-658 (**1960**).

———— Konfiguratsionnyi indeks elektronnoi struktury normal'nykh atomov. *Izv. Moskov. Timir. Sel'skokh Akad.* **40** (3), 161-176 (**1961**).

———— K obosnovaniiu pravila posledovatel'nogo zapolneniia $(n+l)$-grupp. *Zh. Eksper. i Teoret. Fiz.* **41** [2(8)], 465-466 (**1961**).

———— $(n+l)$ rules and their significance for the theory of D. I. Mendeleev Periodic System. *20th Intern. Congr. Pure Appl. Chem.*, Report **D38**, (Moscow) **1965**.

———— *Raspredelenie atomnykh elektronov i pravilo posledovatel'nogo zapolneniia $(n+l)$-grupp.* Moskva: Atomizdat, **1968**.

———— K teorii periodicheskoi sistemy elementov D. I. Mendeleeva. *Vopr. Istor. Estestvozn. Tekhn.* **4** (29), 3-16 (**1969**).

———— Razvitie nekotorykh teoreticheskikh problem Periodicheskoi Sistemy D. I. Mendeleeva. *Sto let per. zak. khim. elem. Dokl. plen. zased. Iubil. Mendeleev. s'ezda, 10th, 1969.* Moskva: Nauka, **1971**, 54-67.

(24) Therald Moeller. *Inorganic Chemistry. An advanced textbook.* New York: J. Wiley & Sons, **1952** (scheme p. 97).

(25) Richard Lepsius und S. K. Asunmaa. Prinzipielle Betrachtungen im Periodischen System der Elemente. *Naturwissenschaften* **39**, 492 (**1952**); **43**, 324 (**1956**).

(26) Oswaldo Baca-Mendoza. Leyes genéticas de los elementos químicos. Nuevo sistema periódico. *Univ. Nacl. del Cuzco* (Peru), 1-23 (in Spanish); 27-42 (in English), **1953**.

_____ The same title. *Bol. Soc. Quim. Peru (Lima)* **21** (1), 5-23 (**1955**).

_____ Potenciales de ionización y radios iónicos en el Nuevo Sistema Periódico" (in Spanish). *Revista Univers., Univ. Nacl. del Cuzco (Peru)*, **46** (112), 227-291 (**1957**).

(27) S. A. Shchukarev. Periodicheskii zakon D. I. Mendeleeva kak osnovnoi printsip sovremennoi khimii (in Russian). *Zh. Obshch. Khim.* **24** (86), 581 (**1954**).

(28) S. I. Tomkeieff. Atomic sizes and bond types in their relation to the Periodic System and the structure of the atom. *Sci. Progress (London)* **43**, 28-44 (**1955**) (p. 30).

(29) Neville Smith. Letter to the editor: The periodic table. *Chemistry & Industry* 1673-1674, **1955**.

(30) A. S. Ganesan. The normal electron configuration of atoms. *Current Sci. (India)* **24**, 10-11 (**1955**) (scheme p. 11).

(31) J. Keller. Configuración electrónica de los atomos. Regla de la diagonal. *Ciencia (Mexico)* **16**, 4-6; 86-88 (**1956**) (scheme p. 87).

(32) Knud Aage Nissen. *Comments on the representation of the periodic table.* Copenhagen: Niels Bohr Legatet, **1956**.

(33) Emil Petrovici. Citeva observaţii privitoare la caracterul elementelor şi pozitia lor in sistemul periodic. *Bulet. stiint. şi techn., Scoala politechn., (Timisvara, Rumania)*, [N.S.] **1** (15) (1), 335-350 (**1956**) (scheme p. 336/337).

(34) Wojciech Rubinowicz. *Quantentheorie des Atoms.* Leipzig: J. A. Barth, **1959**.

(35) S. Dockx. *Théorie fondamentale du système périodique des éléments.* Bruxelles: Office Internat. de Librairie, **1959**.

(36) F. P. Platonov. Sovremennaia teoriia periodicheskoi sistemy elementov i pravilo zapolneniia elektronnoi sfery po V. M. Klechkovskomu. *Dokl. Moskov. Sel'skokh. Akad. Timir.* **1961** (64), 169-185 (table p. 179).

_____ Stoletie Periodicheskogo Zakona i sistema khimicheskikh elementov D. I. Mendeleeva. *Izv. Moskov. Timir. Sel'skokh. Akad.* **1969** (1), 192-202.

(37) F. Herman and S. Skillman. *Atomic Structure Calculations.* Englewood Cliffs, N.J.: Prentice Hall, **1963**.

(38) A. T. Balaban. Numere magice de electroni şi de nucleoni. *Revista de chim. (Bucuresti)* **14** (3), 158-163 (**1963**).

(39) M. L. Chepelevetskii. Periodicheskaia sistema atomnykh iader. *Zh. Fizich. Khim.* **40** (2), 307-316 (**1966**).

(40) Thomas Chr. Frassares. *Chem. Chronika* **B** 31(11-12), 169-171 (**1966**) (in Greek).

(41) B. V. Tokarev. Regulo de $n+l$ grupoj kaj ĉefaj interrilatoj en Perioda Sistemo de D. I. Mendelejev (in Esperanto). *Scienco Revuo* **16** (3/4), 121-130 (**1966**) (table p. 126/127).

(42) Francisco Farré-Torá. Clasificación cuantica de los elementos químicos. *Univ. Nacl. Auton. de Mexico, Escuela Nacl. Preparatoria, Dept. de Química,* **1966**/67 (table p. 8 and a separate table).

_____ Estructuración de los elementos y clasificación periódica. *Ibidem,* **1967**/68, No. 2.

(43) Samuel Madras. The sequence of orbitals in the periodic table. *Can. Chem. Educ.* **2** (2), 4-5 (**1967**) (scheme p. 5).

(44) K. M. Kedrov. Mendeleevskie formy Periodicheskoi Sistemy elementov. *Sto let per. zak. khim. elem.* Moskva: Nauka, **1969**, 42-56.

_____ The same title. *Vopr. Istor. Estestvozn. Tekhn.* **4** (29), 59-70 (**1969**).

(45) D. Neubert. Double shell structure of the Periodic System of the elements. *Z. Naturforsch.* **25A**, 210-217 **(1970)** (in English).

(46) Frank Dieter Leyh. Modell zur Erläuterung atomarer Elektronenkonfigurationen. *Chem. Schule* **17** (6), 278-284 **(1970)** (scheme p. 279).

(47) V. I. Gol'danskii. The Periodic System of D. I. Mendeleev and problems of nuclear chemistry. *J. Chem. Educ.* **47**, 406-416 **(1970)** (scheme p. 410).

(48) David Layzer. Towards an elementary theory of the periodic table. *U.S. Natl. Bur. Stand., Spec. Publ. No. 353*, 20-36, **1971**.

(49) Saburo Otake. The principle for the formation of the periodic system of elements that was expressed in Gregorovich's table and Klechkovsky theory (in Japanese). *Kagaku No Ryoiki* (Tokyo, Japan) **25** (10), 951-963 **(1971)** (scheme p. 955).

Problem of Inner Transition Elements.

(1) David C. Hamilton. Position of lanthanum in the periodic table. *Amer. J. Phys.* **33**, (8), 637-640 **(1965)**.

(2) H. Merz and K. Ulmer. Position of lanthanum and lutetium in the periodic table. *Physics Letters* (Amsterdam) **26A** (1), 6-7 **(1967)**.

(3) V. M. Chistiakov. Vtorichnaia periodichnost' Birona v pobochnykh a-podgruppakh korotkoi tablitsy Mendeleeva (in Russian). *Zh. Obshch. Khim.* **38** (2), 209-210 **(1968)**.

———— Korotkaia tablitsa Mendzialeeva in magchymae mestsarazmiashchenne liutetsiu (in Belorussian). *Vestsi Akad. Navuk Belaruss. SSR (Minsk), Ser. Khim. Navuk* **1968** (3), 50-56.

I. Supporters of the Uranide and Curide Series.

(1) Niels Bohr. *Drei Aufsätze über Spektren und Atombau.* Braunschweig: F. Vieweg und Sohn, 133, **1922**.

———— Über den Bau der Atome. *Naturwissenschaften* **11**, 606-624 **(1923)**.

———— The structure of the atom. *Nature* **112**, 30 **(1923)**.

———— Om atomernes bygning. *Fys. Tidskr.* (Copenhagen) **21** (1-2), 6-44 **(1923)** (table p. 9).

(2) A. Sommerfeld. *Atombau und Spektrallinien.* Braunschweig: 465-470, **1924**.

(3) A. von Antropoff. Eine neue Form des periodischen Systems der Elemente. *Z. angew, Chem.* **39**, 722-725; 725-728 **(1926)**.

(4) Y. Sugiura, H, C. Urey. The quantum theory explanation of the anomalies in the 6th and 7th periods of the periodic table (in English). *Kgl. Danske Videnskab. Selskab, Math.-Fys. Meddelser* **7**(13), 3-18 **(1926)**.

(5) Vladimir Karapetoff. A chart of consecutive sets of electronic orbits within atoms of chemical elements. *J. Frankl. Inst.* **210** (5), 609-624 **(1930)** (p. 616 and table p. 610/611).

(6) Ta-You Wu, S. Goudsmit. Low states of the heaviest elements. *Phys. Rev.* **43**, 496 **(1933)**.

(7) E. McMillan, P. H. Abelson. Radioactive element 93. *Phys. Rev.* **57**, 1185-1186 **(1940)**.

(8) Laurence S. Foster. Bombardment of uranium with fast and slow neutrons. *J. Chem. Educ.* **17**, 448-449 **(1940)**.

(9) M. Goeppert-Mayer. Rare-earth and transuranium elements. *Phys. Rev.* **60**, 184-187 **(1941)**.

(10) Constantin Bedreag, Place des uranides 93, 94 dans la systématique naturelle des éléments. *Compt. rend.* (Paris) **215**, 537-539 **(1942)**.

———— Structure et systématique naturelle des éléments (in French). *Ann. sci. Univ. Jassy* (Roumania) **28**(1), 143-148 **(1942)**.

———— Die Systematik der Elektronenhülle der Elemente. *Naturwissenschaften* **31**, 490 (**1943**).

———— Die natürliche elektronische Systematik der Elemente (in German). *Bull. Acad. Roumaine, Sect. sci.* **25**, 405-409 (**1943**).

———— Système harmonique des éléments (in French). *Bul. Inst. Politechn. "Gh. Asachi", Iasi* **3**, 317-343 (**1948**).

———— Periodic table of elements from 1 to 100. *Acad. rep. populare Romîne, Iasi, Stud. Cercet. Stiint.* **3**(1-4), 71-81 (**1952**).

———— The energy levels of the elements 90-100. *Ibidem* **5**(1/2), 87-94 (**1954**).

———— Les uranides (93-95) et les trans-curiens (97-100). *Compt. rend.* (Paris) **240**, 1767-1769 (**1955**).

———— The same title. *Comm. acad. rep. popul. Romîne* **5**, 1573 (**1955**).

———— The same title. *Studii Cercet. Ştiint. (Iaşi), Ser. I*, **6**(1/2), 233-238 (**1955**).

———— Aktuelle Systematik der Elemente. *Ibidem* Ser. I, **6** (3/4), 83-88 (**1955**).

———— Uranidele şi omologia lor fisică (in Rumanian). *Analele Ştiint. Univ. "A. I. Cuza", Iasi*, Sect. 1 [N.S.] **1**(1/2), 231-242 (**1955**).

———— The occurance of homologs and the quantum structure of the elements. *Acad. rep. popul. Romîne. Fil. Iaşi, Stud. cercet. ştiint. chim.* **7**(2), 1-4 (**1956**).

———— Uranides and the "actinide" controversy. *Ibidem* **11**(1), 13-18 (**1960**).

———— Représentation physique de la systématique des éléments. *Compt. rend* (Paris) **252**, 1604-1606 (**1961**).

———— Drei *f*-Spalten im System der Elemente. *Naturwissenschaften* **49**, 229 (**1962**).

———— Physical systematics of the elements (in English). *Analele stiint. Univ. "A. I. Cuza", Iaşi*, Sect. I, **8**(2), 451-453 (**1962**).

———— Système physique des éléments. *J. phys. radium* [8] **24**, 27-32 (**1963**).

(11) Kurt Starke. Das Element 93 und seine Stellung im periodischen System. *Z. anorg. allgem. Chem.* **251**, 251-259 (**1943**).

(12) Raymond Daudel. Sur une application de la mécanique broglienne à l'étude des propriétés chimiques des éléments des terres rares et des transuraniens. *Compt. rend.* (Paris) **217**, 396-397 (**1943**).

———— Mouvement scientifique. Les agrégats atomiques. *Rev. sci.* (Paris) **81**, 399-408 (**1943**).

(13) Henry DeWolf Smyth. *Atomic energy for military purposes.* Princeton, N.J.: Princeton Univers. Press, 102, **1945**.

(14) Moise Haissinsky. Les nouveaux éléments chimiques et le système périodique. *Bul. Soc. Chim. France* [5] **16**, 668-678 (**1949**).

———— The position of the cis- and trans-uranic elements in the periodic system: uranides or actinides ? *J. Chim. Soc.* (London), **Suppl. issue 2, 1949**, 241-243.

———— Symposium of chemistry of heavy elements. *Brit. Atom. Energy Res. Establ.*, Oxford, England, **1949**.

———— Sur la classification des derniers éléments du système périodique. *J. chim. phys.* **47**, 415-418 (**1950**).

———— *L'état actuel du sistème périodique des éléments chimiques.* Alençon: Poulet-Malassis, **1951**.

———— La place des éléments transuraniens dans le système périodique. *Experientia* **9**, 117-120 (**1953**).

———— Lanthanides, uranides et curides (in French). *Scientia* (Milan) [6] **93**(52), 119-123 (**1958**).

———— La chimie des cis- et transuraniens comparée a celle des lanthanides et des éléments de transition. *17. Congr. Intern. chim. pur. appl.* (Munich) **1**, 185-202 (**1959**).

———— *Nuclear chemistry and its application* (in Russian). Moskva: I L., **1961**.

———— *The same title* (in English). Reading, Mass.: Addison-Wesley Publ. Co., **1964**.

_____ Comparison of the chemistry of cis and trans uranides with that of lanthanides and of transition metals (in Hungarian). *Magg. Tud. Akad. Kém. Tud. Oszt. Közlem.* **18**, 395-411 (**1962**).

_____ *Progress in inorganic and elements-organic chemistry.* Moskva: I L., **1963**.

_____ and C. K. Jørgensen. Les éléments 4f et 5f. *J. chim. phys.* (Paris) **63**(9), 1135-1138 (**1966**).

M. N. Gaisinskii. O sed'mom periode sistemy khimicheskikh elementov D. I. Mendeleeva. *Vopr. Istor. Estestvozn. Tekhn.* **4**(29), 27-33 (**1969**).

_____ Elektronnaia struktura i khimicheskie svoistva tiazhelykh elementov. *Priroda* (Moskva) **1970**(7), 90-93.

_____ Vliianie 4f- i 5f-elektronov na khimicheskie svoistva elementov. *Sto let per. zak. khim. elem. Dokl. plenarn. zased. Iubil. Mendeleev. s'ezda, 10, 1969.* Moskva: Nauka, **1971**, 68-74.

(15) F. A. Paneth. The making of the elements 97 and 98. *Nature* (London) **165**, 748-749 (**1950**) (uranide series).

_____ La radioactivité artificielle et la chimie. *J. phys. radium* [8] **16**, 735-762 (**1955**).

(16) F. Cap. Zum Problem der Aktinidengruppe. *Experientia* (Basel) **6**, 291-292 (**1950**) (in German).

(17) A. W. Wylie. Some aspects of the chemistry of the heavy elements. *J. and Proc., Royal Austral. Chem. Inst.* **17**, 446-461 (**1950**).

(18) N. V. Sidgwick. *The chemical elements and their compounds.* Oxford: **1950**.

(19) Charles D. Coryell. The place of the synthetic elements in the periodic system. *Record Chem. Progr.* **12**, 55-64 (**1951**).

_____ The periodic table: the 6d-5f mixed transition group. *J. Chem. Educ.* **29**, 62-64 (**1952**).

(20) J. K. Dawson. Electronic structure of the heaviest elements. *Nucleonics* **10** (9), 39 (**1952**).

(21) B. M. Mott. Metallurgical characteristics of the heavy elements. *Research* **6**, 238-246 (**1953**).

(22) V. I. Spitsyn. Sovremennoe sostoianie periodicheskogo zakona D. I. Mendeleeva. *Doklad na VIII Mendeleevskom s'ezde po obshchei i prikladnoi khimii.* Moskva: Izdat. Akad. Nauk SSSR, 3-24, **1959**.

(23) D. N. Trifonov. *Redkozemel'nye elementy i ikh mesto v periodicheskoi sisteme.* Moskva: Izdat. "Nauka", **1966**.

II. Supporters of the Actinide Series.

(1) Henry Bassett. A tabular expression of the periodic relations of the elements. *Chem. News* **65**, 3-4; 19 (**1892**).

(2) A. Werner. Beitrag zum Ausbau des periodischen Systems. *Ber. deut. chem. Ges.* **38**, 914-921 (**1905**).

(3) Thomas M. Lowry. *Historical introduction to chemistry.* London: Macmillan & Co., **1915**, p. 468 (foresees the existence of actinides).

(4) Georg Schaltebrand. Die Gliederung des periodischen Systems der Elemente. *Z. anorg. allgem. Chem.* **115**, 127-130 (**1921**).

(5) Royce H. LeRoy. Teaching the periodic classification of elements. *School Sci. Math.* **27**, 793-799 (**1927**).

_____ The same title. *J. Chem. Educ.* **8**, 2052 (**1931**).

(6) Richard Rudy. Quelques chapiters récents du spectroscopie. *Rev. gén. sci. pur. appl.* **38**, 661-671 (**1927**) p. 671 (accepts actinides according to their electronic configuration).

(7) Charles Janet. *Essais de classification hélicoidale des éléments chimiques.* Beauvais, **1928**.
────── The helicoidal classification of the elements. *Chem. News* **138**, 372-374; 388-393 (**1929**).

(8) V. Romanoff. Le système périodique de Mendeléeff par representation graphique. *Rev. sci.* (Paris) **72**, 661-665 (**1934**).

(9) Jean Perrin. Grain de matière et de lumière. II. Structure des atomes. *Actual. sci. ind.* **191**, 1-52 (**1935**) (p. II, 30). (He predicted the possibility of a second series of rare earth elements).

(10) Thomas Midgley jr. From the periodic table to production. *Chemistry & Industry* **56**, 133-136 (**1937**).

(11) Germán E. Villar. Existe en el sistema periódico una pléyada de 15 elementos radioactivos semejante a la de las tierras raras? *Boletin Fac. Ing. Agrimens. Univ. Montevideo* **3** (5), 231-234 (**1938**) (p. 238).
────── Existe una pléyada de elementos radioactivos en el lugar reservado para el Actinio en el Sistema Periódico? *Anales Asoc. Quím. Argentina* **26**, 126-128 (**1938**).
────── Algunas consideraciones sobre la existencia de una pléyada de elementos en el lugar reservado para el actinio en el Sistema Periódico. *Proc. 8th Amer. sci. congr.* **7**, 39-43 (**1940**).
────── The same title. *Anais Acad. Brasil Cienc.* **12**, 51-57 (**1940**).
────── On a suggested revision of the seventh period of the periodic table. *J. Chem. Educ.* **19**, 286 (**1942**).
────── A suggested revision of the position of thorium in the fourth period (group?) of the periodic table. *Ibidem* **19**, 329-330 (**1942**).
────── Cambio de ubicación del torio en el cuadro periódico. *Bol. Fac. Ing. Montevideo* **2**(3), 191-198 (**1942**).
────── The same title. *Anales Asoc. Quim. Argentina* **31**, 213-221 (1943).
────── En lugar del uranio en la nueva pléyada de "tierras raras". *Bol. Fac. Ing. Montevideo* **2**(7), 721-734 (**1944**).
────── La propriedades del neptunio y del plutonio afirman la hipotesis sobre la existencia de la nueva pléyada de "tierras raras". *Ibidem* **3**(2), 167-176 (**1946**).
────── La nueva pléyada de los "actinidos". *Bol. Soc. Quim. Peru* **13**, 73-93 (**1947**).
────── Properties of americium and europium and their position homologs in the periodic table. *Bol. Fac. Ing. Agrimens. Montevideo* **6**, 191-197 (**1955**).
────── La incorporación de electrones f en las envolventes extranucleares de los Lantánidos y de los Actínidos. *Ibidem* **6**(5), 149-162 (**1957**).
────── Position of the actinides in the periodic table. *Proc. U.N. Intern. Conf. Peaceful Uses Atom. Energy, 2nd, Geneva,* **28**, 385-391 (**1958**).
────── The constitution of the lanthanides and actinides series and a deduced law for the electronic configuration of the series of elements. *Bol. Fac. Ing. Agrimens. Montevideo* **7**, 13-26; 27-34; 35-42 (**1959**).
────── Ajuste propuesto para la disposicion actual de cuadro periódico. *Anales Real Soc. Españ. Fis. Quim.* (Madrid) **56 B**(5), 521-524 (**1960**).
────── Las tierras raras no son elementos de transición d. *Boletin Fac. Ing. Agrimens., Montevideo* **8**(4), 119-135 (**1962**).
────── La estructura electrónica y la posición de los elementos en la clasificación periódica. *Ibidem* **8**(12), 493-510 (**1964**).
────── Nota sobre la clasificación periódica de los elementos. *Anales Real Soc. Españ. Fis. Quim.* (Madrid) Ser. B., **60** (2/3), 103-106 (**1964**).
────── A suggested modification to the periodic chart. *J. Inorg. Nucl. Chem.* **28**(1), 25-29 (**1966**).
────── Orbitales de valencia de los lantanidos y los actinidos. *Anales Real Soc. Españ. Fis. Quim.,* Ser. B. **63**(7/8), 731-737 (**1967**).

_____ El centenario de la tabla periódico. Estado actual, posibilidades de desarrollo y algunos ajustes propuestos. *Revista Real Acad. Cienc. Exact., Fiz. Natural.* (Madrid), **63** (3), 517-532 (1969).

(12) Glenn T. Seaborg and Arthur C. Wahl. The chemical properties of elements 94 and 95. *J. Amer. Chem. Soc.* **70**, 1128-1134 (**1948**) (a 1942 year report is published).

Glenn T. Seaborg. The chemical and radioactive properties of heavy elements. *Chem. Eng. News* **23**, 2190-2193 (**1945**).

_____ The impact of nuclear chemistry. The heavy elements. Plutonium and other transuranium elements. *Ibidem* **24**, 1193-1198; 3160-3161 (**1946**) (p. 1197).

_____ The transuranium elements. *Science* **104**, 379-386 (**1946**).

_____ Place in periodic system and electronic structure of the heaviest elements. *Nucleonics* **5**(5), 16-36 (**1949**).

_____ Transuranium elements; present status. *Nobel lecture 1951*, **V. III** (**1942-1962**), 325-349.

_____ and Joseph J. Katz. *The actinide elements*. New York: **1954**.

_____ Recent research of the actinide elements. *16th Intern. Congr. Pure Appl. Chem., Sect. inorg. chem.* (Paris) *1957*, 289-312 (**1958**).

_____ Investigaciones recientes sobre los elementos "actinidos". *Ion* (Madrid) **18**, 136-142 (**1958**).

_____ The transuranium elements. *Endeavour* **18**, 5-13 (**1959**).

_____ Transplutonium elements. *Chemistry* (Washington) **32**(5), 1-15 (**1959**).

_____ Recent work with the transuranium elements. *Proc. Natl. Acad. Sci.* (Washington) **45**(4), 471-482 (**1959**).

_____ Poslednie dostizheniia v oblasti transplutonievykh elementov. *Atomnaia energiia* **6**, 21-33 (**1959**).

_____ *Man-made transuranium elements*. Englewood Cliffs, N.J.: Prentice-Hall, **1963**.

_____ and Arnold R. Fritsch. Synthetische Elemente. *Umschau Wissensch. Techn.* **64**(11), 321-324; (12), 372-376 (**1964**).

_____ Progress beyond plutonium. *Chem. Eng. News* **44**(25), 76-89 (**1966**).

(13) C. C. Kiess, Curtis J. Humphreys, and Donald D. Laun. Preliminary description and analysis of the first spectrum of uranium. *J. Res. U.S. Natl. Bur. Stand.* **37**, 57-72 (**1946**).

(14) William F. Meggers. Electron configuration of "rare-earth" elements. *Science* **105**, 514-516 (**1947**).

(15) Włodzimierz Rodziewicz. Zmiany rozmeszczenia pierwiastkow w ostatnim periodzie ich ukł adu. *Przeglad Chem.* (Poland) **5**, 182-187 (**1947**).

(16) T. T. Hardwick. The actinide series of chemical elements. *Proc. Conf. Nucl. Chem. Inst., Canada*, **1**, 44-50 (**1947**).

(17) Otto Hahn. *Künstliche neue Elemente*. Berlin: Verlag Chemie, **1948**, 46-48.

(18) G. B. Harvey. The actinide elements and the chemistry of plutonium. *Nucleonics* **2** (4), 30-40 (**1948**).

(19) A. G. Maddock. A second group of f shell elements. *Research* **1**, 690-701 (**1948**).

(20) B. C. Purkayastha. Fission of atomic nuclei. *Nucleonics* **3** (5), 2-21 (**1948**).

(21) E. E. Vainshtein. Redkie zeml'i i ikh mesto v estestvennoi sisteme khimicheskikh elementov. *Priroda* (Moscow) **37** (6), 7-21 (**1948**).

(22) H. J. Eméleus. Some recent advances in radiochemistry. *Nature* **163**, 624-626 (**1949**).

_____ Transuranics and other newly discovered elements. *Sci. Progress* **38**, 609-621 (**1950**).

_____ and A. G. Maddock. Die Chemie der Transurane. *Österr. Chem.-Ztg.* **57**, 153-160 (**1956**).
_____ The transuranic elements. *Chemistry & Industry* **1958**(40), 1276-1279.

(23) Zoltán Szabo. The actinide series and the periodic table. *Phys. Rev.* **76**, 147-148 (**1949**).
_____ and Béla Lakatos. Remarks on the article: "Comparison of the chemistry of cis and trans uranides with that of lanthanides and of transition metals." (in Hungarian). *Magg. Tud. Akad.*, Kém. Tud. Oszt. Kozlem. **18**, 413-418 (**1962**).

(24) Robert E. Connick. Oxidation states of the rare-earth and actinide elements. *J. Chem. Soc.* (London), **Suppl. issue 2, 1949**, 235-241.

(25) Ricardo de Carvalho-Ferreira. Sobre a posição, no sistema periódico, dos elementos con numero atomico superior a 88. *Anais Assoc. Quím. Brasil* **8**, 105-111 (**1949**).

(26) J. W. Spronsen. De twaalf nieuwe element. *Chem. Weekblad* **47**, 55-60 (**1951**).

(27) V. M. Klechkovskii. $(n+l)$-gruppy v posledovatel'nom zapolnenii elektronnykh konfiguratsii atomov. *Dokl. Akad. Nauk SSSR* **80**(4), 603-606 (**1951**).
_____ K kharakteristike tipa elektronnoi konfiguratsii atomov. *Ibidem* **135**, 655-658 (**1960**).
_____ Konfiguratsionnyi indeks elektronnoi struktury normal'nykh atomov. *Izvest. Timir. Sel'skokh. Akad.* **40** (3), 161-176 (**1961**).

(28) Reino W. Hakala. The periodic law in mathematical form. *J. Phys. Chem.* **56**, 178-181 (**1952**).

(29) Richard Lepsius und S. K. Asunmaa. Prinzipielle Betrachtungen im Periodischen System der Elemente. *Naturwissenschaften* **39**, 490-491 (**1952**).

(30) Robert Delhez. La question des actinides. *Ind. Chim. Belge* **18**, 1189-1197 (**1953**).

(31) V. I. Gol'danskii. *Novye elementy v periodicheskoi sisteme D. I. Mendeleeva.* Moskva: Izdat. Akad. Nauk SSSR, **1953**.
_____ The same title. Atomizdat, **1964**.

(32) Simon Z. Roginskii. *Mendělejevova periodiská soustava prvků ve světle posledních výzkumnů.* Praha: Státni naklad. techn. lit., **1954**.

(33) A. A. Grinberg, B. V. Ptitsyn, F. M. Filinov, and V. N. Lavrent'ev. Preparation of uranium hexacarbonyl. *Trudy Radiev. Inst. im. V. G. Khlopina, Khim. i geokhim. Akad. Nauk SSSR* **7**, 14-16 (**1956**).

(34) S. V. Markevich. Periodicheskaia sistema elementov D. I. Mendeleeva v svete noveishikh izsledovanii (in Russian). *Vestsi Akad. Navuk Belaruss. SSR* (Minsk), Ser. Fiz.-Tekh. Navuk **1957**(2), 5-19.

(35) Peter Graf. Die Transuranelemente. *Chimia* (Bern) **11**, 57-70 (**1957**) (in German).

(36) V. I. Semishin. Poslednie elementy periodicheskoi sistemy. *Trudy Moskov. Inst. Khim. Mashinostr.* **12**, 107-118 (**1957**).

(37) S. Dockx. *Théorie fondamentale du système périodique des éléments.* Bruxelles: Office Intern. de librairie, **1959**.

(38) Christian Klixbüll Jórgensen. Les électrons d et f dans les complexes des groups de transition. *J. chim. phys.* **56**, 889-896 (**1959**).
_____ Absorption spectra of actinide compounds. *Molec. Phys.* **2**, 96-108 (**1959**).

(39) B. B. Cunningham. Comparative chemistry of the lanthanide and actinide elements. *17th Intern. Congr. of pure appl. chem.*, (Munich), **1**, 64-81 (**1959**).

(40) V. M. Vdovenko. *Khimiia urana i transuranovykh elementov.* Moskva-Leningrad: Izdat. Akad. Nauk SSSR, **1960**.

(41) Mario A. Rollier. Gli elementi attinidici. *Chim. Ind. (Milan)* **43**, 775-781 (**1961**).

(42) D. McWhan. *Thesis.* Berkeley: UCRL-9695, **1961**.

(43) A. V. Lapitskii. Soveshchanie po stroeniiu Periodicheskoi Sistemy Mendeleeva. *Izv. Akad. Nauk SSSR* (9), 1719-1720 (**1961**).

(44) K. W. Bagnall. The transuranium elements. *Sci. Progress* (London) **52**, 66-83 (**1964**).

(45) Dieter Nebel. Zur Position der Actiniden- und Transactinidenelemente im Periodensystem der Elemente. *Z. f. Chem.* **10**, 251-260 (**1970**).

SUBDIVISION IIIB: ELECTRON CONFIGURATION TABLES

CLASS 4. TABLES WITH ELECTRONIC CONFIGURATION DISPOSITION OF ELEMENTS 2, 6, 10, AND 14 (BLOCKS OR SUBPERIODS).

Type IIIA4-1.

(1) Georg Schaltenbrand. Darstellung des periodischen Systems der Elemente durch eine räumliche Spirale. *Z. anorg. allgem. Chem.* **112**, 221-224 (**1920**) (table p. 222).

(2) C. J. Monroe and W. D. Turner. A new periodic table of the elements. *J. Chem. Educ.* **3**, 1058-1065 (**1926**) (table p. 1059).

(3) A. Slingervoet Ramondt. Een ruimte-model van het periodiek stelsel der elementen. *Chem. Weekblad* **25**, 496-498 (**1928**) (table p. 497).

(4) Charles Janet. *Essais de classification hélicoidale des éléments chimiques.* Beauvais: Avril, **1928** (table p. IV-1).
———— *La classification hélicoidale des éléments chimiques.* Ibidem: Novembre, **1928** (table III-9).
———— *Concordance de l'arrangement quantique, de base, des électrons planétaires des atomes, avec la classification scalariforme, hélicoidale, des éléments chimiques.* Ibidem: Novembre, **1930** (table II-5, III-3).
Cited by A. Quintana y Mari. La clasificación helicoidal de los elementos químicos. *Quím. e Industr.* (Barcelona) **8**(94), 287-292 (**1931**) (table IV, V).

(5) N. Efremov. Geochemische Gesetzmässigkeiten in der Zusammensetzung verschieden Magmen und ihr Zusammenhang mit den Gesetzmässigkeiten des Periodischen Systems der chemischen Elemente. *Fortschr. Mineral.* **29/30**, 82-84 (**1951**) (table p. 82).

Type IIIB4-1.

(1) C. J. Monroe and W. D. Turner. A new periodic table of the elements. *J. Chem. Educ.* **3**, 1058-1065 (**1926**) (table p. 1062).

(2) A. Slingervoet Ramondt. Een ruimte-model van het periodiek stelsel der elementen. *Chem. Weekblad* **25**, 496-498 (**1928**) (table p. 497).

(3) Charles Janet. *Essais de classification hélicoidale des éléments chimiques.* Beauvais: Avril, **1928** (table IV-5).
———— *La classification hélicoidale des éléments chimiques.* Ibidem: Novembre, **1928** (table IV-13, V-14).

___ *Considération sur la structure du noyau de l'atome.* Ibidem: Decembre, **1929** (table IV-6).

___ The helicoidal classification of the elements. *Chem. News* **138**, 372-374; 388-393 (**1929**) (table p. 388, 389).

Cited by A. Quintana y Mari. La clasificación helicoidal de los elementos químicos. *Quím. e Industr.* (Barcelona) **8** (94), 287-292 (**1931**) (table p. 290).

Subtype IIIA4-1a.

(1) Richard Lepsius und S. K. Asunmaa. Prinzipielle Betrachtungen im periodischen System der Elemente. VIII. Beziehungen zwischen Atomkern und Atomhülle. *Naturwissenschaften* **41**, 221-227 (**1954**) (table p. 225).

(2) Paul A. Giguère. The "new look" for the Periodic System. *Chem. in Canada* **18**(12), 36-39 (**1966**) (table p. 37/38).

___ The same title. *Can. Chem. Educ.* 3(1), 8-10; 12 (**1967**) (table p. 9).

Subtype IIIB4-1a.

(1) Charles Janet. *La classification hélicoidale des éléments chimiques.* Beauvais: Novembre, **1928** (table VI-16).

___ The helicoidal classification of elements. *Chem. News* **138**, 372-374; 388-393 (**1929**) (tables pp. 390, 392).

Cited by A. Quintana y Mari. La clasificación helicoidal de los elementos químicos. *Quím. e Industr.* (Barcelona) 8(94), 287-292 (**1931**) (table p. 288).

(2) Richard Lepsius und S. K. Asunmaa. Prinzipielle Betrachtungen im periodischen System der Elemente. VIII. Beziehungen zwischen Atomkern und Atomhülle. *Naturwissenschaften* **41**, 221-227 (**1954**) (scheme p. 227).

Subtype IIIC4-1a.

(1) N. Efremov. Geochemische Gesetzmässigkeiten in der Zusammensetzung vershieden Magmen und ihr Zusammenhang mit den Gesetzmässigkeiten des periodischen Systems der chemischen Elemente. *Fortschr. Mineral.* (Stuttgart) **29/30**, 82-84 (**1951**) (table p. 82).

Subtype IIIB4-1b.

(1) Jean Cueilleron. Nouvelle classification des éléments. *Compt. rend.* (Paris) **222**, 742-744 (**1946**) (table p. 743).

Type IIIB4-2.

(1) Carl A. Zapffe. A periodic chart for metallurgists. *Trans. Amer. Soc. Metals* **18**, 239-270 (**1947**) (table p. 248).

(2) Michael J. Plichta. Classification simplification—in an elementary manner. *Marquette Engineer* **35**(2), 7-10; 14; 24 (**1961**) (table p. 8/9).

___ and Sister Christopher Scherr and Daniel T. Haworth. A spiral, periodic, electronic chart. *School Sci. Math.* **65**, 473-476 (**1965**).

Type IIIC4-2

(0000) Frank Austin Gooch and Claude Frederic Walker. *Outlines of inorganic chemistry.* New York, London: Macmillan Co., V. II, 3-11, **1905** (table p. 9).

(000) F. Kirchhoff. Über eine Modifikation des Bohrschen Atommodells. *Z. physik. Chem.* **93**, 623-633 **(1919)** (table p. 626/627).

(00) Luigi Rolla e Giorgio Piccardi. Potenziale di ionizzazione e sistema periodico degli elementi. *Gazz. chim. ital.* **56**, 512-531 **(1926)** (scheme p. 522 and fragments p. 523).
Luigi Rolla. Proprietà atomiche e potenziale di ionizzazione. *Giorn. chim. ind. ed appl.* **8**, 266-267 **(1926)** (scheme p. 267).
_____ and G. Piccardi. Propriétés chimiques et potenciels d'ionisation. *Chim. Ind.* (Paris) **16**, 531-543 **(1926)** (scheme p. 539 and fragments p. 540).
_____ On the ionization potentials of the rare earth elements in relation to their position in the Periodic System. *Phil. Mag.* [7] **7**, 286-301 **(1929)** (scheme p. 295 and fragments p. 297).

(0) Magnad Saha. On the detailed explanation of spectra of the metals of the second group. *Phil. Mag.* [7] **3**, 1265-1274 **(1927)** (separate table).
_____ Über ein neues Schema für den Atomaufbau. *Physik. Z.* **28**, 469-473 **(1927)** (fragment p. 470 and scheme p. 473).

(1) Roy Gardner. A method of setting out the classification of the elements. *Trans. Proc. New Zealand Inst.* **60**, 31 **(1929)** (no table).
_____ A table of electronic configurations of the elements. *Nature* **125**, 146 **(1930)** (table).

(2) J. Gillis. Het periodiek systeem der elementen naar von Rijsselberghe. *Natuurw. Tijdschr.* (Belg.) **17**, 218-220 **(1935)** (table p. 219).

(3) Joseph A. Babor. A periodic table based on atomic number and electron configuration. *J. Chem. Educ.* **21**, 25-26 **(1944)** (table p. 25).

(4) W. Finke. Das periodische System der Elemente nach Quantenzahlen in flächenhafter Darstellung. *Z. Physik* **122**, 230-232 **(1944)** (table p. 231).

(5) José-Ignacio Bolívar. Nueva forma de clasificación periódica de los elementos. *Ciencia* (Mexico) **6**, 157-161 **(1945)** (table p. 160/161).

(6) William J. Wiswesser. The Periodic System and atomic structure. *J. Chem. Educ.* **22**, 314-322 **(1945)** (scheme p. XXI and 318).

(7) T. Grjébine. Classification des éléments d'après le nombre d'électrons sur la couche en formation, en fonction de la masse. *Bull. Soc. Chim. France* **1948**, 473-476 (table p. 475).

(8) A. P. Znoiko. Periodicheskii zakon atomnykh iader. II. Khimicheskie analogi elementov v periodicheskoi sisteme elementov atomnykh iader. *Dokl. Akad. Nauk SSSR* **68**(6), 1021-1024 **(1949)** (table p. 1024/1025).

(9) G. M. Murashov. Kvantovo-mekhanicheskaia forma periodicheskoi sistemy elementov. *Zh. Obshch. Khim.* **19**(3), 399-403 **(1949)** (table p. 400).

(10) Werner Gustav Kraudelat. Sistema periódico dos elementos baseado na estrutura dos atomos. *Anais Assoc. Quim. Brasil* (Rio de Janeiro) **8**, 125-140 **(1949)** (table p. 130/131).
_____ Una expressão matemática das propriedades periódicas dos elementos. *Anais Acad. Brasil Cienc.* (Rio de Janeiro) **35**(4), 521-526 **(1963)**.

(11) M. A. Catalán. Periodic table. *J. Cabrera. Fisica general*. Zaragoza, Spain, **1950**. *Cited* by Charlotte E. Moore. Atomic energy levels. *Circular, Natl. Bur. Stand.*, **V. II**, 467, XXX **(1952)**; **V. III**, 467. XXXVII **(1958)**.

(12) G. Scheibe, Fr. Baumgärtner und M. Genzer. Die Darstellung der Entstehung des Periodensys-

tems und der chemischen Bindung in der Chemischen Abteilung des Deutschen Museums. *Angew. Chem.* **67**, 502-515 **(1955)** (table p. 509).

(13) W. R. Walker and G. C. Curthoys. A new periodic table based on the energy sequence of atomic orbitals. *J. Chem. Educ.* **33**, 68-70 **(1956)** (table p. 68).

(14) Fritz Seel. *Atombau und chemische Bindung.* Stuttgart: F. Enke, **1956** (table p. 7).
———— *Atomic structure and chemical bonding.* London: Methuen & Co.; New York: J. Wiley & Sons, **1963** (table p. 8).

(15) Mircea V. Ionescu. Asupra clasificării periodice moderne a elementelor. *Acad. Rep. Popul. Romîne, Fil. Cluj., Stud. Cercet. Chim.* (Bucharest) **7**(1/4), 7-21 **(1956)** (table p. 10/11).

(16) H. C. Longuet-Higgins. The "Aufbauprinzip" as a basis for classification of the elements. *J. Chem. Educ.* **34**, 30-31 **(1957)**.

(17) Jack W. Eichinger jr. The electron chart. *J. Chem. Educ.* **34**, 70-71 **(1957)** (table p. 71).
———— Teaching electron figuration. *Ibidem* **34**, 504-505 **(1957)** (scheme p. 505).

(18) Lu Shao-chi. Improved table of Periodic System of elements (in Chinese). *Hau Hsueh Tung Pao* (Peiping) **1957** (7), 9-13 (table p. 12).

(19) A. T. Balaban. Numere magice de electroni și de nucleoni. *Revista Chim.* (Bucharest) **14**(3), 158-160 **(1963)** (table p. 158).

(20) Elena Bogdan. O propunere de aranjere a elementelor în sistem periodic (in Rumanian). *Analele Stiint. Univ. "Al. I. Cuza", Iasi* [N.S.] **10**(1), 1-16 **(1964)** (separate table p. 4/5).

(21) Earl D. Smith. A periodic diagram for the chemical elements based on relative electron orbital energies. *Fast Journ.* **8**(4), 3-4 **(1965)**.

(22) W. F. Luder. The atomic-structure chart of the elements. *Can. Chem. Educ.* **5**(3), 13-16 **(1970)**.

Subtype IIIC4-2a.

(1) O. Ramírez-Torres. Stability of the *1s* orbital and the chemical behavior of hydrogen. New position for hydrogen and helium in five-block periodic table. *J. Chem. Educ.* **32**, 450-455 **(1955)** (table p. 450).
———— The same title. *Rev. Soc. Brasil Quim.* **40**, 156-158 **(1955)**.

Subtype IIIC4-2b.

(1) A. E. Shtandel'. O ratsional'nom izobrazhenii Periodicheskoi Sistemy elementov. *Zh. Obshch. Khim.* **19**, 981-982 **(1949)** (separate table).
———— Periodicheskaia Sistema elementov v svete sovremennykh predstavlenii o stroenii atoma. *Uch. zap. Novosib. Inst. Sov. Kooper. Torg.* **1960** (1), 112-116.

Subtype IIIC4-2c.

(0) Edward G. Mazurs, **1958**. Not published.

(1) Rose Aynard. Pour une nouvelle classification des éléments de la matière. *Compt. rend.* (Paris) **248**, 2165-2167 **(1959)** (table p. 2166).

(2) Yves Normand. Une nouvelle présentation de la table de Mendeleiev. *Nucleus* (Paris) **1962** (4), 310-311 (table p. 310).

(3) V. K. Grigorovich. *Elektronnoe stroenie i termodinamika splavov zheleza.* Moskva: Nauka, **1970**.

Subtype IIIC4-2d.

(1) E. U. Condon and G. H. Shortley. *The theory of atomic spectra*. Cambridge, England: The Univers. Press, **1951** (table p. 333).

Subtype IIIC4-2e.

(1) C. H. Douglas Clark. Spectroscopy and valency. I. The periodic groups of atoms and ions. *Proc. Leeds Phil. Lit. Soc., Sci. Sect.* **2**, 336-346 (**1931**) (separate table p. 338/339).

(2) Bolesław Modrzejewski. Układ okresowy pierwistków a budowa atomu. *Wiadomości Chem.* (Poland) **3**, 49-66; 119-133 (**1949**) (separate table).

Type IIIC4-3.

(00) Charles R. Bury. Langmuir's theory of the arrangement of electrons in atoms and molecules. *J. Am. Chem. Soc.* **43**, 1602-1609 (**1921**) (table p. 1607).

(0) V. Romanoff. Le système périodique de Mendeléeff par représentation graphique. *Rev. sci.* (Paris) **72**, 661-665 (**1934**) (table p. 662).
_____ Représentation synoptique du système périodique de Mendeléeff et de la théorie de Bohr. *Compt. rend. 17. Congr. chim. ind.* (Paris) **II**, 2059-2062 (**1937**) (table p. 2060).

(1) David T. Gibson. The periodic table and electronic configuration. *Chemistry & Industry* **1948**, 11-12 (table p. 12).

(2) J. I. Bolívar, B. Bucay, J. Keller y M. M. Madrago G. Fundamentación matemática y desarrollo de la clasificación periódica natural de los elementos. *Cienc. Invest.* (Buenos Aires) **13**, 70-74 (**1957**) (table p. 73).
Jaime Keller T. The same title. II. Corroboración geométrica. *Revista Soc. Quím. México* **2**(3), 161-166 (**1958**) (tables pp. 162, 164).

(3) Edward G. Mazurs, **1958**. Not published.

(4) F. P. Platonov. Sovremennaia teoriia Periodicheskoi Sistemy elementov i pravilo zapolneniia elektronnoi sfery po V. M. Klechkovskomu. *Dokl. Moskov. Sel'skokh. Akad. Timir.* **1961** (64). 160-185 (separate table p. 182/183).

Subtype IIIC4-3a.

(1) Gil Chaverri-Rodríguez. The periodic table of elements. *J. Chem. Educ.* **30**, 632-633 (**1953**).
_____ *Tabla periódica, estructura electrónica y enlace química*. Costa Rica: Ciudad Univ., Ser. 100, **1962** (table p. 7).
_____ Estructura electrónica y tabla periódica. *Conf. Interamer. Enseñanza Quim.*, 1st, Buenos Aires, **1965**, 37-48.
_____ Elektrona strukturo kaj perioda tabelo. *Kem. Intern.* **4**(3), 169-181 (**1968**) (table p. 180/181).

(2) Edward G. Mazurs, **1958**. Not published.

(3) Harriman H. Dash. A quantum table of the Periodic System of elements. *Intern. J. Quant. Chem.* **3**, 335-340 (**1969**) (table p. 338).

Subtype IIIC4-3b.

(1) V. I. Semishin. Periodicheskaia Sistema elementov Mendeleeva i struktura elektronnykh obolochek atomov. *Zh. Obshch. Khim.* **25**, 2375-2380 (**1955**) (table p. 2376/2377).

Subtype IIIC4-3c.

(1) Edward G. Mazurs. A new numeration of periods in the Periodic System and the Kessler principle for the construction of the periodic table. *Can. Chem. Educ.* 4(3), 21-23 **(1969)** (table p. 23).

_____ Ķīmisko elementu periodiskās tabulas izveidošanās (in Latvian). *Technikas Apskats* **16**(57), 4-8 **(1970)** (table p. 7).

(2) Robert Mills. Systematic table of the elements. *Amer. J. Phys.* 40(8), 1169-1170 **(1972)** (table p. 1169).

CLASS 5. SHELL AND SUBSHELL TABLES WITH ELECTRONIC CONFIGURATION DISPOSITION OF ELEMENTS: 2, 6, 10, AND 14.

Type IIIA5-1.

(1) G. Haenzel. Die Polygonfläche und das periodische System der Elemente. *Z. Physik.* **120**, 283-300 **(1943)** (table p. 294).

(2) W. Finke. Bemerkungen zu G. Haenzel: "Die Polygonfläche und das periodische System der Elemente." *Z. Physik* **121**, 586-587 **(1943)** (table p. 587).

(3) L. Talpain. Classification gnomonique des elements. *J. phys. radium* [8] **6**, 176-181 **(1945)** (table p. 178).

Type IIIB5-1.

(1) L. Sibaiya. A circular periodic chart. *Am. J. Phys.* **9**, 122-123 **(1941)** (table p. 122).

(2) Hans Schultze. Über ein neues periodisches System der Elemente. *Naturwissenschaften* **32**, 58-59 **(1944)** (table p. 59).

(3) Alcindo Flores-Cabral. Classifiçăo natural dos elementos. *Inst. agron. do Sul, Pelotas-Rio Grande do Sul* (Brasil), circul. **1**, 1-3 **(1946)** (separate table).

_____ The same title. *Bolet. didat. Esc. agron. Eliseu Maciel, Pelotas* (Brasil) **1**, 3-6 **(1951)** (separate table).

_____ *Elementos de química electrônica.* Ibidem **4**, 1-75 **(1954)** (separate table).

(4) Johannes Mathias Hansteen. Forslag til en oppstilling av grunnstoffenes periodiske system. *Fra. Fysik. Verden* (Norway) **13**, 230-232 **(1952)** (table p. 231).

Type IIIC5-1.

(1) E. Cerasoli. Il sistema periodico e il principio di esclusione di Pauli. *Chim. ind., agr., biol.* (Milan) **17**, 37-43 **(1941)** (table p. 40).

(2) Hans Schultze. Über ein neues periodisches System der Elemente. *Naturwissenschaften* **32**, 58-59 **(1944)** (table p. 58).

(3) Yeou Ta. Une nouvelle représentation du tableau périodique basée sur la structure électronique des éléments. *Compt. rend.* (Paris) **221**, 441-442 **(1945)** (table 441).

_____ Une nouvelle représentation du tableau périodique des éléments. *Ann. phys.* (Paris) [12] **1**, 88-99 **(1946)** (table p. 91).

Subtype IIIA5-1a.

(1) Edward G. Mazurs. Originated **1967. Not published.**

Subtype IIIC5-1a.

(1) Edward G. Mazurs. Originated **1967.** Not published.

Subtype IIIC5-1b.

(1) Milton Kerker. The fable of the atomic theatre. *J. Chem. Educ.* **34,** 32 **(1957).**

Subtype IIIC5-1c.

(1) S. K. Mitra. On the periodic classification of the elements. *Phil. Mag.* [7] **11,** 1201-1214 **(1931)** (table p. 1211).

(2) Oscar Scarpa. Alcune rappresentazioni grafiche del sistema periodico degli elementi. *9. Congr. Intern. quim. pura y aplic.* (Madrid) **2,** 158-164 **(1934)** (table p. 159).

(3) Lewis E. Miller. A periodic chart based on *spdf* distribution. *J. Chem. Educ.* **32,** 198-199 **(1955).**

Type IIIC5-2.

(1) O. M. Corbino. Diagramma rappresentativo degli stati quantici e della formazione degli elementi del sistema periodico. *Rivista nuovo cimento* [N.S.] **5,** LVII-LXIV **(1928)** (table p. LXI).

(2) Pierre van Ryssellberghe. A new periodic table. *J. Chem. Educ.* **12,** 474-475 **(1935)** (table p. 474).

(3) Robert L. Ebel. Atomic structure and the Periodic System. *J. Chem. Educ.* **15,** 575-577 **(1938)** (scheme p. 576, fragmentary table p. 577).

(4) W. F. Luder. An improved periodic table. *J. Chem. Educ.* **16,** 393-395 **(1939)** (table p. 394).
_____ Electron configuration as the basis of the periodic table. *Ibidem* **20,** 21-26 **(1943)** (table p. 25).
_____ Atomic orbitals and valence. Rule of eight. *Ibidem* **22,** 221 **(1945).**
_____ and Saverio Zuffanti. *The electronic theory of acids and bases.* New York: J. Wiley & Sons; London: Chapman & Hall, **1946** (table p. 24).

(5) Fahlenbach. Ein periodisches System der chemischen Elemente in neuer Anordnung. *Umschau Wissensch. Techn.* **44,** 403-404 **(1940).**

(6) W. Finke. Das periodische System der Elemente nach Quantenzahlen in flächenhafter Darstellung. *Z. Physik* **122,** 230-232 **(1944)** (table p. 230).
_____ Über ein symmetrisches Periodensystem der chemischen Elemente nach dem Pauli-Prinzip. *Ibidem* **126,** 106-107 **(1949)** (table p. 107).

(7) Pierre Vallet. Nouvelle présentation du tableau périodique des éléments basée sur leur structure électronique. *Compt. rend.* (Paris) **227,** 58-60 **(1948)** (table p. 59).

(8) Werner Gustav Kraudelat. Sistema periódico dos elementos baseado na estructura dos atomos. *Anais Assoc. Quím. Brasil* (Rio de Janeiro) **8,** 125-140 **(1949)** (table p. 133).

(9) Ricardo de Carvalho-Ferreira. O ensino do sistema periódico dos elementos. *Pernambuco químico* (Brazil) **3**, 16-23 (**1950**) (table p. 19).

(10) J. López-Vigueras. Estudio cuántico de los elementos. *Ion* (Madrid) **14**, 599-607 (**1954**) (table p. 602).

(11) Merle L. Gardiner. *Electron distribution and properties of the elements.* McHenry, Ill.: **1962**.

Subtype IIIC5-2a.

(1) Arrigo Mazzucchelli. Una rappresentazione del diagramma degli elementi secondo Corbino. *Gazz. chim. ital.* **60**, 531-534 (**1930**) (table p. 533).

(2) Emil Petrovici. Citeva observații privitoare la caracterul elementelor și poziția lor in sistemul periodic. *Bulet. stiint. și techn., Scoala politechn. Timisvara* (Rumania) [N.S] **1**(15)(1), 335-350 (**1956**) (scheme on Fig. 1).

(3) Moddie D. Taylor. *First principles of chemistry.* Princeton, N.J.: D.van Nostrand, **1960** (table p. 298).

Subtype IIIC5-2b.

(1) Heinrich Pohl. Eine neue Darstellung des periodischen Systems in Verbindung mit dem Schalenaufbau der Atomhülle. *Universum* (Vienna) **13**(6), 176-180 (**1958**) (table p. 178).

Subtype IIIC5-2c.

(1) Robert A. Steinberg. Correlations between biological essentiality and atomic structure of the chemical elements. *J. Agr. Res.* **57**, 851-858 (**1938**) (table p. 854).

(2) Wojciech Rubinowicz. *Quantentheorie des Atoms.* Leipzig: J. A. Barth, **1959** (table p. 132/133).

Subtype IIIC5-2d.

(1) Edward G. Mazurs. Shortened table of the Periodic System of elements. *Can. Chem. Educ.* **8**, 13-15 (**1973**).

Type IIIB5-3.

(1) Fritz Scheele. Über eine Darstellung des periodischen Systems in Kreisform. *Z. Naturforsch.* **5a**, 11-13 (**1950**) (table p. 12).

Type IIIC5-3.

(1) Edward G. Mazurs. **1965**, Not published.

Subtype IIIC5-3a.

(1) Edward G. Mazurs, **1967**. Not published.

Subtype IIIC5-3b.

(1) Heinrich Pohl. Eine neue Darstellung des periodischen Systems in Verbindung mit dem Schalenaufbau der Atomhülle. *Universum* (Vienna) **13**(6), 176-180 (**1958**) (table p. 179).

(2) F. R. Keßler. Vorschlag einer neuen Darstellung des periodischen Systems. *Physik. Blätter* **17**(6), 270-274 (**1961**) (table p. 272/273).

(3) A. P. Dolgushin. Periodicheskii zakon D. I. Mendeleeva i elektronnoe stroenie atomov. *Priroda* (Moscow) **1969** (4), 58-61 (separate table p. 60/61).

CONCLUSION

(1) Dumas. *Compt. rend.* (Paris) **45**, 709-731 (**1857**).

(2) D. Mendeleev. *Zh. Russk. Khim. Obshch.* **1**, 60-77 (**1869**).

(3) Charles Janet. *La classification hélicoidale des éléments chimiques.* Beauvais: Novembre, **1928**.

(4) L. M. Simmons. *J. Chem. Educ.* **25**, 658-661 (**1948**) (table p. 659).

(5) R. T. Sanderson. An electronic distinction between metals and nonmetals. *J. Chem. Educ.* **34**, 229 (**1957**).

(6) Harriman H. Dash. Position of hydrogen in the Periodic System of elements. *Nature* **202**, 1001-1003 (**1964**).

TABLES NOT CLASSIFIED

(Tables arranged from many different points of view
that do not agree with the Periodic System.)

(1) F. J. Wiik. Forsok till en pä Atomgewigten grundad gruppering af de kemiska Elementerna. *Acta Soc. Sc. Finn.* **10**, 416-437 (**1875**).

(2) M. Zaengerle. *Über die Natur der Elemente.* München, **1882**.

(3) Edmund J. Mills. On the numeries of the elements. *Phil. Mag.* [5] **18**, 393-399 (**1884**); [5] **21**, 151-157 (**1886**) (table p. 156/157).

(4) Gustav Wendt. *Entwickelung der Elemente.* Berlin: Hirschwald'sche Buchhandlung, **1891**.

(5) J. Traube. Die Grundlagen eines neuen Systems der Elemente. *Z. anorg. Chem.* **8**, 77-80 (**1895**).

(6) J. W. Retgers. Über einige Änderungen im periodischen System der Elemente. *Z. physik. Chem.* **16**, 644-654 (**1895**) (tables p. 650, 651).

(7) Fritz Salomon. Über die Grundlagen eines neuen periodischen Systems der chemischen Elemente. *Verhandl. Ver. deut. Ntf. u. Ärzte I. Sitzung d. Naturwiss. Abt.* **69**, 98-99 (**1897**) (no table).
_____ Periodisches System der Elemente. *Z. angew. Chem.* **10**, 523-525 (**1897**) (table p. 523, 524/525).

(8) H. Wilde. Table des éléments, disposée avec les poids atomiques en proportions multiples. *Compt. rend.* (Paris) **125**, 707 (**1897**).

(9) Henry E. Armstrong. The classification of the elements. *Proc. Royal Soc.* (London) **70**, 86-94 (**1902**) (table p. 88-93).

(10) Henri Moissan. Classification des corps simples. *Rev. gén. chim. pure appl.* **7**, 73-82; 97-111 (**1904**) (table p. 108).
_____ The same. Paris: Masson et Cie., **1904**.
_____ The same. *Traité de chimie minérale.* Paris: P. I., **1904** (table p. 31).

(11) G. J. Stokes. A new theory of the periodic law. *Chem. News* **90**, 159 (**1904**).

(12) E. Bradbury. New classification of the elements. *Chem. News* **94**, 157-158 (**1906**) (table p. 157).

(13) Pietro Palladino. Sur le poids absolu des corps élémentaires et dépendance de leurs propriétés chimiques et physiques du poids absolu et de la forme. *Monit. Sci.* (*Docteur Quesneville*) [4] **24**, II, 489-522 (**1910**) (table p. 493).

(14) A. van den Broek. Das Mendelejeffsche "kubische" periodische System und die Einordnung der Radioelemente in dieses System. *Physik. Z.* **12**, 490-497 (**1911**) (table p. 491, 496).

(15) Albert C. Crehore. On the family-tree arrangement of the elements and calculation of atomic weights on the corpuscular ring theory of the atom. *Phys. Rev.* **34**, 241-257 (**1912**) (table p. 249).

(16) F. H. Loring. The cyclic evolution of the chemical elements. *Chem. News* **111**, 157-159; 181-182 (**1915**).

(17) Karl Fehrle. Über die Berechnung des Ortes der chemischen Elemente im periodischen System aus dem Atomgewichte und der Dichte. *Physik. Z.* **19**, 532-533 (**1918**).

(18) Ira D. Garard. A simple rule for the classification of the elements. *J. Chem. Educ.* **3**, 542-546 (**1926**).

(19) J. F. Spencer. The position of the elements of the rare earths in the periodic system. *J. Am. Chem. Soc.* **50**, 264-268 (**1928**).

(20) H. H. Stephenson. A statistical periodic table. *Chem. News* **138**, 129-130 (**1929**).

(21) R. W. Thatcher. A proposed classification of the chemical elements with respect to their function in plant nutrition. *Science* **79**, 463-466 (**1934**).

(22) Marcel Bossu. La classification des éléments. *Nouvelles de la chimie* **20**, 2 (**1936**).

(23) Chester Kennison. *The chemical spiral in metric scale.* **1942**.

(24) G. E. Djounkovsky et S. Kavos. Système périodique dans l'espace. *J. phys. radium* [8] **5**, 53-56 (**1944**).
Stéphan Kavos. *Planche de Stéphan Kavos sur la classification des éléments dans l'espace de Mrs. G. Djounkovsky et S. Kavos.* Paris, **1947**.

(25) K. Wickert. Metallschwund und Periodisches System. *Archiv für Metall.* **1**, 278-281 (**1947**) (table p. 278).

(26) Walter Russel. *The Secret of Light.* New York: Carnegie Hall, **1947** (tables p. 226, 297, 298).
———— and Lao Russell. *Atomic Suicide.* Swannanoa, Waynesboro, Virginia, **1957** (tables p. XXI, XXII, 87).

(27) Wilhelm Klemm. *Anorganische Chemie.* Wiesbaden, T. I, **1948** (table p. 72).

(28) Arrigo Horn. Il sistema periodico degli elementi ed i numeri poligonali. *Chimica* (*Milan*) **5**, 242-246 (**1950**).
———— I lantanidi. *Ibidem* **5**, 328-331 (**1950**) (table p. 328).

(29) Thomas P. Ratigan. *God's Toys. The Atoms.* Seattle, Wash.; **1952**.

(30) Oswaldo Baca-Mendoza. Leyes genéticas de los elementos químicos. Nuevo sistema periódico. *Univ. Nacl. del Cuzco* (*Peru*), 1-23 (in Spanish); 27-42 (in English), **1953** (table I).
———— The same title. *Bol. Soc. Quim. Peru* **21**, 5-23 (**1955**) (separate table p. 24/25).

(31) Francesco Vetere. *Un sistema nuovo per l'uso delle valenze nelle formulazione chimiche.* Benevento, Italia: Tip. Auxiliatrix, **1954** (table p. 15).

(32) Hans-Georg Heim. *Harmonikalperiodisches System.* München, Germany: **1955, 1957.**

(33) D. F. Stedman. Personal letter with two tables. *Natl. Res. Counc.* (Ottawa, Canada) **1958.**

(34) Bertrand A. Landry. A new table of the elements. *Main Currents in Modern Thought* **16** (4), 82-91 (**1960**) (table 3).

(35) F. M. Shemiakin. Zakonomernosti v svoistvakh khimicheskikh elementov, obnaruzhivaemye v treugol'noi tablitse s dvoinym vkhodom. *Trudy pervogo Moskov. Medicin. Inst.* **17**, 252-256 (**1962**) (separate table).

(36) D. N. Trifonov. *Granitsy i evoliutsiia Periodicheskoi Sistemy.* Moskva: Gos. Izd. Lit. po At. Nauke i Tekhn. **1963** (table between p. 128/129).

(37) Lars Gunnar Sillén. Elektronhöljet hos atomjoner. *Svensk Kem. Tidskr.* **77** (3), 123-131 (**1965**) (table p. 129).

(38) Tang Wah Kow. An octagonal prismatic periodic table. *J. Chem. Educ.* **49** (1), 59 (**1972**).

(39) J. H. van de Kamer. *A classification of a number of prototypes of organic chemical compounds on the basis of a corresponding arrangement of the Periodic System of the chemical elements.* Not published, **1972**, personal letter.

ARTICLES NOT FOUND IN LIBRARIES

(1) A. Gorbov. *Khimicheskie elementy; ikh prosteishie soedineniia. X. Estestvennaia periodicheskaia sistema elementov.* St. Petersburg: **1910.**

(2) K. Grinakovskii. *O prostranstvennoi modeli periodicheskoi sistemy elementov v sviazi s razvitiem poniatiia o periodichnosti svoistv materii.* Tomsk: **1914.**

(3) V. Sharvin. Novoe izobrazhenie periodicheskoi sistemy khimicheskikh elementov. *Priroda* (Moscow) **1917** (2), 238-239.

(4) E. Tycho. Structure of the atom and Periodic System. *Medd. Kgl. Vetenskapsakad. Nobelinst.* **5** (18), 7 (**1919**).

(5) A. G. Arkhipovich. Periodicheskaia sistema elementov Mendeleeva i struktura materii. *Smela,* **15**, 1 (**1924**).

(6) S. Plesniewicz. *Klasyfikacja perwiastkow chemicznykh.* Warsaw: Polskie Towarystwo Chemiczne, **1931.**

(7) A. F. Reiter. The structural relation of the rare earths to the periodic table. *Proc. Oklah. Acad. Sci.* **14**, 79-80 (**1934**) (the reproduction of the table which was given to the audience was not found).

(8) V. I. Semishin i S. Ia. Starodubtsev. *Stroenie atomov i periodicheskaia sistema.* Moskva: **1944.**

(9) G. Sandev. Spiral'naia tablitsa khimicheskikh elementov na osnove mendeleevskoi periodicheskoi sistemy i ikh poriadkovykh nomerov. *Khim. Industriia* (Sofia) **1950** (9/10), 524-530.

(10) N. P. Agafoshin. *Periodicheskii zakon i periodicheskaia sistema elementov D. I. Mendeleeva.* Moskva: **1951.**
———— Periodicheskaia sistema elementov D. I. Mendeleeva i stroenie atoma. *Khimiia v shkole* **1951** (4), 15-27.

(11) Hachiro Kakigawa. Periodic System. *Chemistry* (Japan) **1954**, 8061.

(12) D. Balarev. Ideal'naia i real'naia periodicheskaia sistema khimicheskikh elementov. *Priroda* (Sofia) **3** (3), 8-14; (4), 16-21 (**1954**).

(13) M. V. Volkenshtein. *Molekuly i ikh stroenie.* III. *Periodicheskaia sistema elementov D. I. Mendeleeva i stroenie atoma.* Akad. Nauk SSSR: **1955**.

(14) O. Wichterle. K voprosu ob uporiadochenii periodicheskoi sistemy. *Prirod. vědy škole* **7**(3), 247-251 **(1957)**.

(15) P. Roger. Periodic classification of the elements. *Information scientific* (Paris) **12** (4), 125-133 **(1957)**.

(16) V. I. Semishin. Poslednie elementy periodicheskoi sistemy D. I. Mendeleeva. *Trudy Moscov. Inst. Khim. Mashinostroen.* **12**, 107-118 **(1957)**.

(17) Shih-Li Chu. The Bohr form of Mendeleev's periodic table. *Tung Chi Ta Hsüeh Hsüeh Pao* (Shanghai) **3**(10), 42-58 **(1958)**.

(18) N. P. Agafoshin. *Sovremennoe sostoianie problemy sistematizatsii khimicheskikh elementov.* Moskva: **1959**.

(19) J. Moser. Dobavlenie k teorii periodicheskoi sistemy elementov. *Godishen. zb. Prirodno-matem. Fak. Univ. Skople* (Yugoslavia) **13**, 45-47 **(1960)**.

(20) R. Goldhammer und R. Flügel. Darstellung des Periodischen Systems. *Physik. Nachr.* **25**, 11 **(1961)**.

(21) N. Birkenstock. Periodisches System der Elementen nach F. R. Keßler. *Praxis der Naturwissensch.* [Teil A. Physik-Chemie] **A11** (8), 60-61, **1962**.

(22) K. V. Nelipa. *Periodicheskii zakon i stroenie atoma.* Tambovsk. Pedag. Inst.: **1962**.

(23) P. Briceno. Isotope of hydrogen H^4 and a new arrangement of elements in the periodic system. *Rev. Fac. Farmac.* (Venez.) **4** (9), 24-54 **(1963)**.

(24) D. Negoiu. The periodic system of elements. *Revista Fiz. Chim.* (Bucharest) **A1** (3), 81-91 **(1964)**.

(25) V. K. Grigorovich. Effect of structural characteristics of internal electron shells on properties of elements. *Izsled. Met. v Zhidkom. i Tverd. Sostoian. Akad. Nauk SSSR, Gos. Kom. po Chern. i Tsvetn. Met. pri Gosplane SSSR Inst. Met.* **1964**, 139-168.

(26) I. Angelova. Periodicheskaia sistema elementov v svete stroeniia atoma. *Biolog. i khimiia* (Belgrad) **7** (3), 16-64 **(1964)**.

(27) I. V. Klipikov. *Periodicheskii zakon, sistema elementov D. I. Mendeleeva i osnova ikh v stroenii atomov.* Kuibishev: **1964**.

(28) Chu Tsia-mo. Periodicity of chemical elements. Deduction of the periodicity of elements from the structure of outer electronic shells. *Taiiuan Huansueuan suebao* **1965** (1/2), 15-22.

(29) D. J. J. van Rensburg. Types of systematic classification of the elements. *Spectrum* (Pretoria, So. Africa) **7** (3), 154-158 **(1969)**.

(30) I. S. Karapetyan. Periodic system and electronic structure of atoms (in Russian). *Sborn. Nauchn. Trudov Erevan. Arm. Gos. Pedagog. Inst., Khim.,* **1970** (1), 123-130.

(31) G. Richter. *Periodensystem der Elemente.* Leipzig: Deut. Verlag Grundstoffind., **1971**.

Appendices

I. OUTLINE OF THE HISTORY OF DISCOVERY OF DISCOVERY OF THE PERIODIC SYSTEM

Prehistory:	de Morveau and others.	1782.	Fig. 1.
Law of Triads:	Dumas, Döbereiner, and others.	1828.	Fig. 2 and 3.
Valence tables:	Lothar Meyer and others.	1864.	Fig. 5.
Law of Octaves:	de Chancourtois, Newlands.	1863, 1865.	Fig. 6-8.
Forerunners of the discovery of the Periodic Law.	Odling, Hinrichs, and Lothar Meyer	1864, 1867, and 1868.	Fig. 9-11.
Discoverer of the Periodic Law.	Mendeleev	1869-1871.	Fig. 12-15; 17-20.

OUTLINE OF THE CLASSIFICATION OF THE TYPES OF PERIODIC TABLES
DIVISION I: SHORT TABLES
(with 8 columns)

	Group A:	Group B:	Group C:
Description of the types.	helices, space lemniscates, space concentric circles, and space squares.	spirals, lemniscates, and concentric circles.	series tables, zigzags, and parallel-line tables.

CLASS 1. TABLES WITH 8 GROUPS AND NO SUBGROUPS (LAW OF OCTAVES)

	Group A		Group B		Group C	
Type 1. Tables where the representative and transition elements are in the same groups without separation into subgroups.	IA1-1 de Chancourtois [Fig. 21]	1863	IB1-1 Wallin [Fig. 22]	1926	IC1-1 Newlands [Fig. 23]	1865
	—		—		IC1-1a Gibbes	1875
	IA1-1b Oppegaard	1948	—		IC1-1b Reed	1885
	—		—		IC1-1c Nechaev	1894
	—		—		IC1-1d Biltz a. Biltz	1928

218 GRAPHIC REPRESENTATIONS OF THE PERIODIC SYSTEM

Type 2. Tables where the transition and inner transition elements are separated from the representative elements in special tabulations.	—	—	IC1-2 Blanshard 1895 [Fig. 24]
	—	—	IC1-2a Masson; Ramsay 1896 [Fig. 25]
	—	IB1-2b Lyon 1928 [Fig. 26]	IC1-2b Rabinowitsch a. Thilo 1930
	—	—	IC1-2c de Boisbaudran 1895

CLASS 2. TABLES WITH TWO SUBGROUPS "a" AND "b"

Type 1. Symmetrical chessboard-like tables.	IA2-1 A. W. Stewart 1919 [Fig. 27]	—	IC2-1 Mendeleev 1870 [Fig. 28]
Type 2. Symmetrical tables with bridge elements.	IA2-2 O. I. Stewart 1928 [Fig. 29]	IB2-2 O. I. Stewart 1928 [Fig. 30]	IC2-2 von Richter 1889 [Fig. 31]
	—	—	IC2-2a Biltz 1902
	—	—	IC2-2b Preyer 1892 [Fig. 32]
Type 3. Symmetrical tables with elements of short periods divided symmetrically into two parts.	—	—	IC2-3 Benedicks 1904 [Fig. 33]
Type 4. Unsymmetrical tables with elements of short periods placed in the subgroups "a" and "b."	—	IB2-4 Baumhauer 1870 [Fig. 34]	IC2-4 Gretschel a. Bornemann 1883 [Fig. 35]
	—	—	IC2-4a Dauvilier 1920
	—	—	IC2-4b Martin; Molinari 1905
	—	—	IC2-4c Herz 1912
	—	—	IC2-4d Morette 1941 [Fig. 36]

Type 5. Unsymmetrical tables with elements of short periods placed exclusively in subgroups "a."	IA2-5 Harkins a. Hall [Fig. 37]	1916	IB2-5 Huth [Fig. 38]	1884	IC2-5 Arnold [Fig. 39]	1885
	—		—		IC2-5a Lothar Meyer [Fig. 40]	1870

CLASS 3. TABLES WITH THREE SUBGROUPS "a", "b", AND "c"

Type 1. Tables with inner transition elements in two series.	IA3-1 Bassett [Fig. 41]	1892	IB3-1 Green a. Jackson [Fig. 42]	1950	IC3-1 Schenk; Faustov [Fig. 43]	1949
	—		—		IC3-1a Chistiakov [Fig. 44]	1964
	—		IB3-1b Agafoshin	1952	IC3-1b Shemiakin [Fig. 45]	1932
Type 2. Tables with inner transition elements in one series.	IA3-2 Vogel [Fig. 46]	1918	—		IC3-2 Mazurs [Fig. 47]	1955
	—		—		IC3-2a Murashov [Fig. 48]	1949
	—		—		IC3-2b Scheele	1949

DIVISION II: MEDIUM TABLES
(with 16 or 18 columns)

CLASS 1. TABLES WITH 16 GROUPS

Type 1. Helix and spiral with equal revolutions, and tables with equal series.	IIA1-1 Hack [Fig. 49]	1910	IIB1-1 Flavitskii [Fig. 50]	1887	IIC1-1 Mendeleev [Fig. 51]	1869
	—		—		IIC1-1a Vaisman	1948
	—		—		IIC1-1b Loew	1897
Type 2. Equal lemniscate and zig-zag.	IIA1-2 Sir Crookes [Fig. 52[	1898	—		IIC1-2 Reynolds [Fig. 53]	1886

CLASS 2. TABLES WITH DISPOSITION OF ELEMENTS: 2, 8, AND 18

Type 1. Step tables with group 0 on one side of the table.	IIA2-1 B. K. Emerson [Fig. 54]	1911	IIB2-1 B. K. Emerson [Fig. 55]	1911	IIC2-1 Mendeleev [Fig. 56]	1871
	—		—		IIC2-1a Loring	1913
Type 2. Lemniscates and zigzags with group 0 on one side of the table.	IIA2-2 Soddy [Fig. 57]	1914	IIB2-2 Kipp [Fig. 58]	1942	IIC2-2 Woodiwiss [Fig. 59]	1906
	—		—		IIC2-2a Deeley	1893
	—		—		IIC2-2b Spring Shchukarev [Fig. 60]	1881; 1965
	—		IIB2-2c Bindel a. Blickle [Fig. 61]	1952	—	
Type 3. Tables symmetrical about a vertical line and with group 0 on the right side of the table.	IIA2-3 Tomkeieff	1954	IIB2-3 Oddo [Fig. 62]	1925	IIC2-3 Carnelley [Fig. 63]	1886
	—		—		IIC2-3a von Antropoff [Fig. 64]	1925
	—		—		IIC2-3b Grigorovich	1963
	—		—		IIC2-3c Schenk	1951
Type 4. Tables with interrupted short periods.	IIA2-4 E. von Stackelberg [Fig. 65]	1911	IIB2-4 Tocher [Fig. 66]	1910	IIC2-4 Mendeleev [Fig. 67]	1869
	—		—		IIC2-4a Grigorovich	1963
Type 5. Tables with group 0 in the middle of the table.	—		—		IIC2-5 Schmidt [Fig. 68]	1918
	—		—		IIC2-5a von den Broek [Fig. 69]	1911
	—		—		IIC2-5b Horsley [Fig. 70]	1900

APPENDICES 221

Type 6. Step tables with group IIa on the right side of the table.	—	—	IIC2-6 Mendeleev [Fig. 71]	1869
Type 7. Mirror-image tables.	—	—	IIC2-7 Wiberg [Fig. 72]	1936
	—	—	IIC2-7a Hackh [Fig. 73]	1914

DIVISION III: LONG TABLES
(with 32 columns)
Subdivision IIIA: Chemical Tables.

CLASS 1. TABLES OF ONE REVOLUTION AND ONE ROW.

Type 1. Tables of one revolution and of one row.	—	IIIB1-1 Opolonick [Fig. 74]	1935	IIIC1-1 Lea	1895
	—	—		IIIC1-1a Mendeleev [Fig. 75]	1889
	—	IIIB1-1b Stoye [Fig. 76]	1954	IIIC1-1b Vincent [Fig. 77]	1969

CLASS 2. TABLES WITH DISPOSITION OF ELEMENTS: 4, 16, 36, AND 64 (cycles)

Type 1. Tables of 4 planes, 4 revolutions, or 4 cycles.	IIIA2-1 Kapustinskii [Fig. 78]	1953	IIIB2-1 Rydberg [Fig. 79]	1913	IIIC2-1 Mazurs [Fig. 80]	1965
Type 2. Table of 4 lemniscates.	IIIA2-2 Bilecki [Fig. 81]	1915	—		—	

CLASS 3. TABLES WITH DISPOSITION OF ELEMENTS: 2, 8, 18, AND 32 (periods)

Type 1. Step tables with group 0 on one side of the table.	IIIA3-1 Stintzing [Fig. 82]	1916	IIIB3-1 Hackh [Fig. 83]	1914	IIIC3-1 Bassett [Fig. 84]	1892
	IIIA3-1a Stedman	1947	IIIB3-1a Janet [Fig. 85]	1928	—	
	IIIA3-1b Gamov	1940	—		—	
	IIIA3-1c Tremlelt	1963	—		—	
	—		—		IIIC3-1d Stareck	1932
	—		—		IIIC3-1e Verschoyle	1908

	IIIA3		IIIB3		IIIC3	
Type 2. Lemniscates and zigzags with group 0 on one side of the table.	IIIA3-2 Schirmeisen [Fig. 86]	1900	IIIB3-2 Janet [Fig. 87]	1928	IIIC3-2 Saz [Fig. 88]	1922
	IIIA3-2a Gooch a. Walker [Fig. 89]	1905	IIIB3-2a Rinck a. Feschotte [Fig. 90]	1962	—	
Type 3. Tables symmetrical about a vertical line with group 0 on the right side of the table.	IIIA3-3 Aucken [Fig. 91]	1951	—		IIIC3-3 Bayley [Fig. 92]	1882
	IIIA3-3a Sugathan a. Menon [Fig. 93]	1956	—		—	
	—		—		IIIC3-3b Hackh [Fig. 94]	1914
	—		—		IIIC3-3c Wagner a. Booth [Fig. 95]	1945
Type 4. Tables with interrupted short and medium periods.	IIIA3-4 Zmaczynski [Fig. 96]	1937	IIIB3-4 Janet [Fig. 97]	1928	IIIC3-4 Werner [Fig. 98]	1905
Type 5. Tables with group 0 in the middle of the table.	—		—		IIIC3-5 Mendeleev [Fig. 99]	1869
	—		IIIB3-5a Nodder	1920	IIIC3-5a Schmidt [Fig. 100]	1911
	—		—		IIIC3-5b Mazurs [Fig. 101]	1957
	—		—		IIIC3-5c Sheehan [Fig. 102]	1961
Type 6. Step tables with group IIa on the right side of the table.	—		—		IIIC3-6 Janet [Fig. 103]	1927
	—		—		IIIC3-6a Mazurs [Fig. 104]	1955
	—		—		IIIC3-6b Mazurs [Fig. 105]	1955

APPENDICES 223

Subdivision IIIB: Electronic Configuration Tables.

CLASS 4. TABLES WITH ELECTRONIC CONFIGURATION DISPOSITION OF ELEMENTS: 2, 6, 10, AND 14 (blocks or subperiods)

Type 1. Symmetrical helices and spirals, and series electronic configuration tables.	IIIA4-1 Schaltenbrand 1920 [Fig. 108]	IIIB4-1 Monroe a. Turner 1926 [Fig. 109]	—
	IIIA4-1a Lepsius a. Asunmaa 1954; Giguère 1966 [Fig. 110]	IIIB4-1a Janet 1928 [Fig. 111]	IIIC4-1a Efremov 1951 [Fig. 112]
	—	IIIB4-1b Cuilleron 1946 [Fig. 113]	—
Type 2. Right-side electronic configuration tables.	—	IIIB4-2 Zapffe 1947 [Fig. 114]	IIIC4-2 Gardner 1930 [Fig. 115]
	—	—	IIIC4-2a Ramírez-Torres 1955
	—	—	IIIC4-2b Shtandel' 1949
	—	—	IIIC4-2c Aynard 1959 [Fig. 116]
	—	—	IIIC4-2d Condon a. Shortley 1951 [Fig. 117]
	—	—	IIIC4-2e Douglas Clark 1931 [Fig. 118]
Type 3. Left-side electronic configuration tables.	—	—	IIIC4-3 Gibson 1948 [Fig. 119]
	—	—	IIIC4-3a Chaverri-Rodríguez 1953 [Fig. 120]
	—	—	IIIC4-3b Semishin 1955 [Fig. 121]
	—	—	IIIC4-3c Mazurs 1969 [Fig. 122]

CLASS 5. SHELL AND SUBSHELL TABLES WITH ELECTRONIC CONFIGURATION DISPOSITION OF THE ELEMENTS: 2, 6, 10, AND 14

Type 1. Symmetrical tables of concentric circles and parallel lines.	IIIA5-1 Haenzel 1943 [Fig. 123]	IIIB5-1 Sibaiya 1941 [Fig. 124]	IIIC5-1 Cerasoli 1941 [Fig. 125]
	IIIA5-1a Mazurs 1967 [Fig. 126]	—	IIIC5-1a Mazurs 1967 [Fig. 127]
	—	—	IIIC5-1b Kerker 1957 [Fig. 128]
	—	—	IIIC5-1c Mitra 1931 [Fig. 129]
Type 2. Right-side shell and subshell tables.	—	—	IIIC5-2 Corbino 1928 [Fig. 130]
	—	—	IIIC5-2a Mazzucchelli 1930 [Fig. 131]
	—	—	IIIC5-2b Pohl 1958 [Fig. 132]
	—	—	IIIC5-2c Steinberg 1938 [Fig. 133]
	—	—	IIIC5-2d Mazurs 1969 [Fig. 134]
Type 3. Left-side shell and subshell tables.	—	IIIB5-3 Scheele 1950 [Fig. 135]	IIIC5-3 Mazurs 1965 [Fig. 136]
	—	—	IIIC5-3a Mazurs 1967 [Fig. 137]
	—	—	IIIC5-3b Pohl 1958 [Fig. 138]

II. ELEMENT BLOCKS OF WHICH THE TABLES CONSIST

Block s.

s^1
1 H (hydrogen)
3 Li (lithium)
11 Na (sodium)
19 K (potassium)
37 Rb (rubidium)
55 Cs (cesium)
87 Fr (francium)

s^2
2 He (helium)
4 Be (beryllium)
12 Mg (magnesium)
20 Ca (calcium)
38 Sr (strontium)
56 Ba (barium)
88 Ra (radium)

Block p.

p^1
5 B (boron)
13 Al (aluminum)
31 Ga (gallium)
49 In (indium)
81 Tl (thallium)

p^2
6 C (carbon)
14 Si (silicon)
32 Ge (germanium)
50 Sn (tin)
82 Pb (lead)

p^3
7 N (nitrogen)
15 P (phosphorus)
33 As (arsenic)
51 Sb (antimony)
83 Bi (bismuth)

p^4
8 O (oxygen)
16 S (sulfur)
34 Se (selenium)
52 Te (tellurium)
84 Po (polonium)

p^5
9 F (fluorine)
17 Cl (chlorine)
35 Br (bromine)
53 I (iodine)
85 At (astatine)

p^6
10 Ne (neon)
18 Ar (argon)
36 Kr (krypton)
54 Xe (xenon)
86 Rn (radon)

Block d.

d^1
21 Sc (scandium)
39 Y (yttrium)
71 Lu (lutetium)
103 Lw (Lr) (lawrencium)

d^2
22 Ti (titanium)
40 Zr (zirconium)
72 Hf (hafnium)
104 Ku (kurchatovium)

d^3
23 V (vanadium)
41 Nb (niobium)
73 Ta (tantalum)

d^4
24 Cr (chromium)
42 Mo (molybdenum)
74 W (tungsten)

d^5
25 Mn (manganese)
43 Tc (technetium)
75 Re (rhenium)

d^6
26 Fe (iron)
44 Ru (ruthenium)
76 Os (osmium)

d^7
27 Co (cobalt)
45 Rh (rhodium)
77 Ir (iridium)

d^8
28 Ni (nickel)
46 Pd (palladium)
78 Pt (platinum)

d^9
29 Cu (copper)
47 Ag (silver)
79 Au (gold)

d^{10}
30 Zn (zinc)
48 Cd (cadmium)
80 Hg (mercury)

Block f.

f^1
57 La (lanthanum)
89 Ac (actinium)

f^2
58 Ce (cerium)
90 Th (thorium)

f^3
59 Pr (praseodymium)
91 Pa (protactinium)

f^4
60 Nd (neodymium)
92 U (uranium)

f^5
61 Pm (promethium)
93 Np (neptunium)

f^6
62 Sm (samarium)
94 Pu (plutonium)

f^7
63 Eu (europium)
95 Am (americium)

f^8
64 Gd (gadolinium)
96 Cm (curium)

f^9
65 Tb (terbium)
97 Bk (berkelium)

f^{10}
66 Dy (dysprosium)
98 Cf (californium)

f^{11}
67 Ho (holmium)
99 Es (einsteinium)

f^{12}
68 Er (erbium)
100 Fm (fermium)

f^{13}
69 Tm (thulium)
101 Md (mendelevium)

f^{14}
70 Yb (ytterbium)
102 No (nobelium)

III. EQUATIONS OF THE PERIODIC TABLE

Equations in connection with the periodic table are presented here.

1) Number of elements in a cycle — L_c (length of a cycle):

$$L_c = 4x^2 = 4(l + 1)^2$$

x = cycle number: 1,2,3,4.
Results: 4, 16, 36, 64.

$x = l+1$; l = subsidiary quantum number: 0,1,2,3.

2) Number of elements in each of the pairs of periods — L_p (length of periods):

$$L_p = 2x^2 = 2(l + 1)^2$$

x = cycle number: 1,2,3,4.
Results: 2, 8, 18, 32.

l = maximum value of subsidiary quantum number.

3) Number of elements in subperiods — L_s:

$$L_s = 2(2l + 1) = 4l + 2$$

l = subsidiary quantum number: 0,1,2,3.
Results: 2, 6, 10, 14.

4a) Number of elements in periods — L_p:

$$L_p = \sum_{z=0}^{z=3} 2(2l + 1)$$

Results: for 1st and 2nd period ($l = 0$): $L_p = 2(0+1) = 2 \cdot 1 = 2$;

for 3rd and 4th period ($l = 0$ and 1): $L_p = 2(0+1) + 2(2+1) = 2+6 = 8$;

for 5th and 6th period ($l = 0,1,$ and 2): $L_p = 2(0+1) + 2(2+1) + 2(4+1) =$
$= 2+6+10 = 18$;

for 7th and 8th period ($l = 0,1,2,$ and 3): $L_p = 2(0+1) + 2(2+1) + 2(4+1) +$
$+ 2(6+1) = 2+6+10+14 = 32$.

4b) Number of elements in periods — L_p:

$$L_p = 2\left\{\tfrac{t}{2} + \tfrac{1}{4}\left[1-(-1)^t\right]\right\}^2 =$$ [Simmons 1948 (1b)]

$$= \tfrac{1}{2}\left\{t + \tfrac{1}{2}\left[1-(-1)^t\right]\right\}^2 =$$ [Simmons 1947 (1a) and Marson 1956 (4)]

$$= \tfrac{t^2}{2} + \tfrac{2t+1}{4}\left[1-(-1)^t\right] =$$ [Klechkovskii 1951 (2a)]

$$= \tfrac{t^2}{2} + \left(\tfrac{t}{2} + \tfrac{1}{4}\right) - \left(\tfrac{t}{2} + \tfrac{1}{4}\right)(-1)^t = \left(\tfrac{t}{2}\right)^2 + \left(\tfrac{t}{2} + \tfrac{1}{2}\right)^2 - \left(\tfrac{t}{2} + \tfrac{1}{4}\right)(-1)^t$$

t — period number: 1,2,3,4,5,6,7,8.

Results: 2,2,8,8,18,18,32,32.

($t = n+l$; n = principal quantum number, l = subsidiary quantum number.

5) Atomic number of an element when the period terminates — Z_{end}:

$$Z_{end} = \tfrac{1}{6}t(t+1)(t+2) + \tfrac{1}{4}(t+1)\left[1-(-1)^t\right] = \quad \text{[Simmons 1948 (1b) and Hakala 1952 (3)]}$$

$$= \tfrac{1}{6}(t+1)^3 - \tfrac{1}{6}(t+1) + \tfrac{1}{4}(t+1)\left[1-(-1)^t\right] \quad \text{[Klechkovskii 1956 (2e)]}$$

Results, 2, 4, 12, 20, 38, 56, 88, and 120,
which means: He, Be, Mg, Ca, Sr, Ba, Ra,   —

6) Atomic number of an element when a period starts — Z_{beg}:

$$Z_{beg} = \tfrac{1}{6}(t^2-1) + \tfrac{1}{4}\left[1+(-1)^t\right] + 1 \quad \text{[Klechkovskii 1954 (2d)]}$$

Results: 1, 3, 5, 13, 21, 39, 57, and 89,
which means: H, Li, B, Al, Sc, Y, La, Ac.

7) Atomic number of an element when a new cycle starts, i.e. when a new kind of subperiod (s, p, d, or f) is added to the periods — Z_c:

$$Z = \tfrac{1}{3}l(2l+1)(2l+2) + 1 = \quad \text{[Klechkovskii 1953 (2b)]}$$

$$= \tfrac{1}{6}(2l+1)^3 + \tfrac{1}{6}(5-2l) = \quad \text{[Klechkovskii 1954 (2c)]}$$

$$= \tfrac{4}{3}l^3 + 2l^2 + \tfrac{2}{3}l + 1 = (2l^2+1)\left(\tfrac{2}{3}l+1\right)$$

Results: 1, 5, 21, and 57,
which means: H, B, Sc, La.

A following different formula can be given:

$$Z_c = 1 + 4 \sum_{l=0}^{l=3} l^2 \quad \text{[Mazurs 1968]}$$

Results: for s: $Z_c = 1 + 4l_0^2 = 1 + 4\cdot 0^2 = 1$;
for p: $Z_c = 1 + 4(l_0^2 + l_1^2) = 1 + 4(0^2 + 1^2) = 5$;
for d: $Z_c = 1 + 4(l_0^2 + l_1^2 + l_2^2) = 1 + 4(0^2 + 1^2 + 2^2) = 21$;
for f: $Z_c = 1 + 4(l_0^2 + l_1^2 + l_2^2 + l_3^2) = 1 + 4(0^2 + 1^2 + 2^2 + 3^2) =$
$= 1 + 4(1 + 4 + 9) = 1 + 4\cdot 14 = 57$.

1a) L. M. Simmons. A modification of the periodic table. *J. Chem. Educ.* **24**, 590 **(1947)**.

b) The display of electronic configuration by a periodic table. *Ibidem*, **25**, 658 a. 661 **(1948)**.

2a) V. M. Klechkovskii. (n + l)-gruppy v posledovatel'nom zapolnenii elektronnykh konfiguratsii atomov. *Dokl. Akad. Nauk SSSR* **80** (4), 604 **(1951)**.

b) K formulirovke pravil zapolneniia elektronnykh urovnei. *Ibidem*, **92**, 923 **(1953)**.

c) O pervykh elektronakh s dannym *l* v neitral'nom atome. *Zh. Eksper. Teoret. Fiz.* **26** (6), 760 **(1954)**.

d) O pravilakh, formuliruemykh pri pomoshchi (n + l)-gruppirovki kvantovykh urovnei. *Izv. Moskov. Timir. Sel'skokh. Akad.* **2** (6), 212 **(1954)**.

e) O nachale i okonchanii zapolneniia elektronami nekotorykh kvantovykh urovnei. *Dokl. Moskov. Sel'skokh. Akad. Timir.* **23**, 200 **(1956)**.

3) Reino Hakala. The periodic law in mathematical form. *J. Phys. Chim.* **56**, 178-181 **(1952)**.

4) L. M. Marson. Mathematical chemical periodicity. *Nature* (London) **177**, 1179-1180 **(1956)**.

Alphabetical Author Index

Name.	Year.	Type or section.	Reference number
	A.		
Abbott, David	1966.	IC2-2	(22)
	1966.	IIC2-4	(61)
Abelson, P. H.	see McMillan, E.		
Acera, Luis Hurtado	see Hurtado-Acera.		
Adams, Elliott Quincy	1911.	IIIC3-3	(3)
Addison, C. C.	1948.	IIC2-4.	(48)
Agafoshin, N. P.	1951.	not found	(10)
	1952.	IB3-1b	(1)
	1956.	IB2-4	(5)
	1959.	not found	(18)
Ageev, N. V.	1949.	IIC2-5	(6)
Akhmetov, N. S. and Vozdvizhenskii, G. S.	1965.	IC2-4	(45)
Akhumov, E. J.	1947.1961.	IIIC3-3	(10)
Albanskii, V. L.	1950.1951.	IC2-4	(30)
Ali, S. M.	1966.	IC3-1	(8)
Allen, William M.	1970.	IIC2-4	(65)
Amsterdamski, Stefan	1961.	IC3-1	(5)
Anderson, J. S.	see Eméleus, Harry J.		
Angelova, I.	1964.	not found	(26)
Anonymous author	1869.	triads	(13)
Anonymous author	1869.	valence tables	(5)
Anonymous author	1947.	IC2-3	(13)
Anonymous author	1947.	IA3-1	(2)
Antropoff, Andreas von	1925.1926.1927.	IIC2-3a	(1)
	1926.	IIIC3-3b	(2)
	1926.	uranides	(3)
	1937.	preface	(12)
	1937.	IIIC1-1	(2)
Arkhipovich, A. G.	1924.	not found	(5)
Armstrong, Henry E.	1902.	not classif.	(9)
Arnold, Carl	1885.1903.	IC2-5	(1)
Astakhov, K. V.	1971.	IC2-4	(50)
	1971.	IIC2-4	(67)
	1971.	IIIC3-1e	(7)
	1971.	IIIC3-4	(29)
Asunmaa, S. K.	see Lepsius, Richard		
Aucken, I.	1951.1952.	IIIA3-3	(1)
Austin, George T. and Austin, Helen F.	1963.	IIC2-4	(58)
Austin, Helen F.	see Austin. George T.		
Auwers, O. v.	1948.	IIIC3-6	(9)
	1948.	introd. IIIB	(20)
Aynard, Rose	1959.	IIIC4-2c	(1)

B.

Babor, Joseph A.	1944.	IIIC4-2	(3)
Baca-Mendoza, Oswaldo	1953.1954.1955.1957.	IIIC3-1e	(5)
	1953.1955.1957.	introd. IIIB	(26)
	1953.1955.	not classif.	(30)
Bacher, Robert F. and Goudsmit, Samuel A.	1932.	IIC2-6	(4)
Bagnall, K. W.	1964.	actinides	(44)
Baialovich, Ivan	1962.	IIIC3-4	(18)
Bakker, C. J.	1940.	IC2-5	(12)
Balaban, A. T.	1963.	introd. IIIB	(38)
	1963.	IIIC4-2	(19)
Balarew, D.	1921.	IC2-4b	(4)
	1922.	IIIC3-6a	(00)
	1954.	not found	(12)
Baldwin, J.	see Chissick, S. S.		
Balenović, Z.	see Iveković, H.		
Banerjee, Pares Chandra	1945.	IC2-5	(17)
Barraza, Ortega G., Calero A. De Lope, and Francisco Farré-Torá	1964.	IIIC3-6	(15)
Base, Daniel	see Simon, W.		
Baskerville, Charles	1909.	IC2-4b	(2)
Bassett, Henry	1892.	IA3-1	(1)
	1892.	IIIC3-1	(1)
	1892.	actinides	(1)
Batschinski, A. J.	1903.	IIC1-1	(5)
Bauer, Edmond	1922.	IC2-5	(5)
Baumhauer, Heinrich	1870.1873.	discovery	(8)
	1870.1873.	IB2-4	(1)
Baumgärtner, Fr.	see Scheibe, G.		
Baur, Emil	1911.	little per.t.	(3)
	1911.	IC2-5	(3)
Bayley, Thomas	1882.1898.	IIIC3-3	(1)
Bazarov, A.	1887.	IC1-1	(6)
Bedreag, Constantin G.	1916.	IC2-3	(5)
	1924.1925.1926.1927.1928. 1932.1933.1934.1942.1943. 1948.1952.1961.1962.	IIC2-4	(22)
	1942.1943.1948.1952.1954. 1955.1956.1960.1961.1962.		
	1963.	uranides	(10)
Beebe, William Sully	1882.	valence tables	(7)
Benedicks, Carl	1904.	IC2-3	(1)
Berthollet, Claude Louis	1787.	prehistory	(2)
Beutel, Ernst	1913.	IIA1-2	(2)
Bilecki, Alois	1913.	introd. IIIB	(2)
	1915.	IIIC1-1a	(3)
	1915.	IIIA2-2	(1)
Bilibin, Ia.Ia.	1939.	IIB2-4	(6)
Biltz, Heinrich	1902.	IC2-2a	(1)
Biltz, Heinrich, and Biltz, Wilhelm	1928.	IC1-1d	(1)
Biltz, Wilhelm	see Biltz, Heinrich		
Bindel, Ernst	1958.	IIIA2-1	(2)
Bindel, E. and Blickle, A.	1952.	IIB2-2c	(1)

Birkenstock, N.	1962.	not found	(21)
Bjerrum, Niels	1953.	IC2-2	(18)
Blanchard, Arthur A., and Wade, Frank B.	1914.	IC2-3	(2)
Blanshard, C. T.	1895.	IC1-2.	(1)
Blickle, A.	see Bindel, E.		
Blokh, M. A.	1934.	preface	(10)
Blomstrand, C. W.	1869.1870.	valence tables	(4)
Bogdan, Elena	1964.	IA1-1	(4)
	1964.	IB2-4	(6)
	1964.	IIIC4-2	(20)
Bohr, Niels	1922.1923.	IIIC3-3	(5)
	1922.1923.	uranides	(1)
Bohr, N. and Coster, D.	1923.	IIIC3-3	(5)
Boisbaudran, Lecoq de	1895.1897.	IC1-2c	(1)
Boirbaudran, Lecoq de, and Lapparent, A.de	1891 (citation)	octaves	(1c)
Bokii, G. B.	1942.	IC2-5	(14)
	1942.	IIC2-4	(40)
Bolin, Iwan	1924.	IC2-4	(9)
Bolívar, José Ignacio	1945.	IIIC4-2	(5)
Bolívar, J. I., Bucay, B., Keller, J., and Madrago G. M.	1957.	IIIC4-3	(2)
Bommer, H.	see Klemm, W.		
Booth, Harold Simmons	see Wagner, Henry A.		
Bornemann, G.	see Gretschel, H.		
Bossu, Marcel	1936.	not classif.	(22)
Bourgerel, G.	1920.	IIC1-1	(6)
Bowden, S. T.	1947.	IC1-2	(5)
Bradbury, E.	1906.	not classif.	(12)
Brauner, Bohuslaw	1902.	IC2-1	(4)
	1908.	little per.t.	(2)
Briceno, P.	1963.	not found	(23)
Broek, van den A.	1907.	IC1-1b	(3)
	1911.	IIC2-5a	(1)
	1911.	not classif.	(14)
	1914.	IC2-4b	(3)
Brown, Harold P.	1940.	IIIB3-4	(2)
	1940.	IIIC3-5	(7)
Bucay B.	see Bolívar, J. I.		
Büchner, E. H.	1915.	IC2-2a	(3)
Bury, Charles R.	1921.	IIIC4-3	(00)
Bustos, Carlos Lopez	see Lopez-Bustos, Carlos		
Butler, John A. V.	1927.	IC2-4	(11)
	1927.	IIC2-3	(15)

C.

Cabral, Alcino Flores	see Flores-Cabral, Alcindo		
Cáceres, Toribio	1911.	IIC2-4	(14)
Campbell, J. A.	1949.	introd. IIIB	(21)
Cap, F.	1950.	uranides	(16)
Carnelley, Thomas	1879.1884.	IC1-1	(3)
	1886.	IB2-4	(2)
	1886.	IIC2-3	(1)

Carrière, Émile, and Guiter, Henri	1943.	IIC2-6	(5)
Carroll, Benjamin, and Lehrman, Alexander	1948.	introd. IIIB	(19)
Carvalho-Ferreira, Ricardo de	1949.	actinides	(25)
	1950.	IIIC5-2	(9)
Caswell, A. E.	1929.	IIB2-4	(5)
Catalán, N. A.	1950.	IIIC4-2	(11)
Centnerszwer, M.	1926.	IC2-5	(8)
Cerasoli, E.	1941.	IIIC5-1	(1)
Chaikhorskii, A. A.	1970.1971.	IIIC3-1	(9)
Chancourtois, A. E. Béguyer de	1862.1863.	octaves	(1a,b,c,d)
	1863.	IA1-1	(1)
Chaudron, Georges	1940.	IIC2-4	(39)
Chauvierre, Marc	1919.	IIA2-2	(3)
	1919.	IIC2-4	(17)
Chaverri-Rodriíguez, Gil	1953.1962.1965.1968.	IIIC4-3a	(1)
	1957.1962.	preface	(18)
Chepelevetskii, M. L.	1966.	introd. IIIB	(39)
Chiang, Ming Chien	see Hsueh, Chin Fang		
Chicherin, B. N.	1888.	IC1-1	(7)
	1888.	IIC1-1	(3)
	1888.	IIC2-4	(4)
Chissick, S. S. and Baldwin, J.	1965.	IIIC3-4	(22)
Chistiakov, V. M.	1964	IC3-1a	(1)
	1964.	IIIC3-4	(20)
	1968.	inner trans. elem.	(3)
Chu, Shih-Li	1958.	not found	(17)
Chugaev, L. A.	1913.	IC2-1	(5)
Clark, C. H. Douglas	1931.1932.1934.1937.1938.	IC2-4	(14)
	1931.	IIIC4-2e	(1)
	1934.	IIC2-4	(31)
Clark, John D.	1933.	IIB2-1	(3)
	1950.	IIIB3-1	(4)
Clauson, Jennie E.	1952.1954.	IIIA3-1b	(3)
Cleator, P. E.	1950.	IIC2-3a	(3)
Clifford, A. A.	1959.	IC3-2	(2)
	1960.	IIIC3-4	(16)
Condon, E. U., and Shortley, G. H.	1951.	IIIC4-2d	(1)
Coninck, Oechsner de	1902 (citation).	triads	(10)
Connick, Robert E.	1949.	actinides	(24)
Cooke, Josiah P.	1854.	triads	(7)
Cooper, D. G.	1958.	IIC2-3a	(5)
Cooper, W. R.	1924.	IC2-4c	(3)
Corbino, O. M.	1928.	IIIC5-2	(1)
Coryell, Charles D.	1951.1952.	IIIC3-3	(14)
	1951.1952.	uranides	(19)
Coster, D.	see Bohr, Niels		
Courtines, M.	1925.	IIA2-1	(2)
	1925.	IIC2-5b	(5)
Crehore, Albert C.	1912.	not classif.	(15)
Crookes, Sir William	1886.1887.	IIC1-2	(2)
	1891.	priority (of Chancourtois)	(6)

	1898.	IIC1-1	(4)
	1898.	IIA1-2	(1)
Cueilleron	1946.	IIIB4-1b	(1)
Cunningham, B. B.	1959.	actinides	(39)
Curthoys, G. C.	see Walker, W. R.		
Cuthbertson, C., and Metcalfe, E. Parr (suggestion of Porter, A. W.)	1908.	IIC2-5b	(2)

D.

Dahmen, W.	1927.	preface	(9)
Dash, Harriman H.	1964.	conclusion	(6)
	1969.	IIIC4-3a	(3)
Daudel, Raymond	1943.	IIC2-4	(42)
	1943.	introd. IIIB	(12)
	1943.	uranides	(12)
Dauvillier, A.	1921.1922.	IC2-4a	(1)
Dawson, J. K.	1952.	uranides	(20)
Deeley, R. M.	1893.	IIC2-2a	(1)
Delhez, Robert	1953.	actinides	(30)
	1954.	IC2-4	(33)
	1954.1963.	preface	(16)
Delimarskii, Iu. K.	1969.	IB2-5.	(6)
Delimarskii, Iu. K. and Zarubitskii, O. G.	1969.	IB2-5	(6)
De Lope, Carelo A.	see Barraza, Ortega G.		
DeVault, Don	1944.	introd. IIIB	(13)
Diatkina (Dyatkina), M. E.	see Syrkin, Ia. K.		
Djounkovsky, G. E., and Kavos, S.	1944.	not classif.	(24)
Dmitriev, A. K.	1937.	IIC2-2	(6)
Dobrocvetov, E. N.	1948.	IIC1-1b	(3)
Döbereiner, Johann Wolfgang	1817.1829.	triads	(1a,b)
Dockx, S.	1959.	introd. IIIB	(35)
	1959.	actinides	(37)
Dolgushin, A. P.	1969.	IIIC5-3b	(3)
Druce, J. G. F.	1925.	preface	(8)
Dubpernell, George	1946.	IIIC3-1	(4)
Dumas, Jean Baptiste	1828.1851.1857.1859.	triads	(0, 5)
	1857.	conclusion	(1)
Dushman, Saul	1915.1916.1917.	IC2-2b	(2)
Dwight, A. E.	1960.	IIIB3-4	(5)

E.

Ebel, Robert L.	1938.	IIIC5-2	(3)
Efremov, N.	1951.	IIIA4-1	(5)
	1951.	IIIC4-1a	(1)
Eichinger, Jack W., jr.	1957.	IIIC4-2	(17)
Elsen, G.	1930 (citation).	triads	(7)
Eméleus, H. J.	1949.1950.1958.	actinides	(22)
Eméleus, Harry J., and Anderson, J. S.	1938.	introd. IIIB	(9)

Emeléus, H. J., and Maddock, A. G.	1956.	actinides	(22)
Emerson, B. K.	1911.1928.	IIA2-1	(1)
	1911.1928.	IIB2-1	(1)
	1911.	IIC2-3	(12)
	1911.	IIC2-4	(13)
	1928.	IIC2-2	(5)
Emerson, Edgar I.	1944.	IC2-5	(16)
	1944.	IIB2-4	(7)
	1944.	IIIC3-4	(7)
Emu, Ceka	1912.	IIC2-4	(15)
Erdmann, H.	1902.	IIB1-1	(3)
Estok, George K.	1956.	IC2-4	(36)
	1956.	IIC2-4	(55)
	1956.	IIIC3-4	(14)
Exhibition of Science at South Kensington (England)	1951.	IIB2-1	(5)

F.

Fahlenbach	1940.	IC2-4	(23)
	1940.	IIIC5-2	(5)
Fajans, Kasimir	1913.	IC2-4	(5)
	1914.	IC2-2	(8)
	1915.1919.	IC2-3	(3)
Farré-Torá, Francisco	1966/67.1967/68.	IIIC3-6	(15)
	1966/67.1967/68.	introd. IIIB	(42)
	see Barraza, Ortega G.		
Faustov, A. P.	1949.	IC3-1	(1b)
Fawsitt, C. E.	1931.	IC2-2	(16)
Fedorov, E. S. (Fedaroff)	1881.	IC1-1	(5)
Fehrle, Karl	1918.	not classif.	(17)
	1923.	IC2-3	(8)
Ferreira, Ricardo de Carvalho	see Carvalho-Ferreira, Ricardo de		
Fersman, A. E.	1933.	IIC1-1b	(2)
	1936.	IIC2-4	(34)
	1936.	IIC2-5b	(8)
Feschotte, Pierre	see Rinck, Émile		
Fickers, B. A.	1955.	IIC2-4	(54)
Filinov, F. M.	see Grinberg, A. A.		
Fillinger, Harriett H.	1932.	IC2-4	(15)
Finke, W.	1943.	IIIA5-1	(2)
	1944.	IIIC4-2	(4)
	1944.1949.	IIIC5-2	(6)
Fizicheskii Entsiklopedicheskii Slovar'	see Kratkaia Khimicheskaia Entsiklopediia		
Flavitskii, Flavian	1887.1888.1896.	IIB1-1	(1)
Flores-Cabral, Alcindo	1946.1951.1954.	IIIB5-1	(3)
Flügel, R.	see Goldhammer, R.		
Fornoff, Frank J.	see Hazlehurst, Thomas H.		
Foster, Laurence S.	1937.1939.1946.	IIC2-4	(36)
	1940.	uranides	(8)
	1949 (citation).	IIA2-1	(4)
Fourcroy, Antoine François de	1787.	prehistory	(2)

Frassares, Thomas Chr.	1966.	IIIC3-6	(21)
	1966.	IIIB3-4	(7)
	1966.	IIIC3-4	(25)
	1966.	introd. IIIB	(40)
Frémy, E.	1865.	triads	(12)
French, Sidney J.	1937.	IC1-2b	(2)
	1942.	IC2-4	(24)
	1943.	IC1-2	(4)
Friend, J. Newton	1925.	IC2-2	(13)
	1961.	IC2-4	(42)
Fritsch, Arnold R.	see Seaborg, Glenn T.		

G.

Gaisinskii, M.	see Haissinsky, Moise		
Gamov, Georg	1940.	IIIA3-1b	(1)
Ganesan, A. A.	1955.	introd. IIIB	(30)
Garaiev, Z. Sh.	1969.	IIIB3-1a.	(5)
Garard, Ira D.	1926.	not classif.	(18)
Gardiner, Merle L.	1962.	IIIC5-2	(11)
Gardner, Roy	1929.1930.	IIIC4-2	(1)
Garratt, A. J.	1951 (citation).	IIB2-1	(5)
Garrett, A. E.	1909.	preface	(4)
Geauque, H. A.	1925.	IIC2-2	(4)
Genzer, M.	see Scheibe, G.		
Getman, Frederick H.	1913.	IC2-4c	(2)
Ghiorso, Alberto, and Seaborg, Glenn T.	1956.	IIIC3-3	(18)
Gibbes, Lewis R.	1875.	IA1-1	(3)
	1875.	IC1-1a	(1)
Gibbs, Oliver Wolcott	1845.	triads	(3)
Gibson, David T.	1948.	IIIC4-3	(1)
Giguère, Paul A.	1966.1967.	IIIA4-1a	(2)
Gillis, J.	1935.	little per. t.	(14)
	1935.	IIIC4-2	(2)
Gladstone, J. H.	1853.	triads	(6)
Glockler, George, and Popov, Alexander	1951.1952.	IIIC3-3	(13)
Gmelin, Leopold	1843.	triads	(2)
Goeppert-Mayer, M.	1941.	uranides	(9)
Gol'danskii, V. I.	1952.	IC2-5	(21)
	1953.1964.	actinides	(31)
	1964.	IC2-4	(43)
	1968.1969.1970.	IIIC3-4	(27)
	1970	introd. IIIB	(47)
Gol'danskii, V. I., and Polikanov, S. M.	1969.	IIIC3-4	(27)
Goldhammer, D. A.	1896.1897.	IC1-1c	(2)
Goldhammer, R., and Flügel, R.	1961.	not found	(20)
Golgotiu, Tiberiu, Linde, Julieta, and Luca, Angela	1960.	IC1-2b	(4)
Gooch, Frank Austin, and Walker, Claude Frederic	1905.	little per. t.	(1)
	1905.	IIIA3-2a	(1)
	1905.	IIIC4-2	(0000)

Gorbov, A.	1910.	not found	(1)
Gordon, Gilbert	1960.	IIC2-4	(56)
	1960.	IIIA3-1b	(4)
Gorter, C. J., and Rutgers, A. J.	1933.	IC2-4	(16)
Goudsmit, Samuel A.	see Bacher, Robert F. and also Wu, Ta-You		
Graf, Peter	1957.	actinides	(35)
Graves, Stuart	1929.	IC2-4	(12)
Green, Frank O., and Jackson, Bernard G.	1950.	IB3-1	(2)
	1950.	IC3-1	(2)
Green, Jack	1953.	IC1-1	(14)
Gretschel, H., and Bornemann, G.	1883.	IC2-4	(1)
Griff, Helen K.	1964.	IIIB3-4	(6)
Grigorovich, V. K.	1963.1964.1965.1966.	IC3-1	(6)
	1963.1966.	IIC2-3b	(1)
	1963.1966.	IIC2-4a	(1)
	1963.	IIIC3-4	(19)
	1964.	not found	(25)
	1966.	IC1-2	(9)
	1910.	IIIC4-2c	(3)
Grimm, Albert	see Prandtl, Wilhelm		
Grimm, H. G.	1922.	IIC2-5	(2)
Grimsehl, Ernest	1959.	IC3-1	(4)
Grinakovskii, K.	1914.	not found	(2)
Grinberg, A. A., Pitsyn, B. V., Filinov, F. M., and Lavrent'ev, V. N.	1956.	actinides	(33)
Grjébine, T.	1948.	IIIC4-2	(7)
Groshans, J. A.	1884.	IIC2-4	(2)
Guenther, W. B.	1970.	IC1-2	(11)
Guides, Edu, Inc.	1958.	IC2-4d	(3)
Guiter, Henri	see Carrière, Émile		
Guzman, J.	1937.	IIC1-1	(8)

H.

Hack, Karl	1910.	IIA1-1	(1)
	1910.	IIB1-1	(6)
	1926.1934.	IIB2-4	(3)
Hackh, Ingo W. D.	1914.	IIB1-1	(7)
	1914.1915.1918.1919.1929.	IIC2-7a	(1)
	1914.1918.1919.1929.	IIIB3-1	(1)
	1914.	IIIC3-3b	(1)
	1918.1919.	IIIC1-1a	(4)
Haenzel, G.	1943.	IIIA5-1	(1)
Hähnel, Sigge	1949.	introd. IIIB	(22)
Hahn, Otto	1948.	actinides	(17)
Haissinsky, Moise	1930.1961.1964.	IC2-4	(13)
	1949.1950.1951.1953.1958. 1959.1961.1962.1963.1964. 1966.1969.1970.1971.	uranides	(14)
Haissinsky, M., and Jørgensen, C. K.	1971.	IIIC3-3	(25)
	1966.	uranides	(14)

Hakala, Reino	1948.1952.	IIIC3-6e(10)	(10)
	1948.1952.	introd. IIIB	(18)
	1952.	actinides	(28)
	1952.	equations	(3)
Hall, R. S.	see Harkins, William D.		
Hamilton, David C.	1965.	inner trans. elem.	(1)
Hansteen, Johannes Mathias	1952.	IIIB5-1	(4)
Hardt, Horst-Dietrich	1966.1969.1970.	IIIC3-4	(24)
Hardwick, T. T.	1947.	actinides	(16)
Harkins, William D.	1916.	IA2-5	(1)
	1917.	IC2-3	(4)
Harkins, William D., and Hall, R. E.	1916.	IC2-3	(4)
	1916.	IA2-5	(1)
	1916.	IIA2-2	(2)
	1916.	IIB2-4	(2)
Hartog, P. J.	1889 (citation).	octaves	(1c)
Harvey, G. B.	1948.	actinides	(18)
Haughton, Samuel	1888.	IIC1-2	(3)
Havens, William (of the Columbia University staff)	1949.	IIIB3-1	(3)
Haworth, Daniel T.	see Plichta, Michael J.		
Hayek, E.	1952.	IC2-4	(31)
Hazlehurst, Thomas H.	1941.	introd. IIIB	(11)
Hazlehurst, Thomas H., and Fornoff, Frank J.	1943.	IIC2-4	(41)
Heald, Milton T.	1954.	IIC2-4	(53)
Hecht, Selig	1947.	IIA2-1	(4)
Heim, Hans Georg	1955.1957.	not classif.	(32)
Heist, Gerhardt	1969.	IIC2-4	(63)
Heribert, Herbert	1930.	IC2-2	(15)
Herman, F., and Skillman, S.	1963.	introd. IIIB	(37)
Herz, W.	1912.	IC2-4c	(1)
Hicks, W. M.	1914.	IIC2-2	(3)
Hinrichs, Gustavus Detlef	1867.1869.	forerun.	(2)
	1867.	IB2-5	(0)
	1869.	IIC2-2b	(0)
Höltje, Robert	1940.	IC2-4	(22)
Hopkins, Arthur John	1911.	IIC2-3	(11)
Hopkins, B. Smith	1923.1939.	preface	(6)
	1933.1936.	IC2-4	(17)
Horie, S.	1954.	IA3-1	(3)
Horn, Arrigo	1950.	not classif.	(28)
Horsley, G. F.	1893.	IIC2-5b	(1)
Horstmann, A.	1885.	IIC2-4	(3)
Howe, Jas. Lewis	1900.	IC2-2	(5)
	see Venable, F. P.		
Hsueh, Chin Fang, and Chiang, Ming Chien	1937.	IIIC1-1a	(6)
Hubbard, Henry D.	1934.	IC2-4	(18)
Hübner, Gerhard	1951.	IIB2-4	(8)
	1951.1953.	IIC2-7a	(2)
Hull, Albert W.	1922.	IC2-2b	(3)
Hull, William Q.	1952.	IIC2-4	(50)
Humphreys, Curtis J.	see Kiess, C. C.		

ALPHABETICAL AUTHOR INDEX 237

Hurtado-Acera, Luis	1951.	IC3-1	(3)
Huth, Ernst	1884.	IB2-5	(1)
	1884.	IC2-5	(0)

I.

Incolla, Hector Vicente	see Mojos, Ana Maria		
Ionescu, Mircea V.	1956.	IC1-2	(6)
	1956.	IIIC3-4	(15)
	1956.	IIIC4-2	(15)
Irvin, K. Gordon	1938.1939.	IIB2-1	(4)
Istrati, C. I., and			
Longinescu, G. G.	1913.	IIC2-3	(13)
Iveković, H., and			
Balenović, Z.	1960.	preface	(21)

J.

Jackson, Bernard G.	see Green, Frank O.		
Janek, A.	1944.	IIIC3-6a	(0)
Janet, Charles	1927.1928.1929.1930.1931.	IIIC3-6	(1)
	1927.1929.1930.	introd. IIIB	(4)
	1928.1929.1931.	IIIB3-1a	(1)
	1928.	IIIB3-2	(1)
	1928.1929.1930.1931.	IIIB3-4	(1)
	1928.1930.1931.	IIIA4-1	(4)
	1928.1929.1931.	IIIB4-1	(3)
	1928;1929.1931.	IIIB4-1a	(1)
	1928.1929.	actinides	(7)
	1928.	conclusion	(3)
Jantsch, G., and			
Klemm, W.	1934.	little per.t.	(9)
Jawein	1881 (citation).	IIC1-1	(1)
Jørgensen, C. K.	1959.	actinides	(38)
	see Haissinsky, M.		

K.

Kahanovicz, M.	1928.	IIIC3-4	(3)
Kakigava, Hachiro	1954.	not found	(11)
Kamer, J. H. van den	1972	not classif.	(39)
Kapustinskii, F.	1951.	IC1-1	(13)
	1951.	IIIC3-3	(15)
	1953.	IC2-4	(32)
	1953.	IIIA2-1	(1)
Karapet'iants, M. Kh.,			
and Kreshkov, A. P.	1969.	IC2-4	(46)
	1969.	IIIC3-4	(28)
Karapetoff, Vladimir	1930.	IIIC1-1a	(5)
	1930.	uranides	(5)
Karapetyan, I. S.	1970.	not found	(30)
Kavos, Stéphan	1947.	not classif.	(24)
	see Djounkovsky, G. E		
Kedrov, B. M.	1953.	IC2-5	(22)
	1969.	discovery	(3)
	1969.	introd. IIIB	(44)
Keller, J. (Jaime T.)	1956.	Introd. IIIB	(31)
	see Bolívar, J. I.		
	1958.	IIIC4-3	(2)

Kemble, W. F., and Underhill, C. R.	1909.	IIC1-2	(5)
Kennison, Chester	1942.	not classif.	(23)
Kerker, Milton	1957.	IIIC5-1b	(1)
Keszler, F. R.	1961.	IIIC5-3b	(2)
Keszthelyi, Lajos	1955.	IC2-4	(35)
Kiess, C. C., Humphreys, Curtis J., and Laun, Donald D.	1946.	actinides	(13)
Kipp, Friedrich	1942.	IC2-3	(11)
	1942.	IIB2-2	(1)
Kirchhoff, F.	1919.	IIIC4-2	(000)
	1920.	IIC1-1a	(0)
Kiss, Julius	1909.	IC2-2	(7)
Klechkovskii, V. M.	1951.1952.1961.1968.1969.	IIIC3-6	(11)
	1951.1960.1961.	actinides	(27)
	1951.1952.1953.1954.1956. 1957.1960.1961.1965.1968. 1969.	introd. IIIB	(23)
	1951.1953.1954.1956.	equations	(2)
Klemenc, A.	1953.	IIIC3-1	(6)
Klemm, Wilhelm	1929.1930.1932.1937.1938. 1942.1969.	little per.t.	(9)
	1948.	not classif.	(27)
	1969.	IIIC3-3	(23)
	see Jantsch, G.		
Klemm, W. and Bommer, H.	1937.	little per.t.	(9)
Klipikov, I. V.	1964.	not found	(27)
Kohlweiler, Emil	1920.	IC1-1a	(3)
Korff, A. S.	1928.	IIIC3-6	(3)
Kow, Tang Wah	1972.	not classif.	(38)
Kratkaia Khimicheskaia Entsiklopediia	1964.	IC2-5	(24)
Kraudelat, Werner Gustav	1949.1963.	IIIC4-2	(10)
	1949.	IIIC5-2	(8)
Kraus, Charles A.	1927.	IIC2-5b	(6)
Kreshkov, A. P.	see Karapet'iants, M. Kh.		
Krishen, Anoop	see Verma, Mulk Raj		
Kunz, Jacob	1912.	IIB2-1	(2)
Kurbatov, V. Ia.	1925.	IB1-1	(0)
Kwasnik, Walter	1943.	IC2-5	(15)

L.

Ladenburg, R.	1920.	IC2-4	(6)
Lakatos, Béla	see Szabo, Zoltán G.		
Landry, Bertrand A.	1960.	not classif.	(34)
Langhammer, Günter	1949.	preface	(14)
Langmuir, Irwing	1919.	IIIC3-1e	(2)
Lapitski, A. V.	1961.	actinides.	(43)
Lapparent, A. de	see Boisbaudran, Lecoq de		
Laun, Donald D.	see Kiess, C. C.		
Lautié, Raymond	1938.	IIIC3-5	(5)
	1939.	IIIC3-6	(6)
Lavoisier, Antoine Laurent	1787.1789.	prehistory	(2,3)

Lavrent'ev, V. N.	see Grinberg, A. A.		
Layzer, David	1971.	introd. IIIB	(48)
Lea, M. Carey	1895.	IIIC1-1	(1)
Lebedev, I. A.	1972.	IC3-2b	(2)
Lehrman, Alexander	see Carroll, Benjamin		
Lenssen, E.	1857.	triads	(9)
Lepsius, Richard, and Asunmaa, S. K.	1952.	IIIC3-4	(11)
	1952.1954.1956.	IIIC3-6	(12)
	1952.	actinides	(29)
	1952.1956.	introd. IIIB	(25)
	1954.	IIIA4-1a	(1)
	1954.	IIIB4-1a	(2)
LeRoy, Royce H.	1927.	IIIC3-6	(2)
	1927.1931.	actinides	(5)
	1931.	IIIC3-5	(3)
Leyh, Frank Dieter	1970.	IC2-4	(48)
	1970.	introd. IIIb	(46)
Linde, Julieta	see Golgoţiu, Tiberiu		
Loew, E.	1897.	IIC1-1b	(1)
	1897.	IIIB1-1	(0)
Longinescu, G. G.	see Istrati, C. I.		
Longuet-Higgins, H. C.	1957.	IIIC4-2	(16)
Lopez-Bustos, Carlos	1952.1954.	IIC2-4	(51)
López-Vigueras, J.	1954.	IIIC5-2	(10)
Loring, F. H.	1913.1920.1923.	IIC2-1a	(1)
	1913.1942.	IIIC1-1a	(2)
	1915.	not classif.	(16)
	1922.1923.	IIIC3-1e	(3)
	1943.	IC2-3	(12)
Losanitsch, S. M.	1906.1909.	IIC2-3	(9)
Loung, Pai Yen	1965.	IC2-4	(44)
	1965.	IC3-1b	(3)
	1965.	IIC2-4	(60)
	1965.	IIC2-4a	(2)
	1965.	IIIC3-4	(23)
Lowry, Thomas M.	1915.	actinides	(3)
Luca, Angela	see Golgoţiu, Tiberiu		
Luder, W. F.	1939.1943.1945.	IIIC5-2	(4)
	1943.	preface	(13)
	1970.	IIIC4-2	(22)
Luder, W. F., and Zuffanti, Saverio	1946.	IIIC5-2	(4)
Lyon, Darvin O.	1928.	IB1-2b	(1)
	1928.	IIB2-4	(4)
	1928.	IIC2-4	(28)
	1928.	IIC2-6	(3)
	1928.	IIIC3-3	(6)

M.

Maddock, A. G.	1948.	actinides	(19)
	see Eleméus, H. J.		
Madelung, E.	1936.	introd. IIIB	(8)
Madrago G., M.	see Bolívar, J. I.		
Madras, Samuel	1967.	introd. IIIB	(43)
Mahler, Karl	1927.	IC2-4	(10)

	1927.	IIC2-4	(27)
Margary, Ivan D.	1921.	IIC2-3	(14)
Markevich, S. V.	1957.	actinides	(34)
Marson, L. M.	1956.	equations	(4)
Marson, L. M., and Zucchi, U.	1955.	IIIC3-3	(17)
Martin, Geoffrey	1904.1905.	IC1-1	(8)
	1905.	IC2-4b	(1a)
Mashentsev, A. I.	1954.	IC3-1b	(2)
Masson, Orme	1896.	IC1-2a	(1a)
Mast, W. C.	see Wrigley, A. N.		
Mathur, Prem Behari	1954.	IC1-2a	(3)
Matthes, Franz	1961.	IC2-4	(41)
Mazurs, Edward G.	1955.1956.1957.1966.1970.	preface	(17)
	1955.1956.	IC3-2	(1)
	1955.1956.	IIIC3-6a	(1)
	1955.1956.	IIIC3-6b	(1)
	1957.	IIIC3-5b	(1)
	1958.	IIIC4-2c	(0)
	1958.	IIIC4-3	(3)
	1958.	IIIC4-3a	(2)
	1958/1973	IIIC5-2d	(1)
	1965.	IIIC2-1	(1)
	1965.	IIIC5-3	(1)
	1967.	IIIA5-1a	(1)
	1967.	IIIC5-1a	(1)
	1967.	IIIC5-3a	(1)
	1969.1970.	IIIC4-3c	(1)
Mazzucchelli, A.	1930.	IIIC5-2a	(1)
	1935.	IC2-5	(9)
McCutcheon, T. P.	see Wrigley, A. N.		
McCutchon, K. B.	1950.	IC1-2a	(2)
McMillan, E., and Abelson, P. M.	1940.	uranides	(7)
McWhan, D.	1961.	actinides	(42)
Meggers, William F.	1947.	actinides	(14)
Meissner, W. Walter	1931.	IC2-3	(9)
	1931.	IIC2-4	(30)
Mendeleev, D. I.	1869.1870.1871.1872.1877. 1879.1880.	discovery	(1,4,5,6,7,10,11,12)
	1869.	IC1-1	(2)
	1869.1870.1871.1872.1875. 1877.1879.1880.	IC2-1	(0,1)
	1869.1871.1880.1881.1889. 1899.	IIC1-1	(1)
	1869.	IIA2-1	(00)
	1869.	IIC2-2a	(0)
	1869.	IIC2-4	(1)
	1869.	IIC2-6	(1)
	1869.	IIIC3-5	(1)
	1869.	conclusion	(2)
	1871.1880.1899.	priority	(1,4,7)
	1871.1872.1877.1879.1880.	IIC2-1	(1)
	1889.1892.1897.	IIIC1-1a	(1)
	1889.	introd. IIIB	(1)
	1906.	IIC2-3	(8)
Mendoza, Oswaldo Baca	see Baca-Mendoza, Osvaldo		

Menon, T. C. K.	see Sugathan, K. K.		
Menshutkin, B. N.	1934. (citation)	discovery	(1a)(5)
		IIC2-6	(1)
	1937.	IC2-5	(10)
Merz, H., and Ulmer, K.	1967.	inner trans. elem.	(2)
Metcalfe, E. Parr	see Cuthbertson, C.		
Meyer, Lothar	1864.	valence tables	(1)
	1868.	forerun.	(3)
	1868.	IIC2-6	(0)
	1870.	discovery	(8)
	1870.1880.	IC2-5a	(1)
	1872.	IA1-1	(2)
	1880.	priority	(3)
	1880.	IC1-1	(4)
	1880.	IC2-1	(2)
	1880.	IIC1-1	(2)
	1892.1893.	IC2-4	(2)
Meyer, R. J.	1914.	little per.t.	(4)
	1914.	IC2-2	(9)
Meyer, Stefan	1899.1918.	IIC2-3	(5)
	1909.	IIC2-3a	(0)
	1918.	IIC2-5b	(4)
Midgley, Thomas, jr.	1937.	IIIC3-1	(3)
	1937.	IIIC3-1e	(4)
	1937.	actinides	(10)
Mikhailenko, Ia. I.	1940.	IIIC3-3	(9)
Miller, Lewis E.	1955.	IIIC5-1c	(3)
Mills, Edmund J.	1884.1886.	not classif.	(3)
Mills, Robert	1972.	IIIC3-6	(18)
	1972.	IIIC4-3c	(2)
Mitra, S. K.	1931.	IIIC5-1c	(1)
Modrzejewski, Bolesław	1949.	IIIC4-2e	(2)
Moeller, Therald	1952.	introd. IIIB	(24)
Moissan, Henri	1904.	preface	(3)
	1904.	not classif.	(10)
Mojos, Ana Maria and Incolla, Hector Vicente	1969.	IC1-1	(15)
	1969.	IC2-4	(47)
	1969.	IIC2-4	(64)
Molinari, E.	1905.	IC2-4b	(1b)
Monckman, James	1907.	IIC2-4	(11)
Monroe, C. J., and Turner, W. D.	1926.	IIIC3-5	(2)
	1926.	IIIA4-1	(2)
	1926.	IIIB4-1	(1)
Moore, Charlotte E.	1952.1958 (citation)	IIIC4-2	(11)
Morette, A.	1941.	IC2-4d	(1)
Morozov, Nikolai	1907.	IC2-5	(2)
Morveau, Louis Bernard Guyton de	1782.1787.	prehistory	(1,2)
Moser, J.	1960.	not found	(19)
Mott, B. M.	1953.	uranides	(21)
Müller, Robert M.	1944.	IC2-2	(17)
	1944.	IC2-2b	(5)
Murashov, G. M.	1949.	IC3-2a	(1)
	1949.	IIIC4-2	(9)

N.

Napol'skii, S. A.	1948.	IIIB3-1a	(3)
Naquet, A.	1864.	valence tables	(2)
Nebel, Dieter	1970.	IIC2-4	(66)
	1970.	IIIC3-3	(24)
	1970.	actinides	(45)
Nechaev, N. P.	1894.	IC1-1c	(1)
Negoiu, D.	1964.	not found	(24)
Nekrassov, B. W.	1936.	IC2-4	(20)
	1936.	IC2-4c	(4)
	1936.	IIC2-3	(16)
Nelipa, K. V.	1962.	not found	(22)
Nernst, Walther	1892.1893.1895.	IC2-2	(2)
	1893.	IIC2-3	(3)
	1893.	IIC2-4	(7)
Nesmeyanov, A. N.	1960.	IC2-4	(37)
Neubert, D.	1970.	IIIC3-6	(17)
	1970.	introd. IIIB	(45)
Newlands, John Alexander Reina	1863.1864.	triads	(10)
	1864.1865.1866.1875.1878.	octaves	(2)
	1864.1865.1866.1878.	IC1-1	(1)
	1878.1884.1890.	priority	(2,5)
Nikol'skii, K. V.	1934.	introd. IIIB	(7)
Nissen, Knud Aage	1956.	IIIC3-6	(14)
	1956.	introd. IIIB	(32)
Noddack, Ida	1934.1935.	IIC2-4	(33)
Noddack, Walter	1937.	IIC2-4	(35)
Nodder, C. R.	1920.	IIIB3-5a	(1)
Normand, Yves	1962.	IIIC4-2c	(2)
Norrish, R. G. W.	1922.	little per.t.	(6)
	1922.	IIC2-4	(20)
	1922.	IIC2-4a	(0)

O.

Oberhauser, B. F.	1946.	IC2-4	(26)
	1946.	IIC2-4	(46)
Oddo, Giuseppe	1920.1925.1931.	IIC2-5a	(2)
	1925.	IIB2-3	(1)
Odling, William	1857.1861.	triads	(8)
	1864.	forerun.	(1)
	1864.	IIC2-6	(00)
Opolonick, Nicolas	1935.	IIIB1-1	(1)
Oppegaard, A. G.	1948.	IA1-1b	(1)
Oswald, Marcel	1937.1938.	IC1-2	(3)
Otake, Saburo	1971.	introd. IIIB	(49)

P.

Pachmann, Eduard	1966.	IC1-2	(10)
	1966.	IIC2-4	(62)
Palladino, Pietro	1910.	not classif.	(13)
Palmaer, Wilh.	1917.	IC1-1	(10)
	1924.1925.	IC2-5	(6)
Palmer, Charles Skeele	1890.1893.1897.	IIC2-4	(5)

Paneth, Fritz A.	1922.	IC2-3	(7)
	1923.1924.1930.	IC2-4	(8)
	1923.1928.1930.1942.	IIC2-4	(21)
	1950.1955.	uranides	(15)
Partington. I. R.	1920.	IB2-5	(3)
Pascal, Paul	1949.	IIC2-6	(6)
Pauli, W.	1954/55.1955 (citation)	IIIB2-1	(1)
Pauling, Linus	1939.1940.	introd. IIIB	(10)
Pavolini, T.	1957.	IC2-2	(20)
Payne, E. C.	1937.	IC2-4	(21)
	1938.	IA3-2	(2)
Pearce, D. W.	1935.	little per.t.	(12)
Perrin, Jean	1935.	IIIC3-5	(4)
	1935.	actinides	(9)
Petriichuk, D. I.	1966.	IC3-1	(7)
Petrovici, Emil	1956.	IC1-2	(7)
	1956.	IIIC5-2a	(2)
	1956.	introd. IIIB	(33)
Pettenkofer, M.	1850.1858.	triads	(4)
Pfeiffer, Paul	1920.1924.	IIC2-4	(19)
Piccardi, G.	see Rolla, L.		
Pitter, A. V.	1965.	IIIC3-4	(21)
Piutti, Arnaldo	1913.1914.	IIC2-2a	(2)
	1925.	IB2-5	(4)
Platonov, F. P.	1961.	IIIC4-3	(4)
	1961.1969.	introd. IIIB	(36)
Plesniewicz, S.	1931.	not found	(6)
Plichta, Michael J.	1961.	IIIB4-2	(2)
Plichta, Michael J., Scherr, sister Christopher, and Haworth, Daniel T.	1965.	IIIB4-2	(2)
Pohl, Heinrich	1958.	IIIC5-2b	(1)
	1958.	IIIC5-3b	(1)
Polikanov, S. M.	see Gol'danskii, V. I.		
Popov, Alexander I.	see Glockler, George		
Porter, A. W.	see Cuthbertson, C.		
Prandtl, Wilhelm, and Grimm, Albert	1924.	IIC2-4	(24)
Preyer, W.	1892.1893.	IC1-1b	(2)
	1892.1893.	IC2-2b	(1)
	1893.	IIC1-2	(4)
Ptitsyn, B. V.	see Grinberg, A. A.		
Purkayastha, B. C.	1948.	actinides	(20)

Q.

Quam, G. N., and Quam, Mary Battell	1934.	preface	(11)
Quam, Mary Battell	see Quam, G. N.		
Quill, Laurence L.	1938.	IIIC3-3	(8)
Quintana y Mari, A.	1931 (citation).	IIIB3-1a	(1)
		IIIB3-4	(1)
		IIIC3-6	(1)
		IIIA4-1	(4)
		IIIB4-1	(3)
		IIIB4-1a	(1)

R.

Rabinowitsch, Eugen, and Thilo, Erich	1930.	IC1-2b	(1)
	1930.	IIC2-4	(29)
	1930.	introd. IIIB	(5)
Radik, Iu. M.	1901.	IIC2-4	(9)
Radulescu, Dan	1912.	IC2-4	(3)
Ragno, Michele	1938.	IIC2-3	(17)
	1938.	IIC2-4	(37)
Ramírez-Torres, O.	1955.	IIIC4-2a	(1)
Ramondt, A. Slingervoet	1928.	IIIA4-1	(3)
	1928.	IIIB4-1	(2)
Ramsay, Sir William	1896.1900.1902.	IC1-2a	(1b)
	1900.	IIA2-1	(0)
	1903.	IIB1-1	(4)
Rang, P. J. F. (F.)	1893.1895.	IIC2-4	(6)
Ransom, Dorothy	1952.	IIIC3-1	(5)
Ratigan, Thomas P.	1952.	not classif.	(29)
Rayleigh, Lord	1911 (citation).	IIB1-1	(2)
Redfern, J. P., and Salmon, J. E.	1962.	IIC2-4	(57)
Reed, C. J.	1885.1895.	IC1-1b	(1)
Regny, P. Vinassa de	see Vinassa-de Regny, P.		
Reinmuth, Otto.	1928.1929.	IC1-1b	(4)
Reiter, A. F.	1934.	the table not found	(7)
Remy, Henrich	1931.	IIC2-5	(4)
	1931.	introd. IIIB	(6)
Renaud, Paul	1945.	IIC2-4	(43)
Rensburg, van D. J. J.	1969.	not found	(29)
Renz, Carl	1922.	little per.t.	(7)
	1922.	IIIA3-2a	(2)
Retgers, J. W.	1895.	not classif.	(6)
Reuber, Rudolf	1954.	IIIC3-4	(12)
Reychler, A.	1897.1899.	IC2-2	(4)
Reynolds, J. Emerson	1886.1895.	IIC1-2	(1)
Rhodes, Jack Claude	1933.	IIIC3-4	(4)
Rice, William E.	1956.	IA3-2	(3)
Richards, Theodore Williams	1898.	IIC2-3	(4)
Richter, G.	1971.	not found	(3)
Richter, Victor von	1869.1870 (citation).	discovery	(9c)
	1869 (citation).	IC1-1	(2)
	1870 (citation).	IC2-1	(1)
	1889.	IC2-2	(1)
	1889.	IIC2-3	(2)
Rinck, Émile, and Feschotte, Pierre	1962.	IIIC1-1a	(7)
	1962.	IIIB3-2a	(1)
Rixon, F. W.	1933.	IIC1-1	(7)
Rodebush, Worth H.	1925.	IIC2-4	(25)
Rodziewicz, Włodzimierz	1947.	IC2-4	(27)
	1947.	IIIC3-3	(11)
	1947.	actinides	(15)
Roger, P.	1957.	not found	(15)
Roginskii, Simon Z.	1954.	IC2-4	(34)
	1954.	actinides	(32)

Rolla, Luigi	1926.	IIIC4-2	(00)
Rolla, Luigi, and Piccardi, Giorgio	1926.1929.	IIIC4-2	(00)
Rollier, Mario A.	1961.	actinides	(41)
Romanoff, V.	1934.	IIIB3-2	(2)
	1934.1937.	IIIC3-4	(5)
	1934.1937.	IIIC3-6	(4)
	1934.1937.	IIIC3-6b	(0)
	1934.	actinides	(8)
	1934.1937.	IIIC4-3	(0)
Roscoe, H. E., and Schorlemmer, C.	1900.	IIC2-3	(6)
Ruben, Samuel	1964.1965.	IIC2-5	(8)
Rubinovicz, Wojciech	1959.	introd. IIIB	(34)
	1959.	IIIC5-2c	(2)
Rudorf, George	1900.1904.	preface	(2)
	1900.	IC2-1	(3)
	1903.	IIC2-3	(7)
	1904.	IC2-2	(6)
Rudy, Richard	1927.	actinides	(6)
Russell, Lao	see Russell, Walter		
Russell, Walter	1947.	not classif.	(26)
Russell, Walter and Lao	1957.	not classif.	(26)
Rutgers, A. J.	see Gorter, C. J.		
Rydberg, J. R.	1913.1914.	IC2-4	(4)
	1913.1914.	IIC2-2	(2)
	1913.1914.	IIIB2-1	(1)
Ryss, J. G.	1948.	IIC2-4	(47)
Rysselberghe, Pierre van	1935.	little per.t.	(13)
	1935.	IIIC5-2	(2)

S.

Sabatier, Paul	1890.	IIC2-2b	(2)
Saccardo, Pietro	1955.	IB2-4	(4)
Sadikov, V. S.	1940.	IIIC3-6	(7)
Saha, Maghnad (Magh Nad)	1927.	IIIC4-2	(0)
Salmon. J. E.	see Redfern, J. R.		
Salomon, Fritz	1897.	not classif.	(7)
Sanderson, R. T.	1954.1956.	IC2-5	(23)
	1957.	conclusion	(5)
	1964.	IC1-2	(8)
Sandev, G.	1950.	not found	(9)
Sanford, Ferfando	1911.	IC1-1	(9)
Sarkisov, E. S.	1948.1950.	IIC2-2	(7)
Satou, Sumio	1955.1959.1966.	IC1-2a	(4)
	1955.1959.1966.	IIIC3-1e	(6)
Saz, Eugenio	1922.1925.1931.1934.	IIIC3-2	(1)
Sborgi, Umberto	1925.	IC2-2a	(4)
	1951/52.	IIIC3-3	(16)
Scarpa, Oscar	1934.	IIC2-1	(3)
	1934.	IIIC5-1c	(2)
Schaltenbrand, Georg	1920.	IIIA4-1	(1)
	1921.	IIIC3-3	(4)
	1921.	actinides	(4)
Scheele, Fritz	1949.	IC3-2b	(1)
	1949.	IIIC3-3b	(4)

	1950.	IIIB5-3	(1)
Scheer, Roderich	1955.	IC2-2	(19)
	1955.	IIC2-3a	(4)
Scheibe, G., Baumgärtner, Fr., and Genzer, M.	1955.	IIIC4-2	(12)
Schenk, Peter W.	1949.1951.	IC3-1	(1a)
	1951.	IIC2-3c	(1)
Scherer, George A.	1949.	IIIA3-4	(2)
Scheringa, K.	1911.	IC2-5	(4)
Scherr, sister Christopher,	see Plichta, Michael J.		
Schirmeisen, Karl.	1900.	IIIA3-2	(1)
Schlesinger, Gert G.	1966.	IIIB3-1a	(4)
Schmidt, Curt	1911.1917.	IIIC3-5a	(1)
	1917.	preface	(5)
	1917.	IC2-2	(11)
	1918.	IIC2-5	(1)
Schorlemmer, C.	see Roscoe, H. E.		
Schultze, Hans	1944.	IIIB5-1	(2)
	1944.	IIIC5-1	(2)
Science Service	1947.	IC2-4d	(2)
Seaborg, Glenn T.	1945.1946.1949.1959.1969.		
	1971.	IIC2-4	(44)
	1969.1971.	IIIC3-3	(18)
	see Ghiorso, Alberto		
Seaborg, Glenn T., and Fritsch, Arnold R.	1963.	IIIC3-3	(18)
Seaborg, Glenn T., and others	1942.1945.1946.1948.1949. 1951.1954.1958.1959.1963. 1964.1966.	actinides	(12)
Sears, George (Geo) W.	1924.1933.	IC2-2b	(4)
Seel, Fritz	1956.1963.	IIIC4-2	(14)
Selinov, I. P.	1951.	IC2-5	(20)
Sell, Octavia S.	1955.	IIIA3-2	(3)
Semishin, V. I.	1955.	IIIC4-3b	(1)
	1957.	actinides	(36)
	1957.	not found	(16)
	1959.1969.	preface	(20)
	1962 (citation).	IIC2-4	(9)
Semishin, V. I., and Starodubtsev, S. Ia.	1944.	not found	(8)
Seubert, Karl	1895 (citations).	octaves	(1c)
		forerun.	(3)
		discovery	(1a)
		IIC2-1	(1)
		IIC2-6	(0) (1)
		IIIC3-5	(1)
Shao-chi, Lu	1957.	IIIC4-2	(18)
Sharvin, V.	1917.	not found	(3)
Shchukarev, S. A.	1937.	little per.t.	(15)
	1937.	IC2-5	(11)
	1954.	introd. IIIB	(27)
	1960.	IC2-4	(39)
	1962.	IC2-2	(21)
	1965.	IIC2-2b	(3)
Sheehan, W. F.	1961.	IIIC3-5c	(1)

Shemiakin, F. M.	1932.	IC3-1b	(1)
	1962.	not classif.	(35)
Shireby, D.	1966.	IIIA3-4	(3)
Shortley, G. H.	see Condon, E. U.		
Shtandel', A. E.	1949.1960.	IIIC4-2b	(1)
Sibaiya, L.	1941.	IIIB5-1	(1)
Siborg, G.	see Seaborg, Glenn T.		
Sidgwick, N. V.	1950.	IC2-4c	(5)
	1950.	uranides	(18)
Sidle, A. B.	1966.	IIIA3-1b	(5)
Silbermann, T.	1916.	IIC2-6	(2)
Sillén, Lars Gunnar	1965.	not classif.	(37)
Simmons, L. M.	1947.1948.	IIIC3-6	(8)
	1947.1948.	introd. IIIB	(17)
	1947.1948.	equations	(1)
	1948.	conclusion	(4)
Simon, W., and Base, Daniel	1909.	IIB1-1	(5)
Sisler, Harry H., and Vanderwerf, Calvin A.	1943.	motto	
Skillman, S.	see Herman, F.		
Smith, Earl D.	1965.	IIIC4-2	(21)
Smith, John D. Main	1924.	preface	(7)
	1924.1927.	IC2-5	(7)
	1927.	little per.t.	(8)
	1927.	IC1-2	(2)
Smith, Neville	1955.	IIIC3-6	(13)
	1955.	introd. IIIB	(29)
Smiz, Eduard	1919.	IC1-1	(11)
Smyth, Henry DeWolf	1945.	uranides	(13)
Snijder, H. G. S.	1939.	IIC2-4	(38)
Société Chimique de France	1940.	IC2-5	(13)
Soddy, Frederick	1914.	IIA2-2	(1)
Sokoloff, V. P.	1954.	IIB2-4	(9)
	1954.	IIC2-4	(52)
Sommerfeld, Arnold	1924.	uranides	(2)
	1925/26.	introd. IIIB	(3)
	1930.	little per.t.	(10)
Sommerfeldt, Ernst	1934.	IC2-3	(10)
	1934.	IIC2-5b	(7)
Spedding, Frank H.	1951.	IIIC3-3	(12)
Spence, R.	1949.	IC2-4	(29)
Spencer, J. F.	1928.	not classif.	(19)
Spice, J. E.	1960.	IIIC3-3	(19)
Spitsyn, V. I.	1959.	uranides	(22)
Spring, W.	1881.1886.	IIC2-2b	(1)
Spronsen, J. W. van	1951.1968 (citation).	octaves	(1d)
	1951 (citation).	IA1-1	(1)
	1951.	IIIC3-4	(10)
	1951.	actinides	(26)
	1964 (citation).	forerun.	(1)
	1966 (citation).	octaves	(2)
	1968 (citation).	forerun.	(2)
	1969.	preface	(22)

Stackelberg, Eduard von	1911.	IIA2-4	(1)
	1911.	IIC2-4	(12)
	1911.	IIC2-5b	(3)
Stackelberg, Mark von	1925 (citation).	IIC2-3a	(1)
	1929.	IIC2-3a	(2)
	1929.	IIIC3-3b	(3)
Staigmüller, H.	1901.	IIC2-4	(10)
Staff of Chicago Museum of Science and Industry	1946.	IIA2-1	(3)
	1949.	IIIA3-1b	(2)
Stareck, J. E.	1932.	IIIC3-1d	(1)
	1932.	IIIC3-3	(7)
Starke, Kurt	1943.	uranides	(11)
Starodubtsev, S. Ia.	see Semishin, V. I.		
Stedman, D. F.	1947.	IIIA3-1a	(1)
	1947.	IIIB3-1a	(2)
	1958.	not classif.	(33)
Steele, B. D.	1901.	IIIC3-1	(2)
Steinberg, Robert A.	1938.	IIIC5-2c	(1)
Steinmetz, Charles P.	1918.	IB3-1	(1)
Stephenson, H. H.	1929.	not classif.	(20)
Stewart, Alfred W.	1919.	IA2-1	(1)
	1919.	IC2-2	(12)
	1919.1925.	IIC2-4	(18)
	1920.	little per.t.	(5)
Stewart, O. I.	1928.	IA2-2	(1)
	1928.	IB2-2	(1)
	1928.	IC2-2	(14)
Stintzing, Hugo	1916.	IIIA3-1	(1)
	1916.	IIIB3-1	(2)
Stokes, B. J.	1916.	IIIC3-4	(26)
Stokes, G. J.	1904.	not classif.	(11)
Stoney, G. Johnstone	1888.1902.	IIB1-1	(2)
Stoye, K.	1954.	IIIB1-1b	(1)
Strache, H.	1908.	IC1-1a	(2)
Strack, H.	1952.	IIIB3-4	(3)
Strong, Frederick C. III	1959.	IIIB3-4	(4)
Sugathan, K. K., and Menon, T.C.K.	1956.	IIIA3-3a	(1)
Sugiura, Y., and Urey, H.	1926.	uranides	(4)
Swinne, Richard	1920.	IIIC3-4	(2)
	1926.	IIC2-4	(26)
Syrkin, Ia.K.	1934.	IC2-4	(19)
Syrkin, Ia.K., and Diatkina, M. B.	1946.1950.	IC2-5	(18)
	1946.1950.	IIC2-4	(45)
Szabo, Zoltán G.	1949.	actinides	(23)
	1969.	IIC2-5b	(9)
Szabo, Zoltán G., and Lakatos, Béla	1952.1954.1957.	IIC2-5b	(9)
	1962.	actinides	(23)

T.

Ta, Yeou	see Yeou Ta		

Talpain, L.	1945.	IIIA5-1	(3)
Tansley. L. Beaumont	1920.1921.	IB2-4	(3)
Taube, Mieczyslaw	1967.	IIIC3-3	(22)
Taylor, Moddie D.	1960.	IIIC5-2a	(3)
Taylor, Wendell H.	1941 (citation).	IA1-1	(3)
	1941 (citation)	IC1-1a	(1)
	1949 (citation)	octaves	(2)
Ternström, Torolf	1963.1964.	IIIC3-1	(8)
Teudt, Heinrich	1919.	IC1-1	(12)
Thatcher, R. W.	1934.	not classif.	(21)
Thénard	?	triads	(11)
Thilo, Erich	see Rabinovich, Eugen		
Thomsen, Julius (Tomsen, Iu.)	1895.1897.	IIIC3-3	(2)
Tikhomirov, S. V.	1959.	IIIC3-1	(7)
Tilden, William A., Sir	1898 (citation).	octaves	(2)
	1910.	IIC2-3	(10)
Tocher, J. F.	1910.	IIB2-4	(1)
Tokarev, B. V.	1966.	IIIC3-6	(16)
	1966.	introd. IIIB	(41)
Tomkeieff, S. I.	1954.1955.1956.1958.	IIB2-1	(6)
	1954.	IIA2-3	(1)
	1954.	IIC2-3	(18)
	1955.	introd. IIIB	(28)
Traube, J.	1895.	not classif.	(5)
Tremlelt. R.	1963.	IIIA3-1c	(1)
Trifonov, D. N.	1963.	not classif.	(36)
	1966.	uranides	(23)
	1970.	IC2-4	(49)
Tsia-mo, Chu	1965.	not found	(28)
Tsuchida, Ryutaro	1953.	preface	(15)
Turner, W. D.	see Monroe, C. J.		
Tycho, E.	1919.	not found	(4)

U.

Ulmer, K.	see Merz, H.		
Underhill, C. R.	see Kemble, W. P.		
Urbach, Franz	1921.	IC2-4	(7)
Urey, H.	see Sugiura, Y.		

V.

Vainshtein, E. E.	1948.	actinides	(21)
Vaisman, I. A.	1948.	IIC1-1a	(1)
Valcárcel, Antonio García	1945.	IC2-4	(25)
Vallet, Pierre	1948.	IIIC5-2	(7)
Vanderwerf, Calvin A.	see Sisler, Harry H.		
Vdovenko, V. M.	1960.	actinides	(40)
Veimarn, P. P. fon	1915.	IIC2-4	(16)
Venable, F. P.	1895 (citation).	IA1-1	(3)
	1895 (citation).	IC1-1a	(1)
	1895.1896.1898.1904.	IC2-2	(3)
	1896.	preface	(1)
Venable, F. P., and Howe, Jas. Lewis	1898.	IC2-2	(3)
Ventriglia, Ugo	1948.	IC2-4	(28)

Verma, Mulk Raj, and Krishen, Anoop	1952.	IIIA3-2	(2)
Verschoyle, W. Denham	1908.	IIIC3-1e	(1)
Vetere, Francisco	1954.	not classif.	(31)
Vigueras, J. López	see López-Vigueras, J.		
Villar, German E.	1938.	IIIC3-5	(6)
	1938.1957.	IIIC3-6	(5)
	1938.1940.1942.1943.1944. 1946.1947.1955.1957.1958. 1959.1960.1962.1964.1966. 1967.1969.	actinides	(11)
	1960.	IC2-4	(38)
	1962.1969.	IIIC3-4	(17)
	1964.1966.1967.1969.	IIC2-4	(59)
	1964.	IIIC3-3	(20)
Vinassa-de Regny, P.	1927.	IIC2-5a	(3)
	1933.	IC1-1a	(4)
Vincent, H. P.	1969.1971.	IIIC1-1b	(1)
Vogel, J. G.	1941.	IIC1-1	(9)
Vogel, Rudolf	1918.	IA3-2	(1)
Volkenshtein, M. V.	1955.	not found	(13)
Vozdvizhenskii, G. S.	1960.	IC2-4	(40)
	see Akhmetov, N. S.		

W.

Wade, Frank B.	see Blanchard, Arthur A.		
Waechter, Fr.	1878.	valence tables	(6)
Wagner, Henry A., and Booth, Harold Simmons	1945.	IIIC3-3c	(1)
Walden, P.	1908.	discovery	(2)
Walker, C. F.	see Gooch, F. A.		
Walker, James	1891.	IIC2-1	(2)
	1899.	IIC2-4	(8)
Walker, W. R., and Curthoys, G. C.	1956.	IIIC4-2	(13)
Wallin, B. H.	1926.	IB1-1	(1)
Walter, Reinh.	1950.	IC1-2b	(3)
	1950.	IIC2-5	(7)
Weissenberger, G.	1916.	IC2-2	(10)
	1916.	IC2-3	(6)
Wells, P. V.	1918.	IB2-5	(2)
Wendt, Gustav	1891.	not classif.	(4)
Werner, Alfred	1905.	IIIC3-4	(1)
	1905.	actinides	(2)
Wheeler, R. F.	1955.	IIIC3-4	(13)
Wheeler, T. S.	1947.	IC2-5	(19)
Wherry, Edgar T.	1924.	IIC2-4	(23)
White, Harvey Elliott	1934.	IIC2-4	(32)
Wiberg, Egon	1936.	IIC2-7	(1)
Wickert, K.	1947.	not classif.	(25)
Wiik, F. J.	1875.	not classif.	(1)
Wilde, H.	1897.	not classif.	(8)
Williamson, A. W.	1864.	valence tables	(3)
Wichterle, O.	1957.	not found	(14)
Wiswesser, William J.	1945.	IIIC4-2	(6)
	1945.	introd. IIIB	(14)

Wolff, F. von.	1928.	IB2-5	(5)
Woodiwiss, Geo	1906.	IIC2-2	(1)
	1906.	IIC2-7	(0)
Worley, F. P.	1923.	IIC2-5	(3)
Wrigley, A. N., Mast, W. C., and McCutcheon, T. P.	1949.	IIC2-4	(49)
	1949.	IIIC3-4	(8)
Wu, Ta-You, and Goudsmit, S.	1933.	uranides	(6)
Wurtz	1870/71.	triads	(14)
Wylie, A. W.	1950.	uranides	(17)

Y.

Yagoda, Herman	1935.	little per.t.	(11)
Yeou Ta	1945.1946.	IIIC5-1	(3)
	1946.	introd. IIIB	(15)
Yi, Pao-Fang	1947.	introd. IIIB	(16)

Z.

Zaengerle, M.	1882.	not classif.	(2)
Zapffe, Carl A.	1947.	IIIB4-2	(1)
	1969.	forerun.	(2)
Zarubitskii, O. G.	see Delimarskii, Iu. K.		
Zenghelis, C.	1906.	IC2-2a	(2)
Zikmund, Miroslav	1959.	preface	(19)
Zimens, Karl Erik	1948.	IIC2-5	(5)
	1950.	IIIC3-4	(9)
Zmaczynski, Emil V.	1937.	IIIC3-3c	(0)
	1937.	IIIA3-4	(1)
	1937.	IIIC3-4	(6)
Znoiko, A. P.	1949.	IIIC4-2	(8)
Zucchi, U.	see Marson, L. M.		
Zuffanti, Saverio	see Luder, W. F.		

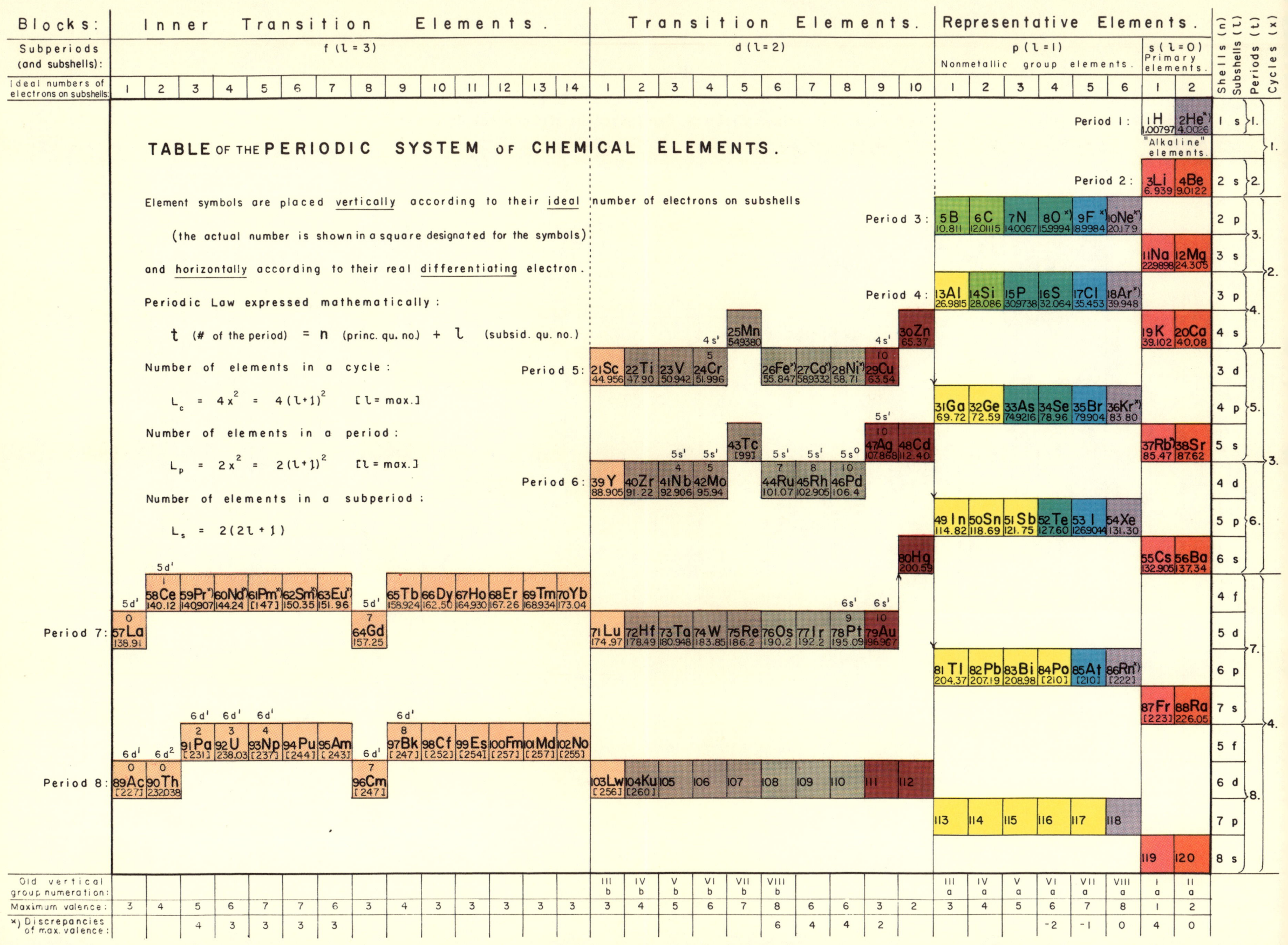

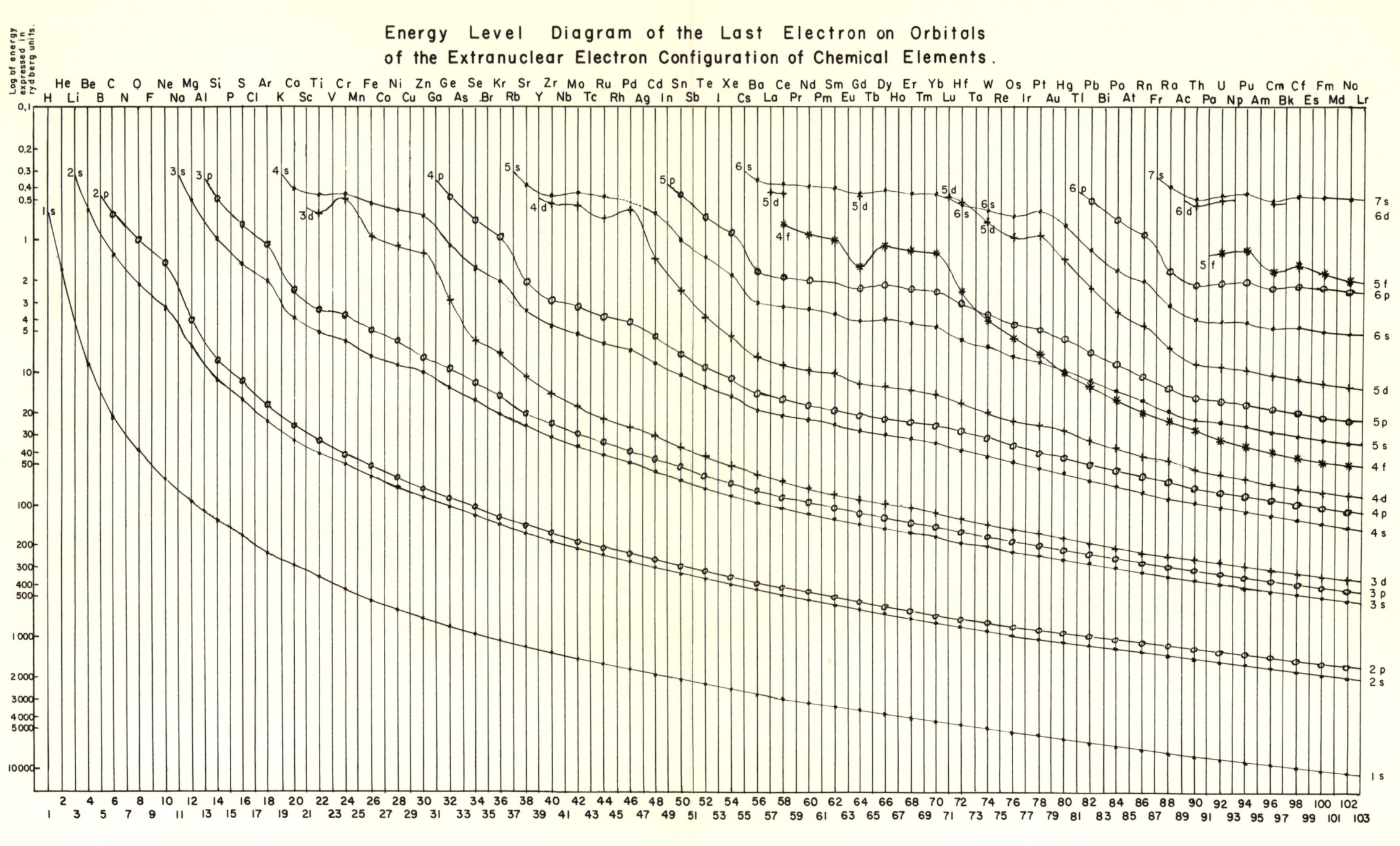

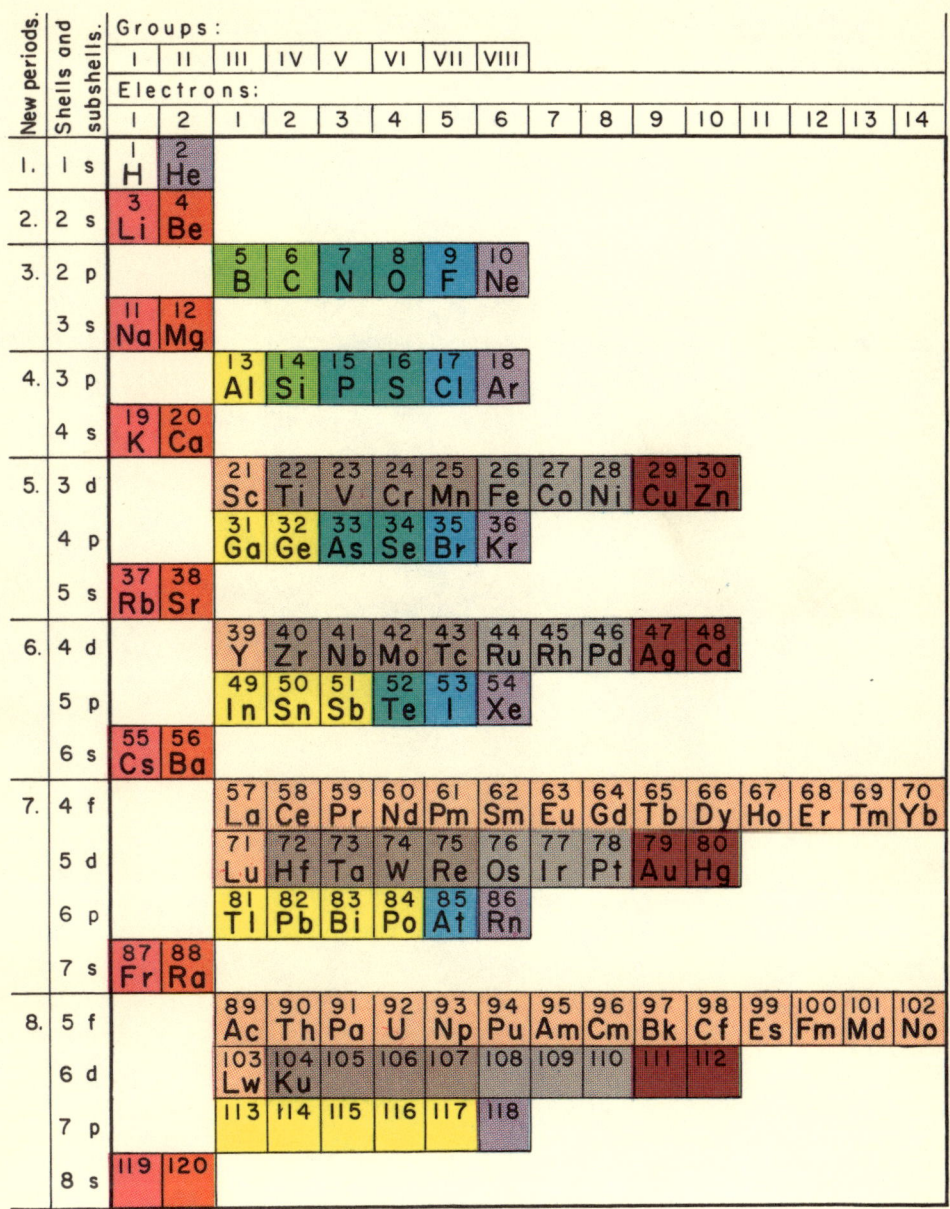

FIG. 134. Mazurs 1958/1973. Subtype IIIC5–2d.